Perspectives on Global Change

The TARGETS Approach

Edited by

Jan Rotmans
and Bert de Vries
National Institute of Public Health and the Environment (RIVM), The Netherlands

PUBLISHED BY THE PRESS SYNDICATE OF THE UNIVERSITY OF CAMBRIDGE
The Pitt Building, Trumpington Street, Cambridge CB2 1RP, United Kingdom

CAMBRIDGE UNIVERSITY PRESS
The Edinburgh Building, Cambridge CB2 2RU, United Kingdom
40 West 20th Street, New York, NY 10011-4211, USA
10 Stamford Road, Oakleigh, Melbourne 3166, Australia

© Cambridge University Press 1997

This book is in copyright. Subject to statutory exception
and to the provisions of relevant collective licensing agreements,
no reproduction of any part may take place without
the written permission of Cambridge University Press.

First published 1997

Printed in the United Kingdom at the University Press, Cambridge

A catalogue record for this book is available from the British Library

ISBN 0 521 62176 3 hardback

CONTENTS

Foreword		xi
Preface		xiii
1	**Global change and sustainable development**	**1**
	Jan Rotmans, Marjolein B.A. van Asselt, and Bert J.M. de Vries	
1.1	Introduction	3
1.2	Global change	4
1.3	Sustainable development	9
1.4	Integrated assessment	11
1.5	This book	13

Part One   The TARGETS model

2	**Concepts**	**15**
	Jan Rotmans, Bert J.M. de Vries, and Marjolein B.A. van Asselt	
2.1	Introduction	17
2.2	The conceptual model and the basic concepts	18
2.3	Pressure-State-Impact-Response (PSIR)	22
2.4	Integration	25
2.5	Presentation	29
2.6	Transitions	30
3	**The TARGETS model**	**33**
	Jan Rotmans, Marjolein B.A. van Asselt, Bert J.M. de Vries,	
	Arthur H.W. Beusen, Michel G.J. den Elzen, Henk B.M. Hilderink,	
	Arjen Y. Hoekstra, Marco A. Janssen, Heko W. Köster, Louis W. Niessen,	
	and Bart J. Strengers	
3.1	Introduction	35
3.2	Integrated assessment modelling	35
3.3	Aggregation, calibration and uncertainty	39
3.4	Description of the TARGETS1.0 submodels	45
3.5	Submodel linkages in the TARGETS1.0 model	51
3.6	Future work	53
4	**The Population and Health submodel**	**55**
	Louis W. Niessen, and Henk B.M. Hilderink	
4.1	Introduction	57
4.2	Health transitions	58
4.3	Modelling population and health	62
4.4	Model description	65
4.5	Calibration and validation	76
4.6	Conclusions	78

v

5	**The energy submodel: TIME**	**83**
	Bert J.M. de Vries, and Marco A. Janssen	
5.1	Introduction	85
5.2	Energy issues	87
5.3	Position within TARGETS	89
5.4	The energy demand module	92
5.5	The fuel supply modules	95
5.6	The electric power generation module	100
5.7	Calibration	102
5.8	Conclusions	106

6	**The water submodel: AQUA**	**107**
	Arjen Y. Hoekstra	
6.1	Introduction	109
6.2	Water policy issues	110
6.3	An integrated approach to water policy issues	112
6.4	Model description	114
6.5	Calibration and validation	130
6.6	Conclusions	134

7	**The land and food submodel: TERRA**	**135**
	Bart J. Strengers, Michel G.J. den Elzen and Heko W. Köster	
7.1	Introduction	137
7.2	Model description	139
7.3	Calibration	153
7.4	Conclusions	158

8	**The biogeochemical submodel: CYCLES**	**159**
	Michel G.J. den Elzen, Arthur H.W. Beusen, Jan Rotmans, and Heko W. Köster	
8.1	Introduction	161
8.2	The global carbon and nitrogen cycles and related feedbacks	163
8.3	Model description	168
8.4	Calibration	179
8.5	Conclusions	184

9	**Indicators for sustainable development**	**187**
	Jan Rotmans	
9.1	Introduction	189
9.2	Indicators and indices	190
9.3	Linkages between indicators and models	193
9.4	A model-based indicator framework	194
9.5	Practical implementation of TARGETS1.0	196
9.6	Linkage with existing indicator programmes	202

10	Uncertainties in perspective *Marjolein B.A. van Asselt, and Jan Rotmans*	205
10.1	Introduction	207
10.2	Model routes	209
10.3	Framework of perspectives	211
10.4	Methodology	219

Part Two Exploring images of the future

11	Towards integrated assessment of global change *Jan Rotmans, Bert J.M. de Vries, Marjolein B.A. van Asselt,* *Arthur H.W. Beusen, Michel G.J. den Elzen, Henk B.M. Hilderink,* *Arjen Y. Hoekstra, and Bart J. Strengers*	223
11.1	Introduction	225
11.2	Experimental set-up and uncertainty analysis	226
11.3	Economic scenarios	231
11.4	The hierarchist utopia: a reference future	232

12	Population and health in perspective *Henk B.M. Hilderink and Marjolein B.A. van Asselt*	239
12.1	Controversies related to population and health	241
12.2	Population and health uncertainties	242
12.3	Perspectives on population and health	243
12.4	Three images of the future	247
12.5	The plausibility of the projections	250
12.6	Risk assessment	252
12.7	Population, health and global change	256
12.8	Conclusions	261

13	Energy systems in transition *Bert J.M. de Vries, Arthur H.W. Beusen, and Marco A. Janssen*	263
13.1	Introduction	265
13.2	Major controversies and uncertainties	266
13.3	Perspectives on world energy	270
13.4	Simulation results for the three utopias	274
13.5	Uncertainties and dystopias: some more model experiments	284
13.6	Conclusions	289

14	**Water in crisis?**	**291**

Arjen Y. Hoekstra, Arthur H.W. Beusen, Henk B.M. Hilderink, and Marjolein B.A. van Asselt

14.1	Introduction	293
14.2	Questions related to water	293
14.3	Major controversies and uncertainties	295
14.4	Perspectives on water	300
14.5	Water in the future: three utopias	304
14.6	Possible water futures: the broader scope	312
14.7	Risk assessment	314
14.8	Water policy and global change	315

15	**Food for the future**	**319**

Bart J. Strengers, Michel G.J. den Elzen, Heko W. Köster, Henk B.M. Hilderink, and Marjolein B.A. van Asselt

15.1	Introduction	321
15.2	Main issues and uncertainties	322
15.3	Perspectives on land and food	324
15.4	Simulation results for the three utopias	328
15.5	Uncertainties and dystopias: some more model experiments	337
15.6	Risk assessment	341
15.7	Conclusions	343

16	**Human disturbance of the global biogeochemical cycles**	**345**

Michel G.J. den Elzen, Arthur H.W. Beusen, Jan Rotmans, and Marjolein B.A. van Asselt

16.1	Introduction	347
16.2	Controversies and uncertainties	348
16.3	Perspectives on the global carbon and nitrogen cycle and climate system	353
16.4	Three images of the future: utopias	357
16.5	Risk assessment: dystopias	364
16.6	Conclusions	370

17	**The larger picture: utopian futures**	**371**

Bert J.M. de Vries, Jan Rotmans, Arthur H.W. Beusen, Michel G.J. den Elzen, Henk B.M. Hilderink, Arjen Y. Hoekstra, and Bart J. Strengers

17.1	Introduction	373
17.2	Inclusion of feedbacks in integrated experiments	374
17.3	Wishful thinking: three utopian futures	376
17.4	Economic growth in the three utopias	379
17.5	Transitions in utopian futures	384
17.6	Into the 22nd century: towards a sustainable state?	391
17.7	Conclusions	393

18	**Uncertainty and risk: dystopian futures**	**395**
	Bert J.M. de Vries, Jan Rotmans, Henk B.M. Hilderink,	
	Arthur H.W. Beusen, Michel G.J. den Elzen, Arjen Y. Hoekstra,	
	and Bart J. Strengers	
18.1	Introduction	397
18.2	Dystopian tendencies	398
18.3	Systematic exploration of dystopias	403
18.4	Additional explorations: policy timing and overconsumption	408
18.5	Uncertainty and risk	413
18.6	Conclusions	415
19	**Global change: fresh insights, no simple answers**	**417**
	Bert J.M. de Vries, Jan Rotmans, Arthur H.W. Beusen,	
	Michel G.J. den Elzen, Henk B.M. Hilderink, Arjen Y. Hoekstra,	
	Marco A. Janssen, Louis W. Niessen, Bart J. Strengers,	
	and Marjolein B.A. van Asselt	
19.1	Introduction	419
19.2	Synthesis of the results	420
19.3	World in transition	428
19.4	Epilogue	431
References		**435**
Acronyms, units and chemical symbols		**457**
Index		**460**

FOREWORD

The National Institute of Public Health and the Environment (RIVM) is a centre of expertise that provides support to the Dutch government in the development of its National Environmental Policy. The relationship between economic activity in the past and projections for the future, as well as the implications for health and the environment, are quantified by means of simulation models in scenario-based studies. These models deal with a wide range of issues, ranging from local air and soil pollution to climate change. Integration of the models, which are defined 'bottom-up', enables 'integrated assessment' of alternative future environmental policies. Integrated assessment, however, cannot be based on a 'bottom-up' approach alone. As early as the 1980s, RIVM developed the IMAGE model which takes a 'top-down' approach, to describe climate change. This model, and the IMAGE 2.0 version, that was developed subsequently, have made a substantial contribution to the assessment rounds of the Intergovernmental Panel on Climate Change (IPCC). It was felt, however, that issues related to sustainable development, as discussed during the 1992 Rio Conference, should be more integrated. As a first step, RIVM co-operated with Dennis and Donella Meadows of the University of New Hampshire on a re-evaluation of their book Limits to Growth, which was presented to the Club of Rome in 1971. The results of this cooperation were published in 1992 in the book Beyond the Limits. By then it had become clear that more integrated methods were required, which would include interactions on a global scale between human population and health, the economy, energy, water, food and land use, and biogeochemical cycles. Such an approach would be indispensable in operationalising the central theme of the Agenda 21 declaration. In response to this need, a research project - Global Dynamics and Sustainable Development - was started in 1992 within RIVM's Bureau for Environmental Assessment. The resulting model, called TARGETS, is described in this book. Apart from its truly integrated character, it adds several novel dimensions to the work at RIVM which have, so far, received insufficient attention. The two most important of these are the treatment of uncertainties by explicitly formulating cultural perspectives on controversial issues and the development of a visualisation tool which can convey a picture of alternative systems to a wider scientific audience. This was done in cooperation with RIVM's Department of Environmental Information Systems to enable the TARGETS model to be presented in a transparent and interactive way.

Because humankind is confronted with the need to manage increasingly complex systems, it would seem appropriate for us to learn to deal with controversial issues by explicitly including different perspectives on how the world functions and should be managed. As the complexity of these systems grows, it becomes increasingly difficult to acquire knowledge from controlled laboratory experiments. In other words: the scope for 'knowledge rationality' diminishes. There are two sides to the coin here. On the one hand, one has to be more careful in interpreting the results because it is easier to fool or be fooled. But on the other hand, broader problem areas, such as the major issues of global change, can become legitimate subjects of interdisciplinary scientific research. Clearly, the confidence with which results can be presented and used is more limited than in traditional science. But we need no longer refrain from making scientific contributions to the important debates on global change and sustainable development. A tool like the TARGETS model can actually provide a platform for placing many apparently unrelated events in a larger and more meaningful context. TARGETS can also be used to frame the issues more rigorously and consistently. Moreover, science can, in this way, contribute to policy-making by adding value-oriented aspects to the 'knowledge rational' analysis of how to reconcile ends and means and how to deal with questions as to what can be achieved, what should be achieved, and which measures are likely to be most effective. Such a contribution seems rather urgent in view of the transition which humankind is currently going through. With the development and presentation in print of the TARGETS model, RIVM has therefore taken a deliberate step beyond purely 'knowledge rational' studies. The model creates a scientific platform for expressing and clarifying values and convictions about society and nature. The results represent a move away from the clear-cut distinction between the value-oriented policymaker and the science-driven, independent and objective modeller. However, it should not be overlooked that contributions from the social sciences, such as the Cultural Theory of Thompson *et al.*, can provide a framework for motivating the choice of certain assumptions and methods in the context of cultural perspectives. Applying this model will make it possible to group elements within these choices - that are inevitably subjective - into stereotypical (e.g. optimistic or pessimistic) perspectives. In this way the choices will become explicit, more objective and thus more manageable. The model can then be used in an interactive dialogue with policymakers and thus allow the RIVM to achieve its core task: bringing quality into the political debate on the environment and sustainable development.

Nicolaas D. van Egmond
RIVM Director of Environment

PREFACE

In 1992 the National Institute of Public Health and the Environment (RIVM) launched the interdisciplinary research programme 'Global Dynamics and Sustainable Development'. The main objective of this research has been to investigate the concept of sustainable development from a global perspective. A key project within the programme was the development of a global model called TARGETS (Tool to Assess Regional and Global Environmental and health Targets for Sustainability). TARGETS belongs to a class of integrated assessment models which build on a tradition started in the early 1970s with the World3 model used in the Report to the Club of Rome. Despite better information and greater insights into the global system, a model like TARGETS only represents a simplified description of the world, and therefore still has many limitations and deficiencies. However, TARGETS distinguishes itself from previous integrated assessment models in that it deals explicitly with prevailing uncertainties in the form of perspective-based model routes.

It has been our intention for the TARGETS model to be used to assess the interlinkages between social, economic and biophysical processes on a global scale. One of our motivations has been the fact that the increasing rate and complexity of global change processes forces us to go beyond disciplinary boundaries and carry out integrative research, using integrated assessment models. We see the TARGETS model as a tool for experimenting with new concepts and techniques, not as some kind of 'truth machine' that generates predictions. More particularly, we have aimed at to produce fresh insights - not ready-made answers - on issues of global change processes in the context of the Rio Declaration on Sustainable Development and Agenda 21.

This book presents the results achieved so far, which we feel can make a real contribution to the debate on global futures. The model-based explorations add numerical consistency and integrate many scientific disciplines. The integrated perspective provides a consistency which is absent from more narrowly based analyses. And the explicit inclusion of values and beliefs in the form of perspective-based model routes stimulates an open, process-oriented approach. One of the insights gained from the model experiments is that coherent sets of assumptions offer more than one consistent and desirable image of the future. These so-called utopias are vulnerable to criticism, because the assumptions behind them may turn out to be incorrect. Another important insight is that the human-environmental system is characterised by changes on widely differing time-scales.

It is now up to you - the readers - to judge. Whatever your verdict, we sincerely hope that this book will encourage other researchers to participate in helping policymakers and interested lay-people to see the larger whole and we hope it will inspire them to formulate the vision and strategies needed to build a sustainable global future.

The work on the model was carried out within the group Global Dynamics & Sustainable Development, or GLOBO for short. It started in the summer of 1992 and has focused from the outset on the construction and integration of meta-models - simplified expert models - as part of a larger framework. The scientific emphasis was on integrating insights from a variety of disciplines, including demography, health science, economics, environmental science, mathematics and philosophy. As a consequence, a broad spectrum of other research groups both within and outside the GLOBO team have been involved in the project. To ensure that the model set-up and results are communicated clearly and correctly, the simulation language 'M' was developed within RIVM alongside the TARGETS model. This permits easy visualisation in developing and presenting the model. An interactive version of the model is available on CD-ROM[1]. In the course of the project, contributions were made to several other RIVM projects, notably the IMAGE2.1 and UNEP's Global Environmental Outlook (GEO).

It is a difficult task to properly acknowledge all those who have contributed to this book and the research, model development and assessment behind it. We make an attempt here, realising that our list is bound to be incomplete. First of all, we are grateful to the RIVM directorate that initiated and financially supported this interdisciplinary research programme. In particular, we thank Fred Langeweg for his dedicated support; without him TARGETS would not have been possible. In the first and perhaps most difficult phase of the project, Anton van der Giessen and Mirjam Kroeze played a key role. An indispensable team member who took care of many integrative tasks at that time was Marco Janssen. The authors of the various chapters, who, as individual authors but also as a team, are responsible for the contents, form the core group. One team member not mentioned as an author but who played an important role is Esther Mosselman; we thank her for the skill and enthusiasm with which she designed the model interface which is an essential part of the CD-ROM version. In the last, hectic stages important contributions came from Martin Middelburg and Anita Meier. We are grateful to Martin for his dedication in making the graphics and the lay-out and Anita provided the organisational support which is essential to a project like this. Michael Gould helped structure and formulate the book and he and Ruth de Wijs were responsible for making the text more English, more consistent and more readable. Invaluable conceptual contributions, as well as on-line assistance, came from the main developers of the simulation language 'M': Jos de Bruin, Pascal de Vink, Jarke van Wijk and Marco van Zwetselaar. During the

---

[1] Baltzer Science Publishers, P.O. Box 37208, 1030 AE Amsterdam, The Netherlands.
Tel: +31 20 637 0061; Fax: +31 20 632 3651; E-mail: publish@baltzer.nl; internet site http://www.baltzer.nl

project, numerous people in a number of projects and institutes contributed as researchers, research assistants and trainees. A word of thanks for Jodi de Greef for his critical but always human and humorous role as senior scientist in GLOBO, and for Rob Maas, Rob Swart and Rik Leemans of RIVM's Bureau for Environmental Assessment for their critical but supportive attitude towards the project. Special thanks are due to Pim Martens and Koos Vrieze of the Department of Mathematics of Maastricht University; to Frans Willekens and Charles Vlek and their colleagues at the University of Groningen; to Thomas Fiddaman of MIT Sloan School of Management and Ruud van den Wijngaart of the Dutch Environment Ministry VROM; to the members of the IMAGE team for their cooperation; and to the research assistants and trainees Michel Bakkenes, Joost Bakker, Henry van den Bedem, Hessel van den Berg, Petra Costerman Boodt, Anco van Duivenboden, Frank Geels, Marc Geurts, Wander Jager, Linda Kamp, Richard Klugkist, Martijn Root, Renzo van Rijswijk, Eric Verbruggen, Jasper Vis and Detlef van Vuuren for the various ways in which they have contributed to the sustained development of the GLOBO activities.

Finally, we are grateful to colleagues both within and outside RIVM who have critically reviewed our work and in this way contributed to the quality of the research. In particular, we thank Hadi Dowlatabadi, Carlo Jaeger, Donella Meadows and other members of the Balaton Group, Harry van der Laan, Steve Schneider and Mike Thompson, who have been a constant source of inspiration.

Jan Rotmans and Bert de Vries

Global change and sustainable development

1 GLOBAL CHANGE AND SUSTAINABLE DEVELOPMENT

Jan Rotmans, Marjolein B.A. van Asselt and Bert J.M. de Vries

This chapter provides an introduction to the theme of the book by explaining the importance of the three central concepts global change, sustainable development and integrated assessment. The book focuses on five areas: population and human health, energy, water, land and food, and global biogeochemical cycles. The idea of using multiple definitions of sustainable development in an integrated approach to global change is put forward. The construction and use of the TARGETS integrated assessment model (Tool to Assess Regional and Global Environmental and health Targets for Sustainability) is justified in terms of providing a platform for communication within the scientific community that can inform policy debates about likely trends over the next 100 years or so.

1.1 Introduction

With the approach of a new millennium, global change and sustainable development are evolving as key concepts for assessing the future of the planet and of humankind. Over the last few decades, we have become used to the idea that our activities may have serious and irreversible impacts on the environment. Human-induced changes are recognised as having the potential to significantly modify the structure and functioning of the Earth system as a whole. Furthermore, activities at one place on Earth can affect the lives of people around the globe and even jeopardise those of future generations. The use of land, water, minerals and other natural resources by humans has increased more than tenfold during the past two centuries. Future economic development and increases in population will presumably intensify this pressure and the cumulative impacts of human activities, such as agriculture, stock breeding, fishery, industrial production, transport, recreation and domestic activities, are likely to cause major environmental changes, varying from disruption of local ecosystems to disturbances of the biosphere. At the same time we have become aware that progress in human development is becoming increasingly dependent on the state of the environment and may be constrained or even reversed by its future deterioration. This makes the relation between humans and the environment extremely complex, with numerous interactions of many kinds that operate on different scales (Clark and Holling, 1985; Clark and Munn, 1986). This book aims to shed some light on these mutual interactions by unfolding an integrated approach to global change.

1.2 Global change

Awareness of the interconnectedness of human and environmental systems is nothing new (see Friedman (1985) for a comprehensive historical overview). As early as 1864, Marsh provided the first indications of the impact of human beings on the environment (Marsh, 1864). In the early 19th century the concept of the Earth's 'biosphere' was introduced, first in biology by Lamarck and later in geology by Suess. Vernadsky (1945) developed the concept into an umbrella term to describe the integrated living and life-supporting systems on Earth together with its surrounding atmosphere. In this view, humans are seen as merely one of the many species living on Earth. However, in the early 20th century, it was recognised that humankind is a significant agent of change. The concept 'noösphere' was employed to describe global processes in such a way that Earth's systems are seen as being organised by human activity. This concept was first introduced by Le Roy and Teilhard de Chardin in 1927, and was subsequently developed and brought to a wider audience by Vernadsky (1945).

These global concepts were not widely recognised until the second half of the 20th century. The first major interdisciplinary, international review based on these concepts took place at a 1955 symposium entitled 'Man's Role in Changing the Face of the Earth' (Thomas, 1956). Notwithstanding the severe criticism by academics, public officials and business elites (Freeman, 1973; Humphrey and Buttel, 1982; Morrison, 1976) the first Report to the Club of Rome, 'Limits to Growth' (Meadows *et al.*, 1972), can be considered as an important milestone in global thinking and reasoning. In 1974, Molina and Rowland (1974) launched their theory on the depletion of stratospheric ozone, which represented a breakthrough in the awareness that human activities can have serious and irreversible consequences on a global scale. In the early 1980s, the concept of global change began to appear explicitly in scientific reports with increasing frequency. The issue of global climate change prompted a flight of publications. Some of these proved to be of great importance in the development of the concept of global change, in particular the IIASA study 'Sustainable Development of the Biosphere' in 1986 (Clark and Munn, 1986), reports on conferences held in Villach and Bellagio (Jaeger, 1988), the regular overview reports of the Intergovernmental Panel on Climate Change (IPCC) of which the first volume appeared in 1990, and the scientific assessments of ozone depletion (WMO, 1992).

Increasing recognition of global change as a major field of research led, in turn, to a number of international, interdisciplinary research programmes (Price, 1990). The oldest international research programme to promote the co-operation of natural and social scientists is the Scientific Committee on Problems of the Environment (SCOPE), established by the International Council of Scientific Unions (ICSU) in 1969. Other examples of major interdisciplinary research programmes are the World Climate Research Programme (WCRP) initiated in 1979 by the World

Meteorological Organisation (WMO) and the ICSU, the International Geosphere-Biosphere Programme (IGBP) established in 1986, the aim of which is to set priorities for global change research (IGBP, 1988, 1990, 1992), and the Human Dimensions of Global Environmental Change Programme (generally abbreviated to HDP), initiated by the International Social Science Council (ISSC) in 1988 (HDP, 1995).

Notwithstanding the considerable success booked by some of these programmes in integrating social and natural sciences, the research tended to remain disciplinary and dominated by the natural sciences (Price, 1990). Even within the most recent HDP programme, the contributions from the social and the natural sciences tend to diverge. Interdisciplinary research suggests that there is some common conceptual or systemic framework which underpins the entire research framework. It requires a concerted effort to find unifying concepts that will foster and reinforce understanding across disciplines (Zube, 1982). One of the aims of this book, which is the result of co-operative efforts made by scientists from a wide variety of disciplines, is to advance the search for such unifying concepts.

The concept of global change

The various ways in which the term global change is used can cause considerable confusion. Despite the growing interest in this subject, it remains loosely defined (Price, 1989). 'Global change' is frequently used to refer to human-induced changes in the environment. In this restricted definition, global change serves as an umbrella for a whole range of mutually dependent global environmental problems, such as climate change, stratospheric ozone depletion, atmospheric deposition (including acidification), euthrophication, land degradation, desertification and the dispersion of chemicals. In this book the term 'global change' is used to denote the totality of changes evoked by the complex of mutual human-environment relationships. We advocate the use of 'global environmental change' to specifically address the geophysical, biological, chemical and ecological components of global change, and the term '(global) climate change' to refer to transregional climatic changes.

Two types of global environmental change can be distinguished, but not separated: (i) local changes which are of global importance, such as acidification, deforestation, desertification, and other processes leading to changes in the quality of land, in biodiversity and in human health; and (ii) changes occurring throughout the global system, such as climate change and the associated rise in sea level, and stratospheric ozone depletion (Clark and Holling, 1985; di Castri, 1989; Price, 1989). Because the former are local processes that occur in more or less similar forms throughout the world, they are often called 'universal' to distinguish them from the truly 'global' processes. Turner II *et al.* (1990b) define the first type as 'cumulative' and the second as 'systemic' (see *Table 1.1*). One of the major challenges of global change research is to investigate the mutual relationships between these two types of change.

Investigating global change

The global change research described in this book is based on the following key hypotheses:

- the human-environment system is a complex system of interrelated cause-effect chains, in which human disturbances can be disentangled from natural changes;
- global change is by nature universal. Many global trends can be analysed irrespective of aggregational, spatial and temporal differences, using generic descriptions;
- a top-down approach is an adequate strategy for global change research. Disaggregation of global averages which are interesting from a regional point of view is possible using classes and spatial distribution functions.

In line with these hypotheses we distinguish the following characteristics of the processes that constitute global change (Kates *et al.*, 1990; O' Riordan and Rayner, 1991):

- global change comprises those societal and environmental changes which are, either directly or indirectly, human-induced;
- the effects of these changes have all-encompassing implications, affecting present and future global and local societal and ecological structures;
- the rate of change is so rapid that it can be identified within a human lifetime;
- the changes are considered to be irreversible.

Furthermore, we recognise three important dimensions of global change, namely social dynamics, economic dynamics and ecological dynamics. *Social dynamics* is about the social behaviour of a wide range of actors, varying from individuals to institutions at the micro, the meso and the macro level. A key notion here is quality of life, determined by items such as welfare, human health, population growth and

Type	Characteristic	Examples
Cumulative	Impact through distribution worldwide	(a) Groundwater pollution and depletion (b) Species depletion, genetic mutation (biodiversity)
	Impact through magnitude of change (share of global resources)	(a) Deforestation (b) Toxic industrial pollutants (c) Soil depletion on prime agricultural land
Systemic	Direct impact on systems that function globally	(a) Industrial and agricultural emissions of greenhouse gases (b) Industrial and consumer emissions of ozone-depleting gases (c) Land-cover changes and its impacts on albedo

Table 1.1 Two types of global environmental change (Source: Turner II et al., 1990b).

structure, cultural patterns and needs and wants. It is closely linked to human capital (people and their health and skills) and social capital (institutions, cultural cohesion and collective knowledge). *Economic dynamics* is based on the production and consumption patterns for different economic sectors, such as energy, agriculture, industry and services. For each sector the capacity is determined by demand and supply mechanisms, steered by allocation of investments. A basic economic problem is that of efficient allocation of scarce production factors, including labour, capital, land and natural resources. The underlying demand-supply mechanism can be described on different levels, namely in terms of: (i) monetary flows; (ii) physical materials and energy flows; and (iii) communication and decision flows. *Ecological dynamics* is, in the present context, largely determined by the magnitude and nature of human disturbances in relation to the resilience of the biosphere. At different scale levels this can be represented by the disturbance of the element cycles of carbon, nitrogen, phosphorus and sulphur in connection with the water cycle. Also the dissipation of numerous chemical substances is an important determinant of ecological quality. A multiple-stress approach is the best way to evaluate the consequences of human interventions for the functioning of ecosystems.

Clearly, it is impossible to take into account all of the processes subsumed under these dynamics in one single framework. Therefore, the next step is to determine which specific global change issues will be considered. Two criteria which may be relevant in this selection process are: (i) whether the issue is a key one in the totality of global changes, and (ii) whether it is possible to analyse it given the previously outlined hypotheses.

What are the key issues in global change? Any enumeration is incomplete and to a certain extent biased. An intersubjective approach would be to evaluate which issues are commonly addressed in contemporary global change research. To this end we have assessed studies on global change as the starting point for our selection of key issues to be tackled in our research. An overview of the issues addressed in major studies on global change is given in *Table 1.2*. Which studies have been considered? A milestone in global change research is the IIASA report 'Sustainable Development of the Biosphere' (Clark and Munn, 1986). Another study taken into account is the well-known report from the World Commission on Environment and Development entitled 'Our Common Future' (WCED, 1987) which identifies key global concerns. Another monitor of emerging issues is the Worldwatch Institute, with its annual 'State of the World' reports which have appeared since 1984. Since 1986, the World Resources Institute has produced a number of comprehensive surveys on global resources. Turner II *et al.* (1990a) provide an overview of the major human forces that have transformed the Earth over the last three hundred years. The International Conference on an Agenda of Science for Environment and Development into the 21st Century (ASCEND), provided a research agenda for global change. From the perspective of developing countries, the report on the State of the Environment in

	IIASA	WCED	Worldwatch Institute	World Resources Institute	The Earth as transformed by Human Action	ASCEND	State of the Environment in Southern Africa
Global change issues							
Economy	x	x					x
Population	x	x	x	x	x	x	x
Human health			x	x		x	
Urbanisation		x	x	x	x		
Culture					x		
Armed conflicts		x					x
Political processes	x	x		x		x	x
Industry		x			x	x	
Energy	x	x	x	x		x	x
Agriculture	x			x		x	
Food security		x	x	x		x	x
Technology	x				x	x	
Fresh-water resources			x	x	x	x	x
Pollution & waste			x			x	x
Climate change	x		x	x	x	x	x
Forests			x	x	x		x
land cover and land use			x	x	x		
Biodiversity	x	x	x	x			x

Table 1.2 Overview of global change issues addressed in major studies on global change. (The recent GEO-study is not included in this table.)

Southern Africa (Chenje and Johson, 1994) is of interest. A recent overview of environmental issues in both the developed and the developing regions is given in the Global Environmental Outlook (GEO) (UNEP, 1997).

Inspection of *Table 1.2* shows that population developments and climate change are addressed in all studies. Other topics covered by the majority of the studies are urbanisation, political processes, energy, food security, fresh-water resources, forests and biodiversity. Assessment of urbanisation requires insight into the distribution of human settlements. Biodiversity involves the site-specific accumulation of various species. Urbanisation and biodiversity can therefore not be addressed generically. Political processes also do not easily lend themselves to descriptions in generic causal terms. This difficulty is intensified by the fact that existing knowledge of global political and institutional dynamics is scattered. We therefore decided to omit this aspect in our approach to global change. The resulting list of key issues consists of human population, human health, energy use and resources, water use and resources, land use and resources including forests, and global biogeochemical cycles including climate change. Our aim is to describe current knowledge pertaining to these issues and the interlinkages between them in causal terms. This integrative approach may help to clarify the past and present as well as anticipate future global changes.

1.3 Sustainable development

The notion of 'sustainable development' was introduced in 1980 (IUCN *et al.*, 1980). Sustainable development is on the one hand closely allied to the resilience of the environmental system to anthropogenic disturbances, while on the other it involves societal concerns, such as poverty and equity. In this context, an important aspect of global change research is to inform decision-makers about plausible developments, and to present concrete strategies for sustainable development. It took about a decade before the concept of sustainable development became widely known in policy circles. The Brundtland report (WCED, 1987) played a key role in promoting the use of sustainable development as a framework for environmental policy making. Another important milestone in international policy making was the United Nations Conference on Environment and Development (UNCED) in 1992 in Rio de Janeiro (Brazil), followed by a series of UN conferences on sustainability issues. The outcome of this conference, 'Agenda 21' (UN, 1992), is an attempt to translate the concept of sustainable development into an international action programme. An evaluation of the five year period following the succeeding Rio conference is currently under way at the UN's Commission on Sustainable Development.

The concept of sustainable development

Despite the importance of the concept of sustainable development in current policy and scientific debates, there is still no single definition which is shared by all stakeholders. This is because sustainable development is a highly normative, value-loaded notion which can be interpreted and worked out according to various perspectives. Providing a comprehensive overview of all the different interpretations of sustainable development used in policy-making and scientific communities would be a research project in its own right. We restrict ourselves to some dominant uses of this concept to illustrate that several, even contradictory, definitions of sustainable development are viable. For a broader coverage, see for example Redclift (1987), de Vries (1989), Lele (1991) and Serageldin and Steer (1994). Numerous definitions of sustainable development have been given, but the most cited definition is the one given in the Brundtland report (WCED, 1987):

> Sustainable development is development that meets the needs of the present generation without compromising the ability of future generations to meet their own needs.

This definition reasons from the intergenerational principle that no burdens should be inherited by future generations, so that their starting position will at least not be worse than that of past and present generations. The definition reflected in the preamble to Agenda 21 (UN, 1992) is suffused with a similar anthropogenic spirit:

Principle 1:
Human beings are at the centre of concerns for sustainable development. They are entitled to a healthy and productive life in harmony with nature.
Principle 3:
The right to development must be fulfilled so as to equitably meet developmental and environmental needs of present and future generations.

An alternative, more ecocentric definition is:

> Sustainable development is improving the quality of human life while living within the carrying capacity of supporting ecosystems (IUCN et al., 1991).

This definition starts from the continuity principle that holds that ecosystems must be preserved. Some argue for the preservation of all ecosystems, although a less extreme view aims to preserve the resilience and dynamic adaptability of natural life-support systems.

The concept of sustainable development highlights the need to simultaneously address developmental and environmental imperatives, but it is clear from the above that the operational implications of sustainable development are neither unequivocal nor unambiguous. While there is general agreement that economic, social and environmental concerns all matter, the interpretation and valuation of the various aspects vary significantly. A strong emphasis on environmental assets in interpretating sustainable development is reflected in terms such as 'environmental sustainability' (Brown et al., 1991) or 'ecological sustainability' (Dovers, 1990). In this ecological frame of reference, human economic activity and social organisation are considered as subsystems that operate within a larger, but finite, ecosystem. Social and economic objectives for sustainable development are considered to depend on environmental quality (Rees, 1994). In the other extreme point-of-view economic growth overshadows other objectives and sustainable development is seen as 'sustained economic growth'. The goal in this economic context is then to maximise the net welfare of economic activities while maintaining or increasing the stock of economic, ecological and socio-cultural assets over time (Munasinghe, 1993). An economist would readily acknowledge the importance of social and environmental factors, but would interpret them from an economist's perspective. Social concerns tend to be reduced to questions of inequality and poverty reduction, and environmental concerns to questions of natural resource management. Absent are important concerns such as social cohesion, cultural identity and ecosystem integrity (Serageldin and Steer, 1994). In a third, sociological perspective, it is emphasised that the key actors are human beings, whose patterns of social organisation are considered to be crucial for devising viable paths to sustainable development. Economic concerns are interpreted in terms of social arrangements (Cernea, 1994). From this ongoing debate about sustainable development, one can conclude that it is a concept which is more of a guiding principle to be applied heuristically than a scientific concept waiting for a strict definition. It is also clear that human values and beliefs with regard to culture, nature and technology to a large extent determine the specific interpretation people give to sustainable development (de Vries 1989, 1994; Schwarz and Thompson, 1991; WRR 1994)[1]. Rather than try to define sustainable development in an objective way, this book attempts to accommodate multiple definitions. In so doing, we aim to explore the limits to what is understood by the concept of sustainable development.

1.4 Integrated assessment

Integrated assessment (IA) has emerged as a new avenue in the quest to explore complex environmentally-related problems. As an intuitive process it is not new. For instance, thousands of years ago Egyptian farmers were already applying it in the sense that they made use of integrated land management techniques, in particular ingenious crop-farming and methods of irrigation in combination with clever weather forecasting schemes. Since the 1970s, the notion of IA has been used within a broad context, in particular in Europe and North America. However, where in Europe IA has its origins in the population-environment, ecological and acidification research, in North America the focus was mainly on the economic impacts of anthropogenic disturbances (Rotmans and Van Asselt, 1996). During the last decade, IA efforts have increasingly focused on climate change. A comprehensive overview of recent projects on integrated assessment of climate change is given by Rotmans *et. al* (1996).

Methods and tools for IA are still relatively immature. So far, IA has been merely an intuitive process without precise rules and standards of good practice and without a theory which offers guidelines for combining the various disciplinary pieces. There are no theoretically based minimum standards that indicate degrees of quality. As a consequence, IA often lacks credibility both in disciplinary science and in the policy community. Another weakness is that current IA efforts do not include the participation of a wide range of disciplines; in particular social scientists are missing. Notwithstanding its acknowledged weaknesses, proponents of integrated assessment argue that the complexity of the issues demands an integrated approach so that key interactions and effects are not inadvertently omitted from the analysis. The increasing complexity of the problems around us forces us to undertake integrative research, leaving the mechanistic and reductionistic paradigms behind. There is growing consensus that the various pieces of the global change puzzle can no longer be examined in isolation.

---

1 See also CPB (1992) for a long-term study of the global economy in which divergent interpretations of economic driving forces are used in a modelling context.

Integrated assessment endeavours to keep track of how the pieces of the puzzle fit together.

A definition of integrated assessment

While there are a number definitions of IA (Ravetz, 1996), we use the following: *Integrated assessment is an interdisciplinary and participatory process of combining, interpreting and communicating knowledge from diverse scientific disciplines to allow a better understanding of complex phenomena.* IA has two main characteristics: (i) it should provide added value compared to insights derived from research within a single discipline; and (ii) it should offer decision-makers useful information. The term integration has two dimensions: capturing as many as possible of the cause-effect relationships of a phenomenon (vertical integration) and addressing the cross-linkages between different phenomena (horizontal integration). Adopting this interpretation of integrated assessment means that the various causes, mechanisms and impacts of the issue under concern need to be addressed. It is likely that gaps in disciplinary knowledge will be encountered in this process, so IA can help in setting research priorities. One of the objectives of IA is to support public decision-making by developing a coherent framework for assessing trade-offs between social, economic, institutional and ecological determinants and impacts.

Three key aspects of integrated assessment are comprised in the above definition: combination, interpretation and communication processes. The problem with *combining* pieces of knowledge from a variety of sciences is that there is as yet no unifying theory for how to do this. It should be noted that combining is not the same as linking: combination implies a thorough examination of feedback processes. This touches upon an overall difficulty in the IA process, namely that there are no theoretically-based protocols for aggregation and disaggregation in time, space and dynamics. The most important tools for combining disciplinary information and insights in a consistent way are integrated assessment models. IA models should be regarded as an aid in formulating possible images of the future, and not as a means of generating predictions as such. They are not 'truth' machines, but rather serve as heuristic devices in exploring the future. *Interpretation* takes place in iterative processes of understanding and explaining results and insights between the various stakeholders: IA practitioners, scientists working in specific fields and policy-makers. The difficulty with the process of interpretation is that there are no official platforms or forums in which such an exchange of insights can occur. So if there is a dialogue between IA practitioners, subject experts and policy makers on interpretative issues, it takes place on an ad hoc basis. *Communication* in the context of integrated assessment means knowledge transfer from and to a broad audience. This vital part of IA is probably the most complicated and underestimated constituent, for which there are hardly any tools and methods available. There are some attempts under way to develop tools for the communicative processes within integrated assessment in the form of policy exercises (Brewer, 1986; de Vries *et al.*, 1993; Duinker *et al.*, 1993; Jaeger *et al.*, 1990; Toth, 1988; 1992; 1993) and jury panels (Dürrenberger *et al.*, 1996; Jaeger, 1995). Policy exercises are simulation games in which multiple teams play roles and have the task of developing policy strategies which are evaluated using computer models. Jury panels consist of a series of guided discussions on complex issues with a small group of lay-people who are given scientific knowledge, including IA models. These participatory methods are in an embryonic stage of development, but the general direction looks promising.

A framework for integrated assessment
IA is an iterative and participatory process. On the one hand integrated insights from the scientific community are communicated to the decision-making community in particular and to society in general. On the other hand, experiences and lessons learned by decision-makers and visions and values expressed by society form an input for scientific assessment. IA is thus a cyclical process of mutual learning. The role of scientists as stakeholders in this process is to sketch the spectrum of what is possible and plausible in the light of state-of-the-art knowledge. The mutually constructed answers as to what is possible and desirable constitute the major inputs for the communication process. Participatory processes which involve communication between a wide variety of stakeholders should lead to appealing narratives[2]. The narratives from a first IA cycle feedback to problem definition processes in the scientific community and in society at large. They can also serve as a framework for consensus building. The resulting policy actions and their effects also provide feedback for the integrated assessment process. The proposed framework for IA as a cyclical and participatory process comprising various parallel processes is summarised in *Figure 1.1*. We do not argue that current IA efforts, including our own, adequately reflect this iterative procedure. It is a first step, which we hope will contribute to the iterative and participatory research process which should provide guidelines for humanity in its search for an equitable, efficient and sustainable future for planet Earth.

1.5   This book

This is a book on integrated assessment of global change, based on experiments with the TARGETS model. TARGETS is an acronym for Tool to Assess Regional and Global Environmental and health Targets for Sustainability. It also presents a methodology to take account of a variety of perspectives. The book consists of two parts. Part One (Chapters 2 to 10) describes the systems approach and the philosophy

---

2   Other formulations which are used for this process are scenario construction and computer-aided story-telling.

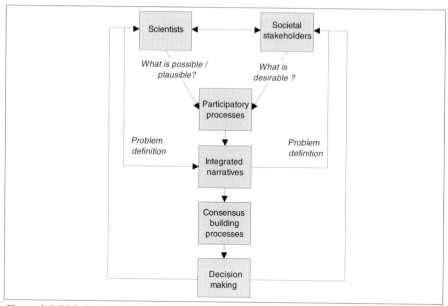

Figure 1.1 Global change research as part of a cyclical and participatory process (Rotmans and van Asselt, 1996).

behind the TARGETS modelling framework, and a detailed description of the submodels and their linkages. Part Two (Chapters 11 to 19) focuses on assessment in which a number of controversies related to global change issues are addressed from different perspectives, using model experiments and perspective-based model routes.

Chapter 2 outlines the basic concepts used in our approach, while Chapter 3 describes the TARGETS model in general terms. The various submodels and their calibration for the world over the period 1900-1990 are described in detail in Chapters 4 to 8. Chapter 9 elaborates on the issue of indicators for sustainable development, while Chapter 10 outlines the methodology of perspective-based model routes. Chapter 11 serves as an introduction to the second part of the book which deals with various aspects of global change controversies. Chapters 12 to 16 focus on model-based assessments of issues relating to population and health, energy, water, food and element cycles and the climate system. Utiopian and dystopian model experiments are used as guidelines. Chapters 17 to 19 synthesise the insights.

We are fully aware that this book does not account to a comprehensive integrated assessment of global change. It merely provides stepping stones which may be useful for traversing the rapids between disciplinary research and integrative approaches. Furthermore, it supplies organised scientific material that indicates what is possible and plausible. One consequence of the adopted definition of integrated assessment as a cyclical, participatory process is that a book like this can never be conclusive. It is nothing more, but also nothing less, than a reflection of lessons learned at a certain stage in an ongoing process.

Part One   The TARGETS model

Concepts

"How complex or simple a structure is depends critically upon the way in which we describe it."
H. Simon, The Sciences of the Artificial (1969)

2 CONCEPTS

Jan Rotmans, Bert J.M. de Vries and Marjolein B.A. van Asselt

Global change is an extremely complex phenomenon, encompassing a wide variety of issues. An adequate approach to such a broad subject demands careful consideration of a host of interactions between people and the environment and a clear understanding of driving forces, be the demographic, social, economic or technological. If we wish to tackle such a complex issue, we need to establish some basic guiding concepts. This chapter proposes an integrated systems approach to a number of key aspects of global change: population, health, energy, land, water and element cycles. We define what we mean by 'system' and 'model' and introduce a conceptual framework for analysing global change. As a mechanism to structure this conceptual framework we use the Pressure-State-Impact-Response (PSIR) approach. We look at two different kinds of integration (vertical and horizontal) and discuss different levels of complexity. Finally, we explain the importance of communicating the results of integrated systems analysis and suggest the value of using different communication methods such as indicators and visualisation.

2.1 Introduction

Most systematic studies of global change have so far focused on subsystems in isolation. However, it is well known that when parts are combined into more complex structures, the resulting system may exhibit quite different properties and behaviour (Gregory, 1981). As mentioned in Chapter 1, there is a growing interest in an integrated approach to global change (for an overview see Parson (1996) and Rotmans *et al.* (1996)). In this book, too, we use an integrated systems approach to address issues of global change and sustainable development. The main argument is that, in studying complex, large-scale global phenomena, there is the need for analysis which concentrates on the interactions between the different subsystems instead of focusing on each subsystem in isolation.

The first step, then, in such a multi-disciplinary analysis is the design of a conceptual framework which enables an adequate description of the dynamics of the human system, the environmental system and their interactions. We give a brief overview of the concepts and tools which are relevant for the design, construction and use of such a framework. Among these are the cause-effect chain and the Pressure-State-Impact-Response or PSIR concept, and the related notion of transitions. Various aspects of integration, both horizontal and vertical, and aggregation are introduced as a heuristic device to arrive at an integrated framework. Finally, some facets of the model are considered: indicators, uncertainties and the use of models to support policy-making.

2.2 The conceptual model and the basic concepts

A conceptual model defines the boundaries of a system, and represents its essential elements and the relationships between them. By expressing these elements and their relationships in a more formal way, using mathematical methods, the conceptual model serves as the starting point for a simulation model. Global change phenomena can be represented by a set of interrelated cause-effect chains, which form a complex system. Its properties are more than the sum of its constituent parts, the subsystems. This is the major argument for using the systems approach to analyse global change and sustainable development. The objective of a systems approach is to analyse the structure of the system as a whole, based on an understanding of the entities which the system is composed of. It also aims to provide insight into the processes, interactions and feedback mechanisms within the system that generate changes in its structure. Such an integrated systems approach is inter- and multi-disciplinary, based on the integration of knowledge gleaned from a variety of scientific disciplines. Our hypothesis is that, in the search for sustainable development strategies, one has to go beyond the analysis of subsystems in isolation, because what is sustainable from the perspective of one subsystem may be unsustainable for the whole system at large or, even worse, may cause a shift in a less sustainable direction in other subsystems.

It is clearly not possible to build a single model capable of addressing all the nuances of global change. The price for building an integrated model of global change phenomena is that many issues have to be treated at a very high level of abstraction: abstractions that, according to critics of integrated assessment modelling, vitiate the value of any insights gained from their development and application. Yet, it is our hypothesis that a simple, transparent and generic conceptual model can be used as a blueprint for a computer simulation model and can serve, supported by quantitative simulations, as an organising framework for structuring the discussion about global change and sustainable development. It creates the possibility to analyse systematically and in a quantitative, dynamic context the linkages and trade-offs between social, economic and environmental changes.

Such a conceptual model should be as generic, or universal, as possible, i.e. it should not be tied to a specific type of problem, situation, or region. This means that the theories and assumptions used for (sub)model construction should be applicable at different levels of spatial aggregation and for different regions in different periods. We make use of meta-models to include the expertise available on region and local-specific dynamics and data is used as much as possible. Nevertheless, the claim of genericity can only partially be fulfilled. In particular, the search for a generic description of land and water-based processes faces serious problems.

The conceptual model we propose consists of two strongly interconnected components: the human subsystem and the environmental subsystem. Each of these has in turn been split up into subsystems which are all interlinked. In line with the major themes associated with global change (Chapter 1), the human system includes

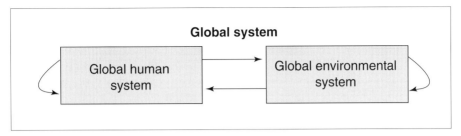

Figure 2.1 A highly aggregated conceptual framework for global change.

the population and health subsystem and the economic subsystem in a broad sense, which includes the provision of food, water, energy and materials. The environmental system comprises element reservoirs and fluxes, both in relation to natural processes and human interferences. The resulting scheme at the highest aggregation level is shown in *Figure 2.1*. We now proceed with a brief description of some basic concepts.

Models

In the absence of complete knowledge and in order to guide their actions, people use simplified images of the world arround them. Such images are constructed from the flow of sensory experiences in combination with rules. One may call these images models of a, usually small, part of the world. Less broadly, a model is defined here as a material, conceptual or formal representation of part of the observable reality, that is, of the system under consideration (de Vries, 1989; Rosen, 1985). Here, we confine ourselves to conceptual models and mathematical models. Conceptual models represent the system's boundaries, the essential entities of the system and their interrelationships. Mathematical models are conceptual models in which entities and interrelationships are formally represented by variables and relationships, often in the form of a set of differential or difference equations.

Systems

In a formal sense, a system is an interpretative representation of a part of reality that is bounded *vis-à-vis* its surroundings and consists of a number of entities (elements, components) that interact with each other. An entity is a part of the system that can be specified by defining its properties. The state of a system at a given moment in time is denoted by the values of relevant properties of its entities. A process is defined as a time-dependent relation, that changes the state of a system (Ackoff, 1971). A subsystem is an element of a larger system which fulfils the conditions of a system in itself, but which also plays a role in the operation of a larger system (Young, 1964). There is no sharp distinction between models and systems. A commonly used definition is that a system is a subjective reflection of the researcher's observations and that there are therefore as many interpretations of a system as there are observers

(Kramer and Smit, 1991). In this view, models and systems are almost identical. We will use the word model if the emphasis is on the conceptual or formal description.

A particularly useful technique for describing systems composed of many interacting entities and feedback loops, is system dynamics (Forrester, 1961, 1968; Goodman, 1974; Randers, 1980). A key distinction is between state variables (stocks, reservoirs) and the rate variables (flows, fluxes) which connect them. Rate variables not only represent phsyical fluxes but also information flows. We will use systems theory in combination with some basic principles of system dynamics to describe the human-environment system on a global scale. We distinguish the following set of variables and parameters to describe a system (see e.g. Morgan and Henrion (1990)):

- *state variables:* represent the state of a system at an arbitrary point in time (stocks, reservoirs);
- *rate variables:* represent processes which relate the various state variables to each other (flows, fluxes);
- *auxiliary variables:* input or help variables and parameters;
- *response variables:* variables and parameters that represent the options available to decision makers to influence the system's behaviour;
- *empirical quantities (observables):* variables and parameters that represent measurable properties of the real-world system being modelled;
- *value quantities:* variables and parameters that represent preferences of the decision makers, researchers or modellers.

These classes are not mutually exclusive. Quantities classified as response variables or state variables may also be empirical quantities or value quantities. Nevertheless, this classification makes sense because it signifies the role a quantity plays in the model, or it indicates the type of uncertainty it is related to. This classification can therefore support a correct treatment of calibration and uncertainty issues. Another useful distinction is between endogenous and exogenous processes. A process is called endogenous when it is fully part of the (sub)system description; it is called exogenous when its enters the model as an input, e.g. as a time-series or an empirically based relationship between two variables. Processes which are exogenous to the system description demarcate the system boundary.

Cause-effect chains
Another organising concept we choose for operationalising global change and sustainable development is the cause-effect chain which is also the basis for the PSIR framework discussed in more detail in section 2.3. This concept reflects the observation in environmental research that material flows in the human-environment system are best understood in terms of interacting chains of causes and effects at the level of compartments (RIVM, 1991). Interconnected cause-effect chains form an organised whole, a complex system the properties of which are more than the sum of

> **System concepts**
>
> *System*
> Representation of a part of reality that forms a coherent whole of entities.
>
> *Reservoir (Stock)*
> A cluster of entities (elements, components) that is considered to be homogeneous.
>
> *Flow (flux)*
> A flux is the quantity of entities being transferred from one reservoir to another in a unit of time. The flux density is defined as the quantity of entities transferred per unit of time per unit of area.
>
> *Source*
> A source is a reservoir from which fluxes of entities originate.
>
> *Sink*
> A sink is a reservoir, which receives fluxes of entities.
>
> *Steady state*
> If all sources and sinks balance and do not change over time, a reservoir is in a steady state.
>
> *Feedback*
> A negative feedback is a process that, after completion of a cycle, suppresses the output signal caused by the original change in the system's state, while a positive feedback reinforces the output signal.
>
> *Circulation time*
> The circulation time is the time it takes to empty the reservoir in the absence of incoming fluxes.
>
> *Response time*
> The response time of a reservoir is the time it takes to reach a new steady state after a sudden change in the system.
>
> *Reference state/Target state*
> The reference state is the 'undisturbed' condition of the system(s); the target state is the 'desirable' state to be achieved.

its constituent parts[1]. From a systems perspective, the interdependencies between the cause-effect chains are most important.

Metamodels

Although some serious attempts are being made to construct an integrated, multi-dimensional model of the Earth's atmosphere, hydrosphere and terrestrial biosphere (Fischer, 1988; Krapivin, 1993), it is conceptually and technically not yet possible

---

1 See for a detailed discussion of some interconnected cause-effect chains: Rotmans *et al.* (1994), Swart and Bakkes (1995).

to link, let alone integrate a variety of complicated, detailed, three-dimensional models. Apart from computational limitations, integration of the relevant expert models will not provide a transparent integrated framework which allows for systematic and comprehensive analysis. It is therefore often more appealing to make use of metamodels for each component of the integrated assessment model. Metamodels, or reduced form models, are highly aggregated, simplified representations of an original model which is more detailed and referred to as the expert model. An important requirement of a metamodel, besides its flexibility and transparency, is that its structure and behaviour has been validated extensively against the original expert model(s) and against empirical data over the relevant range of values[2]. Sometimes, the metamodel consists of only one or a few equations. Such a metamodel then is an aggregate description of a set of often complex processes, based on correlations among key observables. The integrated assessment model then consists of a suite of interlinked metamodels. Although the larger, integrated model is simple at the level of metamodels, it may show complex behaviour and structure through the interrelations and feedback mechanisms which arise by linking the submodels.

2.3 Pressure-State-Impact-Response (PSIR)

The human-environment system can be seen as a set of reservoirs and fluxes connecting the reservoirs, representing interrelated cause-effect chains. In order to structure (sub)systems, the Pressure-State-Impact-Response framework (PSIR) is used. It is similar to the Pressure-State-Response framework developed by the OECD (1993) to structure environment-related indicators. However, the 'state' part of the OECD approach is split into State and Impact in order to accommodate information about changes in the various functions of (sub)systems from changes in the state of a system. The PSIR mechanism can be applied at each level of aggregation. *Figure 2.2* shows the application at the highest aggregation level. At a lower aggregation level, the subsystems are broken down recursively into more detailed subsystems, using the PSIR as organizing principle where possible.

Although we realise that the framework represents only one characterisation of the global change issue, it is based upon a plausible division of the cause-effect chains of global change into the following subsystems:

---

[2] An example of a suite of interlinked metamodels to describe the energy/economy/climate system is the IMAGE 1.0 model, which has been extensively analysed as a mathematical system (Braddock, 1994; Rotmans, 1990). An example of a metamodel for which the validity domain has been extensively tested is PowerPlan, a model designed to simulate electric power generation (Benders, 1996; de Vries *et al.*, 1991)

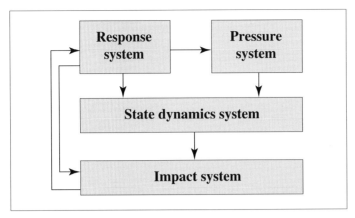

Figure 2.2 The Pressure-State-Impact-Response (PSIR) framework.

- *Pressure:* represents social, economic and ecological driving forces underlying the pressures on the human and environmental system;
- *State:* represents physical, chemical and biological changes in the state of the biosphere, as well as changes in human population and resource and capital stocks;
- *Impact:* represents social, economic and ecological impacts as a result of human and/or natural disturbance;
- *Response:* represents human intervention in response to ecological and societal impacts.

The PSIR framework can be conceived of as a dynamic cycle. Socio-economic developments exert pressure on the environment and, as a consequence, the state of (parts of) the environmental system changes. These changes have impacts on the socio-economic functions of the environment, sometimes indirectly through environmental functions. Next, these impacts elicit a societal response in the form of actions that feed back upon the driving forces, or directly upon the state or the impacts, through adaptation or curative rather than preventive action. For instance, demographic and consumption pressures lead to changes in land use and energy consumption, which in turn cause environmental and economic impacts which may affect the future evolution of both demographic and consumption patterns. Obviously, the PSIR framework is relative: a pressure in one of the (sub)systems may be an impact in another and vice versa. Its application depends on the point in time and space, and the scope of the analysis chosen. The PSIR components are interchangeable and form a dynamically interwoven pattern in such a way that there is no beginning and no end, constituting an inextricable continuum of interconnections.

Example of 'figure of eight'
As an illustration of the PSIR chains and their relative nature, *Figure 2.3* gives an example of the kind of chains discussed in Chapters 17 and 18. The chain is presented as a lemniscate or 'figure of eight'. Population and economy represent the pressures which cause a demand for food and energy. This affects the state of the system: it requires fertiliser which leads to an accumulation of nitrogen (N) in soils and water; it also causes depletion of high-quality oil and gas occurrences and the released carbon dioxide (CO_2) accumulates in the atmosphere. This in turn results in impacts: water quality deteriorates and the global average temperature and sea level are expected to rise. This and the resulting climate change can influence mortality and morbidity which affects life expectancy and hence population growth. It will also affect the economy. There will be responses to this, for example, water and energy prices will go up, which will in turn change demand.

The relative nature of the PSIR chains can be seen from this illustrative example. The demand for food and water are pressures within the respective subsystems; the flows of nitrogen and the carbon dioxide emissions are impacts. For the element cycles submodel these emissions are pressures. Similarly, the change in water quality is an impact which can be responded to with more waste-water treatment; at the same time rising water supply costs are an impact, too. Temperature and sea-level rise are pressure components in the water and the Population and Health submodels.

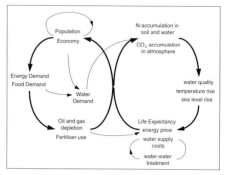

Figure 2.3 An example of a lemniscat or 'figure of eight'.

From a policy perspective, it is useful to distinguish two different levels for the PSIR cycles: the micro and macro level, as shown in *Figure 2.4*. In modelling parts of the world system, one constructs from experiences in the real-world aggregate, abstract classes of physical objects. The collection of elements in such classes are associated with a system state. Changes in state refer to inflows from outside into the system or outflows out of the system, or, if there are subclasses within a class, transitional flows between subclasses. The rate equations behind these changes in the system state are based on gradients within the system which generate driving forces. In natural systems, these gradients are mostly physical. In human systems, the gradients often take the form of a difference between the actual or perceived and the desired characteristics of the system state[3]. These gradients lead to the system's natural response at the micro-level. The driving forces, which lead to additional interference with natural cycles, generate the micro-pressures on the system.

In a macro-context, these driving forces, the changes in the system state and the resulting impacts and responses constitute the PSIR chain. The micro PSIR is the system's autonomous dynamics. The driving forces resulting from people's actions are associated with pressure at the aggregated level: micro-pressures are translated into macro-pressure. Similarly, the system's state and impacts are interpreted and

---

3 Anthropologists as well as sociologists and economists may argue that there is no difference: the human subsystems are also governed by laws.

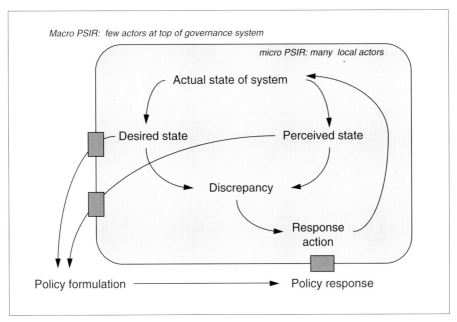

Figure 2.4 The PSIR cycle at both the micro and macro level.

evaluated at the aggregate level of the macro PSIR. It is in this sense that 'policy' can best be understood as a form of collective (government) interference with the autonomous system dynamics of the local level.

2.4 Integration

Since interests have become more interwoven and problems more complex, a trend from problem-specific analyses towards more integrated and more interdisciplinary research is taking place. This has resulted in comprehensive and elaborate computational frameworks, linking (simplified versions of) originally separate models. Although these computational frameworks have proved to be very useful, they often provide a poor basis for exploring and formulating coherent policies because they lack integration. We have endeavoured to perform an integrated analysis, and we have developed concepts for it. Some of these have been introduced in the previous section. Here, we discuss some other aspects of integration as applied in the TARGETS approach.

Horizontal and vertical integration

A first distinction is between vertical and horizontal integration. Vertical integration is based on the causal chain. The idea is to close the PSIR loop, linking a pressure to

Examples of integration

Vertical integration. A vertical integrative approach to the energy subsystem implies the following steps. Past and future activity levels and the resulting demand for heat and electricity are the main pressures. To meet these demands, capital stocks are built up for coal, oil and gas production and for electric power generation. These reflect the state of the energy system. Combustion of fuels results in emissions of, for example, carbon, sulphur and nitrogen oxides which are associated with environmental impacts. Response measures then encompasses decisions with regard to technology and investments, as well as measures addressing consumer behaviour.

Within the energy sector the present and future role of transportation, power generation and other combustion processes is considered. By doing this, the question can be addressed, as to whether the global energy system will be able to cope with the social and economic effects of significant shifts in supply patterns (e.g. penetration of renewables in the market) and changes in the level of demand (e.g. less transportation).

Horizontal integration. This can be illustrated by the example of integration of the various biogeochemical cycles. Horizontal integration here means that the interactions and feedbacks between the various global biogeochemical cycles are taken into account by coupling the various compartments: the atmosphere, terrestrial biosphere, lithosphere (soils), the hydrosphere (oceans; lakes; river basins) and the cryosphere, and subsequently integrating the physical, chemical and biological interactions between the diverse compartments.

Full integration. Full integration for land use for instance, involves the interplay between population growth, economic growth, land management and food supply (pressure factors); the biogeochemical changes in the land system (state dynamics); and the impacts on global and regional climate, and the impacts of changes in food supply conditions on human health, agriculture and economics. By analysing this loop, integrated land and soil management strategies can be explored.

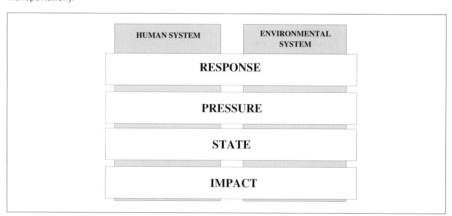

Figure 2.5 Interlinkages between the human and the environmental system.

a state, a state to an impact, an impact to a response and a response to a pressure. The different issues are analysed along the complete cause-effect chain, considering not only the state dynamics of the system, but also the effects of the pressures on the system and the performance of the socio-economic and ecological functions of the system. Horizontal integration addresses cross-linkages and interactions between pressures, states, impacts and responses for the various subsystems distinguished in the integrated model. It is intended to shed light on the interrelationships between common pressures, common state changes and common impacts. Integrating horizontally leads to a focus on interactions between the elements of the pressure vector (such as population and economic growth), of the state vector (concentrations, land-use changes, etc.), of the impact vector (such as malnutrition and vector-borne diseases), or of the response vector (investments, taxes, education, etc.). The horizontally integrated response subsystem represents an imaginary single global governmental actor, who allocates financial resources and performs other interventions through which the human and environment system can be influenced. Total integration means that various pressures are linked to various states, to various impacts and to various responses. In the realm of integrated assessment modelling, many hybrid forms of integration are found. So far, most integrated assessment modelling approaches have focused on vertical integration. Combining this with the distinction between human and environment systems is the basis for the template for the TARGETS approach shown in *Figure 2.5*.

Complexity and interdisciplinarity
A second aspect of integration is that it should bridge what is usually referred to as the domains of natural and social sciences. A useful approach is to distinguish three levels of complexity which differ with respect to the degrees of freedom of the system elements and, partly as a consequence of this, with respect to the nature of our knowledge about them (*Figure 2.6*). The relevance of distinguishing these three levels is that it allows an explicit discussion of the concepts and methods used in integrated assessment, and their differences.

The first level consists of physical reservoirs and flows which correspond partially to observable reality. Model variables, at this level, usually have an explicit

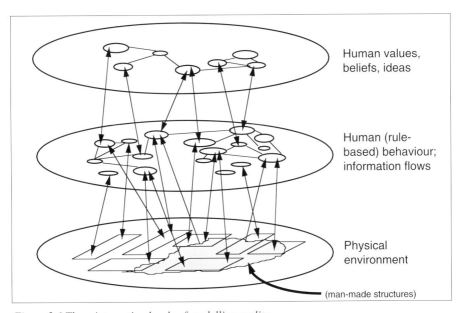

Figure 2.6 Three interacting levels of modelling reality.

and formal correspondence with real-world observable phenomena. At least in principle, the laws of physical, chemical and biological science hold, e.g. conservation of mass and energy. The next level maps the behavioural and informational structures which govern human interference in the underlying physical environment. Such behaviour is described by information-dependent sets of rules. The rules describe actors, varying from individuals to multinational companies and institutions. Models of actors are usually meta-relationships based on correlational analysis of a limited sample of data. The third level comprises values, beliefs and ideas that reflect and rationalise people's behaviour. Policy issues arise at this highest level, so that at this level the design of macro-oriented policies enters the scene. Generally speaking, this level is merely included in models in the form of response variables chosen *ex ante*. The normative dimension and decision-making processes, also at this level, are not – and mostly cannot be – included in quantitative models. We need to search for complementary methods to address those aspects and nuances of human values (Pahl-Wostl *et al.*, 1996).

These three levels of complexity are of course only a simplified representation of a continuous spectrum. Its use, however, may help to communicate that 'strong' science, generating statements on the basis of controlled experiments, only covers a limited domain of the physical environment and an even smaller part of the levels of behaviour and values. Many of the controversial issues related to global change, for example, are rooted in limited understanding or ignorance about certain physical phenomena. This gives rise to quite distinct interpretations of observations about the physical environment, thereby supporting conflicting models of how this part of the world functions. Such uncertainties may be resolved as science proceeds. However, there will always be competing explanations of real-world observations which in turn can be used to support one's behaviour and one's beliefs, values and preferences, especially at higher levels of complexity. This aspect is dealt with in more detail in Chapter 10.

Another facet of this complexity axis is that it allows an explicit discussion of the concepts and methods – as well as their differences – used in the natural sciences and in the social sciences. The former use the techniques of differential and integral calculus to describe physical and chemical processes in environmental compartments, but they also have to deal with uncertainties as soon as the applications are outside the realm of controlled experiments. The latter are used to large uncertainties in describing (human) behaviour and have often employed models from the physical sciences as analogues for the construction of hypotheses (de Vries, 1989). The science of ecology is somewhere in between the first and the second level of *Figure 2.6*, which has given it a great heuristic role in modelling global change (Clark and Holling, 1985). In the last decades, the search for new methods and approaches to bridge the gaps has intensified. Systems dynamics, applied general equilibrium and actor-oriented models and cellular automata and genetic algorithms are some of the tools that have been applied more recently.

2.5 Presentation

User interface

Integratedness and complexity are partly in the eyes of the beholder. We venture that a model like TARGETS may actually make its developers and users more aware of the complexity of global change phenomena. In concrete terms, the models described and used in this book can be presented and interpreted in many different ways, depending on the disciplinary background of the user. The environment system and the corresponding submodels, for instance, can be perceived in terms of environmental compartments, chemical substances, policy themes or from a mathematical perspective. Therefore, in order to address the needs of diverse user groups, the framework should be geared to the users' perception of global change. It is for this purpose that the modelling and visualisation tool 'M' has been developed to facilitate the flexible visualisation of different system representations.

In a broader sense, one would like to use the model as a tool to explore global change phenomena from quite different points of view. This is highly relevant. Issues associated with global change are distinguished from familiar scientific problems in that they are universal in scale and long term in their impact. Moreover, the available data are often inadequate and the phenomena, being novel, complex and variable, are themselves not well understood (Funtowicz and Ravetz, 1993). Hence, there are numerous valid perspectives from which one can describe the human-environmental system and assess future developments. This should be reflected in the model's user interface.

The 'M' model environment

Signifying the need for interactive, easy-to-use and well-documented versions of simulation models, RIVM (Netherland Institute of Public Health and the Environment) has developed a new modelling, simulation and visualisation tool called 'M' (de Bruin *et al.*, 1996). 'M' provides models with 'instrument panels' on which model variables are presented as visual objects such as sliders, graphs or maps. Model inputs can be manipulated directly, and cause immediate recomputation of all output graphs. As all intermediate results are stored, users can gain further insight into the system dynamics by replaying or stepping through former experiments. Context-dependent documentation provides background information and can guide users through the model.

Since models will be used by different users and for different purposes, 'M' gives developers the opportunity to create different instrument panels (so-called 'views'), each providing a different representation of the same underlying model. A view can consist of a flat panel just showing the main scenario inputs and model outputs; it can also be set up as a block diagram in which model variables are organised hierarchically and showing interdependencies. Users can select one or more views and explore these by opening, zooming in and out, and closing components. Users can easily reorganise or mix views and import model variables which are not shown. Scenarios can be saved and exported to other tools.

Indicators

An indicator characterises the status and dynamic behaviour of the system concerned, whereas an index is a multi-dimensional composite made up from a set of indicators. Indicators may serve as vehicles for the communication of model results, according to which response strategies can be mapped out. The use of indicators is necessary because the plurality of model variables forms an opaque web of possible choices, which is too complex for users of the model too handle. To ensure an adequate and structural selection of relevant model options, indicators are indispensable.

We propose to link a set of indicators to the conceptual framework described above. The main advantage of linking a set of indicators to such a framework is that it yields insight into the dynamics of the system concerned. This enables the production of coherent information about linkages between causes and effects (vertical integration) and makes it possible to address the cross-linkages between various issues (horizontal integration). In Chapter 9, a hierarchical framework of indicators with different layers of aggregation, varying from measurable quantities to highly aggregated indices, is presented.

2.6 Transitions

Another interesting concept that can be used to investigate human-environment interactions is the concept of transitions. A transition is defined here as a process during which, in a specific period of time, a phenomenon undergoes a shift from slow to rapid change followed by a return to relative stability. The main aspects of a transition are: (i) it is a shift from one relative equilibrium to another, moving from slow to rapid change, before returning to a different level of stability; (ii) the characteristics of the new equilibrium may differ from those that preceded it; and (iii) stability is relative and does not necessarily mean permanence. For each transition, there is a critical period when society is especially vulnerable to damage. During that period, rates of (autonomous) change are high, and adaptive capacity is often limited, and there is a higher chance that the system will become severely imbalanced. From a modelling perspective, during a transition process one or more state variables start to grow or decline and move away from the initial situation of dynamic equilibrium. Other variables will change too, either as a cause or as a consequence. Positive feedbacks tend to accelerate this process of change, but after some time negative feedbacks become stronger and the system tends towards a new state of dynamic equilibrium. One has to be careful applying this concept – rooted in population biology and demography – to real-world situations. Yet, we feel it can offer an important heuristic function.

In this book, the concept of transitions is used, in combination with the PSIR concept, to describe some aspects of the complex dynamics between population, economy and environment. The major pressure components are within the population and economy. These systems experience a demographic transition, in which a decline in fertility follows the decline in mortality but with a delay, and an economic transition in which human activities are increasingly resource-extensive and knowledge- and information-intensive (Drake, 1993; Ness et al., 1993). These two driving forces lead to a rising demand for food, water and energy, to mention but the most important ones. In the provision of these one can discern transitions, often within broader transitions which belong to the core of past technological developments (Tylecote, 1992). For example, there is well-documented evidence that energy use per unit of economic activity initially rises after which it declines and eventually stabalises. Other characteristics which coincide with growing population and welfare are more widespread access to safe water, an increasing share of animal-based food and rising relative health expenditures. All three may be part of transition processes which are almost finished in some regions of the world and may be underway in other regions. In a broader context, they are all part of the development transition[4]. Environmental problems emerge from these development transitions, which may give rise to response actions which then may become the second stage of what has been called an ecological transition (Baldwin, 1995; Grossman, 1995). The description of changes in the population-economy-environment system as a family of interconnected transitions on a global scale links up with widely observed and accepted transitions within the human system, such as the demographic transition and the epidemiological transition. It may help to gain insight in the complex dynamics behind these changes and to investigate their integrated nature. Experiments with the TARGETS model are performed and presented in Chapters 17 and 18, which aim to reproduce and illustrate the various transitions.

---

4   This is not to say that such a development transition is a rigid pattern for which the developments in the present OECD countries represent the only or even a typical template. Dystopian futures usually involve transition failures (see Chapter 18).

The TARGETS model 3

"In theory, theory and practice are the same. In practice, they aren't."

3 THE TARGETS MODEL

Jan Rotmans, Marjolein B.A. van Asselt, Bert J.M. de Vries, Arthur H.W. Beusen, Michel G.J. den Elzen, Henk B.M. Hilderink, Arjen Y. Hoekstra, Marco A. Janssen, Heko W. Köster, Louis W. Niessen, and Bart J. Strengers

When tackling a subject as complex as global change and sustainable development, it is essential to be able to 'frame the issues'. This was one of the main reasons for developing the TARGETS model, an integrated model of the global system, consisting of metamodels of important subsystems. In this chapter we introduce TARGETS. Building on the previous chapters, we elaborate on the possibilities and limitations of integrated assessment models. Some of the key issues discussed are aggregation, model calibration and validation, and dealing with uncertainty.

3.1 Introduction

One of the main tools used in integrated assessment of global change issues is the Integrated Assessment (IA) model. This chapter introduces such an integrated model, TARGETS, which builds upon the systems approach and related concepts introduced in Chapter 2. Previous integrated modelling attempts either focused on specific aspects of global change, for instance the climate system (IPCC, 1995), or consisted merely of conceptual descriptions (Shaw *et al.*, 1992). We have tried to go one step further, linking a series of cause-effect chains of global change. Although we realise the shortcomings in our current version of the TARGETS model, we felt there was a need to present our model to a wide audience. We first give some advantages and limitations of IA models. Next, we discuss issues of aggregation, calibration, validation and uncertainty. We proceed with a brief description of the five TARGETS submodels which coincides with the PSIR concept and the vertical integration as introduced in Chapter 2. A more detailed description of these submodels is given in Chapters 4 to 8. Then, we discuss the horizontal integration of the submodels and the cross-linkages between them.

3.2 Integrated assessment modelling

Background
Current projects in IA modelling build on a tradition started in the early 1970s by the Club of Rome (Meadows *et al.*, 1972). This first generation of IA models, the so-called global models, focused on resource depletion, population and pollution. Over the past twenty years, numerous global models have been built (Brecke, 1993; Toth

et al., 1989), most of which were rather complicated, highly aggregated and partially integrated. The next generation of IA models addressed specific environmental issues. Examples are the RAINS model developed in the early 1980s (Hordijk, 1991), the IMAGE model (Alcamo, 1994; Rotmans, 1990), the DICE model (Nordhaus, 1992), the PAGE model (Hope and Parker, 1993) and the ICAM1.0 model (Dowlatabadi and Morgan, 1993a; 1993b). The development of a new generation of IA models is now under way. They focus on and benefit from recent findings in such divergent fields as ecosystem dynamics, land-use dynamics and the impacts of climate change on human health and water resources (Rotmans *et al.*, 1996).

It is important to point out that IA models of global change are meant to frame issues and provide a context for debate. They analyse global change phenomena from a broad, synoptic perspective. One of the challenging aspects of building such a model is to find the right balance between simplicity and complexity, aggregation and realism, stochastic and deterministic elements, qualitative depth and quantitative rigour, transparency and adequateness. It is essential to keep in mind the limitations of models like TARGETS and to recognise the kind of issues and questions that can *not* be addressed or are *beyond* the scope of the model.

Value and limitations

Any attempt to fully represent the human and environmental systems and their numerous interlinkages in a quantitative model is doomed to failure. Nevertheless, we maintain that even a simplified but integrated model can provide a useful guide to global change and sustainable development and complement highly detailed models of subsystems that cover only some parts of the phenomena. Among the major advantages of IA models are:

- *exploration of interactions and feedbacks:* explicit inclusion of interactions and feedback mechanisms between subsystems can yield insights that disciplinary studies cannot offer. It can indicate areas of promising new and interdisciplinary research, and also of the potential range and magnitude of global phenomena and of the scale of the interventions needed to counteract or mitigate undesirable aspects;
- *flexible and rapid simulation tools:* the simplified nature and flexible structure of submodels in IA models permit rapid prototyping of new concepts and scientific insights and the indicative simulation and evaluation of long-term scenarios and strategies;
- *coherent framework to structure present knowledge:* by consistently representing and structuring current knowledge, major uncertainties can be identified and ranked. Crucial gaps in current scientific knowledge and weaknesses in discipline-oriented expert models can be identified.
- *tools for communication:* because of their 'umbrella' function these models can be outstanding tools to communicate global change phenomena within the scientific community and between scientists, the public and policy makers and

analysts. Their simplicity enables a transparency which is one of the preconditions for effective communication and debate.

Obviously, IA models also have limitations and drawbacks. Some of these are just the negative side of the above-mentioned advantages; others have to do with current limitations in computer modelling. In our view, the most important ones are:

- *high level of aggregation:* many processes within the human-environment system occur at a micro level, far below the spatial and temporal aggregation level of current IA models. Parameterisations are used to mimic these processes at the scale and aggregation of the model. This may cause serious errors, as is discussed in the next section;
- *inadequate treatment of and cumulation of uncertainties:* by trying to capture the entire cause-effect chain of a problem, IA models are prone to an accumulation of uncertainties. This, together with the variety of types and sources of uncertainty that IA models comprise, makes an uncertainty analysis for integrated frameworks rather difficult;
- *absence of stochastic behaviour:* most IA models assume that real-world processes can be described in terms of continuous, deterministic mathematical equations. In reality, many processes are stochastic by nature. The resulting extreme conditions may exert significant influence on the overall long-term dynamic behaviour of the system; hence they may play a decisive role even though their occurrence has a low probability;
- *limited calibration and validation:* one of the most vexing aspects of modelling a complex, global system is the absence of real-world observations which allow for rigorous model validation. The high level of aggregation, the dynamic, long-term nature of the model and the high level of complexity of the subsystems and their interactions often implies an inherent lack of empirical variables and parameters. If one can identify relevant data, the available sets are often too small and/or unreliable to apply a thorough calibration and validation procedure. We come back to this in the next section.

In designing, constructing and using a model like TARGETS, there are a number of pitfalls. One of them, on the side of the designers, is that familiarity with particular formalisms, e.g. with optimisation techniques, may lead to an 'availability bias' which imposes restrictions on how the problem is formulated and solved. Another pitfall on the side of the user is to consider the IA model as a 'truth machine' rather than as a tool to understand the issues (Wynne and Schackley, 1994). This easily leads to vigorous but rather pointless debates, as the history of the World3-model has made clear (Freeman, 1973; Peccei, 1982; Meadows *et al.*, 1991).

Integrated models of global change attempt to offer an overall picture of those processes that are causally relevant for understanding global change phenomena.

They are by no means comprehensive. After all, there are no entirely reliable models of the underlying processes, and the integration effort inevitably simplifies such models. In our view, the interpretative and instructive value of an IA model is far more important than its predictive capability.

Model set-up

To describe and model the complex global system, we constructed a set of metamodels which have been linked and integrated. This resulted in the **TARGETS** model: **T**ool to **A**ssess **R**egional and **G**lobal **E**nvironmental and Health **T**argets for **S**ustainability. It consists of five submodels: the population and health, the energy submodel, the land and food, and the water submodel, and the submodel describing the biogeochemical element fluxes ('cycles'). These submodels are interlinked and related to the economic scenario generator. Within each subsystem – and submodel – we distinguish pressure, state, impact and response modules. These represent a vertically integrated cause-effect chain. From the point-of-view of horizontal

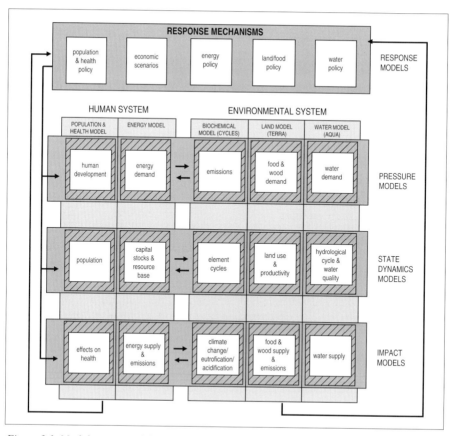

Figure 3.1 Modular set-up of the TARGETS model.

integration, the TARGETS model can be conceived of as pressure, state, impact and response components in each of which the five submodels are linked. Vertical and horizontal integration aspects lead to a representation of the TARGETS model as shown in *Figure 3.1*.

Although the main advantage of the TARGETS model is its integrated character, each submodel has been constructed and can be used independently. Variables which link the submodels are then introduced as exogenous inputs. A first argument for this is that the submodels had to be constructed and implemented in a stand-alone context to allow comparison with other modelling efforts (expert models) in the field. It is also necessary for model calibration and validation and it makes it possible to address submodel-specific issues. A second argument is that the added value of integration can only be evaluated against a background of non-integrated simulation experiments. Moreover, only a thorough understanding of individual submodel behaviour can lead to insight into the consequences of specific links between the various submodels. In Chapter 11 we outline the way in which integrated and non-integrated model experiments have been set up. Detailed simulation experiments with the stand-alone versions of the submodels are presented in Chapter 12 to 16. Before giving a brief description of the various submodels, we deal with a few issues which are relevant to the design, construction and use of a model like TARGETS.

3.3 Aggregation, calibration and uncertainty

Aspects of aggregation
The elements of a system usually have a wide variety of characteristics, among them location in space. They are involved in all kinds of dynamic processes, often with quite different time-scales. An additional problem is that these models consist of a variety of submodels, each of which differs with respect to feasible and desirable levels of aggregation, complexity, and spatial and temporal resolution. Hence, aggregation is a crucial issue in model design. Which classes are distinguished and which spatial and temporal resolution are chosen for model variables?

In general, the answer to the above question is – or should be – based on the purpose of the model. The TARGETS model has been set up as a generic framework so that, in principle, it can be applied at different levels of aggregation. Our first objective is to develop a quantitative, transparent tool to explore the long-term dynamics of the world system and present 'the larger picture'. Hence, we have implemented the model in the first instance for the world as a whole. At this high aggregation level one can easily include the more speculative interactions between the human and environmental system and search for their relevance in the context of global change. At lower levels of aggregation, this is often a tedious task.

The different levels of temporal and spatial aggregation give the integrated model an 'hourglass' structure. For example, economy-energy models usually operate in

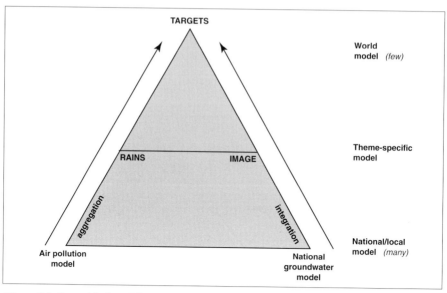

Figure 3.2 Pyramid of models according to different levels of aggregation and integration.

multi-year time steps with national or regional political boundaries. However, atmospheric chemistry models operate in small time steps on a small scale, while climate models have a relatively coarse spatial resolution but run at a fine temporal resolution. Ecological impact models generally require data at fine spatial resolutions but their time resolution varies greatly, from one day to a season or a year. The submodels of the TARGETS model have been set up as generic metamodels which can be calibrated and applied at several levels of temporal and spatial aggregation. Applying them at the global level, as in the TARGETS1.0 model[1], demands a coordinated choice with regard to time-scales, spatial scales and attributes. In the TARGETS1.0 model, both the distributions on a scale below the temporal resolution and the heterogeneities below the spatial resolution level of the model are dealt with by introducing classes and spatial distribution functions.

In *Figure 3.2* different types of models developed by or used at RIVM are categorised along the vertical axis of a pyramid, which indicates the level of aggregation and integration. The TARGETS1.0 model is placed on top of this pyramid, because it has the highest level of aggregation and integration. The high aggregation level makes it an appropriate tool to frame issues regarding global change and sustainable development but it cannot help in formulating what these mean at the regional or local level of a city or a country. Theme-specific integrated

---

[1] Henceforth, we refer to the current, global version of the TARGETS model as the TARGETS1.0 model.

Genericity of metamodels

In setting up the metamodels which provide the building blocks for the integrated model, we have attempted to model the subsystems as generic (or universal) as possible. Genericity, here, means that the model represents the subsystem dynamics in such a way that it is a valid description at different levels of spatial and temporal detail. Hence, concepts, hypotheses and theories used should be applicable at different levels of spatial aggregation and for different regions in different periods. Such a genericity is only possible up to a certain point. One limitation is that aggregate global, slow dynamics may be driven by local, fast processes in a way that cannot be covered in a metamodel – and yet may turn out to be crucial. In fact, the question of genericity of certain relationships is one of the key uncertainties we are faced with.

Because a highly aggregated approach like in the TARGETS1.0 model lacks specific regional/local dynamics, genericity of (sub)models can only partly be realised. This holds for all three levels of complexity shown in *Figure 3.2*. For example, for a commodity like oil in an increasingly free trade context, depletion and technological innovation with regard to exploitation of the resource base can be dealt with adequately at the global level.

However, modelling the import and export flows of oil and their impacts on economic performance requires desaggregation to the level of economically, politically and institutionally relevant actors and dynamics. In our research we have explored the validity domain of several submodels by implementing and parameterising them at lower aggregation levels, e.g. a river basin for water and a country for population and energy. Through these applications we have gained an understanding of the problems one may expect upon aggregating a metamodel derived from local observations to a globally aggregate description.

models such as RAINS and IMAGE, with more spatial and process detail, are better equipped for such questions, but even these models cannot be used to provide detailed spatial descriptions or specific policy proposals. For such purposes, one has to rely on expert models of small populations, environmental compartments and the like which are also the basis for the metamodels. Many of such models are being used for the Dutch National Environmental Outlook (RIVM, 1994). In the process, one may gain scientific quality but lose relevance because the changes in the external variables dominate the dynamics of the modelled system. The actual choices with regard to spatial aggregation in the global TARGETS1.0 model are presented in *Figure 3.3*.

Time

The various subsystems which make up the larger system described by the TARGETS1.0 model are characterised by dynamics with specific time-scales. Economic processes and the related pace of technical change are to a large extent governed by the operational lifetime of the different capital stocks. Within the food and water supply system, similar time-scales are relevant but they are imbedded in the much slower dynamics of processes like soil erosion and groundwater recharge. For the biogeochemical element cycles relevant time-scales range from months for atmospheric processes to hundreds of years for the dynamics of oceans. The time step used within the TARGETS1.0 model is one year, although in some modules a smaller time-step is used. Some physical processes require a time step of one month, such as the hydrological processes in the water submodel (Chapter 6), or of one season, e.g. the biogeochemical processes in the cycles submodel (Chapter 8). The time horizon for the TARGETS1.0 model spans two centuries. All simulation experiments start at the beginning of this century, in the year 1900, which can be thought of as the beginning of the industrial era. The simulations end in the year 2100, which is three to four generations away from people living today. With this time horizon, we look as far ahead as we look back.

Space

In all subsystems but especially in those which deal with the reservoirs and flows of elements and the dynamics of land and water use, there is a distinct spatial heterogeneity. We utilise aggregated data and processes derived from models such as

Spatial categories in the TARGETS submodels

In the land and food submodel (TERRA) spatial heterogeneities are introduced by disaggregation into specific classes for soil, climate and land use. Seven land-cover types have been distinguished. 'Degraded land' is not shown in *Figure 3.3*, because it is negligible in 1900. Next, the land-cover types have been disaggregated further into the following classes (Chapter 7):
- two economic or temperature zone classes (developed and developing);
- three length of growing period (LGP) classes;
- three inherent soil productivity (Q) classes.

In the water submodel (AQUA), a total of ten water reservoirs are distinguished, three of which are groundwater stocks (for more details, see Figure 6.5). The water reservoirs which are most important for the biosphere: fresh surface water, soil moisture, biological water and renewable fresh groundwater, are a small proportion of the total. In the element cycles submodel (CYCLES), we use the same classes as in the TERRA submodel for the terrestrial biosphere. The oceans are modelled as seven separate layers. The atmosphere is represented as a single, uniformly mixed reservoir.

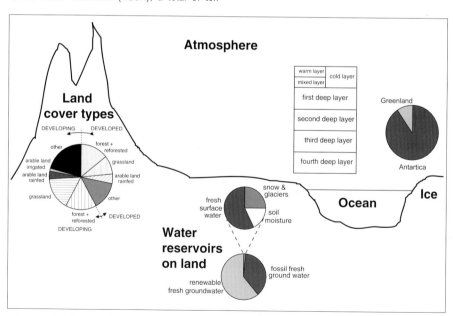

Figure 3.3 Schematic overview of the initial disaggregations used in the TARGETS1.0 model.

the IMAGE2.0 model (Alcamo, 1994) and from the HYDE data base (Klein Goldewijk and Battjes, 1995). Details are given in *Figure 3.3* and Chapters 4 to 8.

Other attributes

One also has to aggregate with respect to reservoir attributes other than location in time and space. Ideally, this should be based on an understanding of the system at the micro-level. For example, if the diet or provision of clean water is below the level of people's aspiration, this induces behaviour to improve the situation. The resulting actions are inherently non-equilibrium processes, which influence the state of the local system and thereby lead to impacts and responses. The crucial question here is how to scale up and down between the macro and micro levels. For example, with regard to demographic developments, there is extensive knowledge of the determinants of fertility and diseases at the individual level. What is usually done is to combine this knowledge in clusters (such as disease clusters and population cohorts) and in aggregated indicators (such as the total fertility rate and disability-adjusted-life-years), which can be used at the level of populations.

In practice, the choice to disaggregate will depend on the model objectives and the kind of questions the model should be able to address. In the TARGETS1.0 model, the human population is disaggregated into five age cohorts and a corresponding number of disease burden classes. Fuel producing capital is represented by four distinct capital stocks, based on the distinction between solid, liquid and gaseous fuels. For water quality, four classes have been formulated. Other causative factors, like food availability, access to safe water and income distribution for the human population, are not related to subclasses. More details are given in Chapters 4 to 8.

Calibration and validation

There are many definitions and interpretations of the terms calibration and validation. Complete calibration and validation of models for a system as large and complex as the Earth is impossible, because the underlying systems are never fully closed (Oreskes *et al.*, 1994). Within the TARGETS1.0 model calibration is defined as a procedure which gauges the most important parameters in such a way that the model simulations come close to the observations.

Validation is defined here as a procedure for testing the adequacy of a mathematical model. There are two types of validation. The first is practical validation. It is done by comparing model outcomes with other model-based research or simulating a period different from the one used for calibration. All submodels have been calibrated for the period 1900-1990 and have, in a limited way, been practically validated using parameter sensitivity analyses (Chapters 4 to 8).

A second type of validation is conceptual validation, which tests whether the concepts and laws used to represent the system under consideration are interpreted and formulated correctly. Conceptual validation of the TARGETS1.0 model requires that each submodel should be scientifically valid in the sense that the model structure and dynamic behaviour over the period 1900-1990, or for a shorter period if there are data limitations, reflect prevailing insights about the modelled subsystems. One way of conceptually validating the submodels is by comparing them with expert models which they are supposed to represent at the metalevel. An example of validating a simple carbon cycle model based on Goudriaan and Ketner (1984), against observational data and more complex two and three-dimensional carbon cycle models is presented in Rotmans and den Elzen (1993b). The CYCLES submodel has been validated by comparing it with the Terrestrial Ecosystem Model (Melillo *et al.*, 1993), the GLOCO model (Hudson *et al.*, 1994), and a general nitrogen and carbon cycle model (den Elzen *et al.*, 1997; Rastetter *et al.*, 1991). The health module of the population and health submodel has been compared with the Harvard incidence-prevalence model (Murray and Lopez, 1994; World Bank, 1993) and the disease module with a similar model for the Netherlands (Barendregt and Bonneux, 1992). For the other submodels, expert models have also been used for comparison and validation.

A second conceptual validation approach is the so-called Strategic Cyclical Scaling (SCS) method proposed by Root and Schneider (1995). This involves continuous cycling between large and small-scale assessments. Such an iterative scaling procedure implies that for a specific submodel the global version is disaggregated and adjusted for a specific region, country or river basin. The new insights are then used to improve the global version, after which implementation for another region, country or river basin follows. Applying the SCS approach at the level of submodels of TARGETS, case-studies have been carried out in order to validate the claim that the generic metamodels cover the crucial processes. With regard to energy, a developing country has been chosen (India) as well as a country which portrays the transition pathway in developed countries (USA). With the population and health submodel, case-studies with parts of the submodel have been done for India (Hutter *et al.*, 1996), China, Mexico (van Vianen *et al.*, 1997) and the Netherlands. For the water submodel, the Ganges-Brahmaputra (Hoekstra, 1995) and the Zambezi (Vis, 1996) basins have been chosen.

Conceptual validation can also be performed by involving experts. First, submodels and their modules have been developed in cooperation with other research institutions and universities. Besides, experts in small-scale, detailed models in a specific region or country have been asked to analyse the model results and to validate regionalised model results for a specific world region against regional data subsets. This has, for example, been done for the fertility component of the population and health submodel in a three-day workshop organised by the Population Research Centres of the University of Kerala (India) and the University of Groningen, in which leading Indian demographers participated (Hutter *et al.*, 1996). In cooperation with the Dutch Ministry of Housing, Physical Planning and Environment (VROM), a workshop was held in which energy experts and policy analysts were asked to give feedback on the energy submodel (de Vries and Janssen,

1996). In the last stage of the project, the submodels and the TARGETS1.0 model as a whole were reviewed by a total of over fifty national and international experts. The TARGETS1.0 model as a whole has been calibrated against historical time-series for major outcomes and in interaction with the stand-alone experiments with the submodels (Chapter 11). There has not yet been a systematic and rigorous conceptual validation.

Uncertainty analysis
Exploring future global change and its consequences for human society is beset with many uncertainties. These may be scientific in nature, arising from incomplete knowledge of key physiological, chemical and biological processes and related to the first level of complexity (Section 2.4). Many are of a socio-economic nature – related to people's behaviour – and reflect inadequate knowledge with respect to the second level of complexity. Finally, uncertainties also enter at the third level of complexity: the level at which norms and values are shaped and reinforced.

Uncertainty analysis should not be confused with sensitivity analysis, although both are essential if one wishes to gain insight into the reliability of models. We adopt the definitions given by Janssen *et al.* (1990). Sensitivity analysis is the study of the influence of variations in model parameters and initial values on model outcomes. Uncertainty analysis is the study of the uncertain aspects of a model and the influence of these uncertainties on model outcomes. Sensitivity analyses are useful to indicate which parameters represent crucial assumptions in the model. In order to indicate reliable confidence bounds, the uncertainty analysis should be comprehensive, which means that as many different sources of uncertainty as possible need to be considered. To this end we use classical uncertainty analysis but we also introduce the idea of model routes to deal with uncertainties arising from disagreement among experts. A perspective-based (or multiple) model route is a chain of biased interpretations of the crucial uncertainties in a model. To invest the model routes with coherence, we use cultural perspectives in order to make choices with regard to controversial model parameters and relations (Chapter 10). This methodology encourages us to make subjective judgements explicit and to consider at least more than one perspective. In this way, differences in future projections can be understood as the outcome of divergent views and valuations, instead of merely low, high and medium values. This approach also facilitates the interpretation of fundamental uncertainties in terms of risk.

3.4 Description of the TARGETS1.0 submodels

Vertical integration
One distinction within the TARGETS framework is between the human system and the environmental – or natural – system. The former mainly focuses on demographic

and health aspects of the human population and the provision of food, water and energy. The latter comprises a variety of flows of natural and man-made substances between the atmosphere, the ocean and the terrestrial biosphere. Of course, the distinction is not sharp. A difficulty is to disentangle the anthropogenic changes from the changes which are part of the natural evolution of environmental (sub)systems. We conceptualise human interventions as superimposed on a 'steady state' of environmental subsystems, ignoring for example long-term evolutionary changes in ecosystems. Ecosystem-related processes only feature in the TARGETS1.0 model as part of a highly aggregated description of aquatic and terrestrial ecosystems, with indicators such as water quality and land-use and land-cover distributions. We now proceed with a description of the TARGETS1.0 model. In our terminology, a model consists of submodels, while submodels in turn have modules as building blocks. Vertically, the TARGETS1.0 model consists of five submodels, each representing the cause-effect relationship for a particular theme of global change, and an economic scenario generator. *Figure 3.4* in section 3.5 shows the different submodels and their interactions.

The Population and Health submodel (Chapter 4)
The objective of the population and health submodel is to simulate changes in morbidity and mortality levels under varying social, economic and environmental conditions. Based on a number of socio-economic and environmental determinants, it simulates the population size and the health of the population in terms of both life expectancy and healthy life expectancy. The submodel consists of three modules: a fertility module, a disease module and a population state module.

- A *pressure* module represents the socio-economic and environmental factors that determine the fertility level, the health risks and the causes of illness and death. The socio-economic pressures are socio-economic status and female literacy level, while the environmental pressures are food and water availability, global climate change and changes in UV-B radiation;
- The *state* module consists of a number of reservoirs, which differ with respect to age, sex and health. Births are determined in the fertility module. The disease module calculates disease-specific morbidity and mortality. In the population state module, the calculated birth and death figures are used to simulate the population size distributed over five age groups and between males and females;
- An *impact* module represents the quantitative and qualitative aspects of demographic developments. The quantitative aspect reflects the size and structure of the total population. We consider the disease-adjusted life expectancy as one of the impacts representing the quality of the population;
- A *response* module includes policy responses regarding fertility behaviour, investments in health care and some other broad policy options. Health services can be allocated among primary and secondary prevention and curative care.

The energy submodel TIME (Chapter 5)

The role of the energy submodel is to simulate the demand and supply of commercial fuels and electricity, given levels of economic activity, and the associated emissions. It consists of five modules: Energy Demand, Electric Power Generation, and three Fuel Supply modules (Solid, Liquid, Gaseous).

- The *pressure* module simulates demand for commercial fuels in five separate economic sectors: residential, commercial/services, industrial, transport and other. Heat and electricity end-use demand are calculated from economic activity levels.
- The *state* module consists of the Fuel Supply modules and the Electric Power Generation module. The key state variables are the capital stocks used to produce energy and the fossil fuel resources. Important features in the Fuel Supply modules are resource depletion, penetration of commercial biofuels and learning-by-doing. The Electric Power Generation Model simulates the generation of electricity by utilising thermal, non-thermal and hydropower generating capital stocks.
- An *impact* module generates yearly emissions of six energy-related gases, CO_2 being the most important one. Land requirements for biofuel production are calculated.
- A *response* module makes it possible to include policy measures which influence energy efficiency and fuel substitution. Among them are the hydropower expansion path and research, development and demostration (RD&D) programmes with respect to biofuel, and non-thermal electricity generation.

The water submodel AQUA (Chapter 6)

AQUA takes into account the functions of the water system that are considered most relevant in the context of global change. Human-related functions considered are the supply of water for the domestic, agricultural and industrial sectors, hydroelectric power generation and coastal defences. Ecological functions taken into account are natural water supply to terrestrial ecosystems and the quality of aquatic ecosystems.

- A *pressure* module describes both socio-economic and environmental pressures on the water system. Total water demand is calculated as a function of population size, economic activity levels, demand for irrigated cropland and water supply efficiencies. The model includes the option of treatment of waste water before discharge;
- The *state* module simulates hydrological fluxes and changes in fresh water quality. The hydrological cycle is modelled by distinguishing ten water reservoirs, some of which are fresh surface and ground water, atmospheric water and oceans (*Figure 3.3*). The water flows between these reservoirs are simulated. Water quality, distinguished in four classes, is described in terms of nutrient concentrations;

- An *impact* module describes the impacts of water system changes on the environment and human society. It describes the performance of the various functions of the water system. The actual water supplies to households, agriculture and industry are calculated, as are the generation of hydroelectric power and the impact of a sea-level rise on the world's coast lines;
- A *response* module enables the user to model human response to negative impacts in the form of water policy measures comprising financial (e.g. water pricing), legislative and managerial measures.

The land and food submodel TERRA (Chapter 7)

The land and food submodel TERRA simulates food supply and demand, and land-use changes in relation to the element fluxes modelled in the CYCLES submodel. It is designed to offer understanding of human pressures on the global land and food system, and of potential impacts of changing food supply conditions on human health.

- A *pressure* module describes the demand for food resulting from the demand for vegetable and animal products, for tropical wood (excluding fuelwood). Demand is calculated as a function of economic activity and population size. Environmental pressures considered in the TERRA submodel are water availability for irrigation and climate change;
- The *state* module simulates changes in the physical state of the Earth's land surface in terms of changing land use and changes in the inputs and outputs of food production as a function of environmental and socio-economic pressures and land policies. The three main modules in the state system are: land-use/land-cover dynamics, erosion and climate change, and food and feed supply. Several land classes are distinguished (*Figure 3.3*);
- The *impact* module describes the impacts of food shortages which are a pressure in the Population and Health submodel. It also calculates changes in the forested and natural grassland areas – which are crude proxies for the loss of natural ecosystems – and loss of arable land through degradation;
- A *response* module gives various policy options: land clearing, expansion of the area of irrigated arable land, increased use of fertilisers and other inputs on rainfed arable land, land or soil conservation, and reforestation.

The element submodel CYCLES (Chapter 8)

The CYCLES submodel describes the cause-effect chain of the global biogeochemical element cycles. It links the anthropogenic pressures in the form of emissions and land and water use to the flows of elements within and between the various compartments. The basic elements C (carbon), N (nitrogen), P (phosphorus) and S (sulphur) are simulated because of their important role in global change phenomena. Some other chemical substances are also explicitly modelled. There is no separate *response* module. The anthropogenic pressures can be counteracted by measures incorporated in the energy, water and land response modules.

- The *pressure* module describes the driving forces underlying anthropogenic interference with the element cycles: emissions and flows of compounds of C, N, P and S from the energy and industrial sector, land-use changes, biomass burning, erosion, fertiliser use, harvesting, and water flow changes;
- A *state* module models the physical, chemical and biological fluxes of the basic elements and other chemicals within and between the atmosphere, terrestrial biosphere, lithosphere (soils), and hydrosphere (fresh surface waters and oceans). The atmosphere, the oceans and the terrestrial biosphere have been disaggregated into specific classes (*Figure 3.3*);
- The *impact* module describes the impacts of the changes in the fluxes of basic elements and other chemical substances on the global environment. It has two modules: for climate and ozone. The most important one, the climate assessment module, simulates the radiative forcing and global-mean temperature changes due to changes in concentrations of greenhouse gases and sulphate aerosols.

The economic scenario generator
Apart from these five submodels, we use a model of the economy. It is merely a transparent mechanism to reproduce exogenous Gross World Product (GWP) trajectories. It is referred to as the 'economic scenario generator'. GWP is taken to be the sum of consumption, value added in industry and services, and the monetary value of food production. Part of industrial output is used to satisfy the investment requirements for the provision of food, water and energy. In this way the scenario generator provides us with a simple money accounting framework.

We discriminate between two capital stocks: industrial capital and service capital. Industrial capital generates industrial output, which is partly reinvested in new industrial capital. The remainder is invested into various sectors: food, water, energy, and health and other services. The investment categories are: irrigation, agricultural inputs (fertiliser in particular), land clearing and conservation, domestic and industrial water supply, waste-water treatment, fossil fuel supply, electricity supply and energy efficiency. Health services investments are taken from service output and subdivided into preventive and curative services. Investments required to satisfy the derived demands for food, water and energy are fully met. Given a presumed relationship between the growth of industrial output and of service output, the remainder is assumed to be for consumption. We use the resulting – and 'maximum allowable' – growth rate in the per capita consumption as one of the indicators of welfare and sustainable development.

There are two reasons why we have opted for such a simple approach towards the economic system. The first is that we wish to avoid major controversies in the field of economic modelling dominating the results of the TARGETS1.0 simulation experiments. The second reason is more mundane: we lacked the expertise and resources to incorporate an economic model which would simulate at least the key factors in long-term economic growth in a satisfactory way. However, the first steps towards cooperation with economic researchers have been made (CPB, 1992; Duchin and Lange, 1994).

Limitations
After this brief description of the submodels, some of the limitations of the TARGETS1.0 model can be seen more clearly. Because economic developments are very important factors in assessing global change, the simple representation of economic processes is a serious limitation. The part of the investment goods, for example, which is dealt with in explicit detail (food, water, energy) is at most one quarter to one third of total world economic output. Important response mechanisms within the human system which determine the overall pattern of economic activities, such as changes in capital and labour productivity and in interest rates, are absent. Another omission is that the impacts of global change on the functioning of the world economy are not modelled explicitly. Feedbacks from for instance climate change on the productivity level in the industrial and service sector may have significant consequences for the overall behaviour of the system. With regard to ecosystems, the model gives at best a rough impression of ecosystem health in the form of indicators such as water quality and the size of the area covered by original vegetation. Ecological processes which may occur in response to changing element fluxes, for example, are not modelled. Hence, the issue of biodiversity is beyond the scope of the present model version. Due to the chosen aggregation level, the causes and impacts of such processes as immigration, urbanisation, wars and refugee movements are implicitly dealt with.

Horizontal integration
One can also focus on the horizontal integration. From that point-of-view, the TARGETS1.0 model can be subdivided into linked pressure, state, impacts and response components. The *pressure* modules are intended to chart the driving forces behind the increasing worldwide pressure on the environment and human society. The *state* modules describe the biogeochemical status of the environmental system and the social and economic status of the human system. The *impact* modules can be divided into three types of interrelated impacts. First, there are the effects of anthropogenic stresses on the environment which affect water availability, water quality, erosivity and climate conditions. A second type of impact is the influence of global change on human health, both direct and indirect. Direct effects relate to changes in disease determinants; indirect effects occur through deterioration of the world food and water supply. A third type of impact are the socio-economic effects of large-scale environmental problems. These show up in rising costs for food, water and energy. They can also take the form of direct losses of land or capital.

goods, which are however not simulated in the present version. Although only partly implemented, there is also an assessment of vulnerability in the form of people and capital at risk and the costs of flood protection for the coastal defence sector.

With regard to the *response* modules, it should be noted that there are many endogenous responses within the model, for instance the decision to invest in electric power plants if electricity demand grows or to increase agricultural inputs if food demand grows. This shows up as a change in costs or prices, which can be viewed as a model-endogenous response which in turn affects the system's behaviour. In a more narrow sense, the response components contain model variables which can be used to simulate exogenous interference with the way in which the system develops in a model experiment. These variables are information variables, unlike the largely physical stock and flow variables within the pressure, state and impact components. The response variables cover a variety of policy-related actions, among them financial incentives, regulation, information programmes and RD&D programmes (Rayner, 1991). Financial incentives are only implicitly included. Regulation encompasses legislation and rules, designed to control the activities of citizens and/or institutions. Important in the TARGETS1.0 model are the abortion legislation, water pricing and imposition of a carbon tax. Public information programmes are designed to alter the behaviour of citizens. Three important response variables relate to the representation of programmes to improve education, especially of women, of mass communication programmes for population policy, and of information campaigns devoted to the efficient use of water. The fourth one, RD&D programmes, involve policy incentives with regard to energy and water efficiency, biomass, and non-thermal electricity production.

3.5 Submodel linkages in the TARGETS1.0 model

Many integrated models use outputs from one submodel in the form of complete time-series as inputs for another submodel. This is a quite limited type of integration. In a model simulation experiment with the TARGETS1.0 model, data flow between the different submodels in each time step, which allows instantaneous simulation of interactions between submodels. The interactions between the submodels are shown in *Figure 3.4*. We briefly describe these interactions here; a more detailed description is given in the Chapters 4 to 8.

Gross World Product (GWP) is exported by the economic scenario generator to all submodels, except CYCLES. Sectoral GWP, or GWP per capita if combined with population size as exported by the Population and Health submodel to all other submodels, determines energy, food and water demand in the Energy, TERRA and AQUA submodel, respectively. In the Energy submodel, two components of GWP, i.e. value added services and value-added industry, are the drivers for energy demand in the five sectors. In AQUA, industrial output is used to determine industrial water demand. In the Population and Health submodel, GWP per capita has an effect on health and life expectancy. The required investments from the Energy, TERRA and AQUA submodel, and the health services demand from the Population and Health submodel, are accounted for in the economic scenario generator.

There are a number of outputs from the Energy submodel to the other submodels. The combustion of fossil fuels generates emissions of CO_2, SO_2, NO_x, N_2O and CH_4 which are inputs for the CYCLES submodel. The land requirements for biofuels are supplied to the TERRA submodel and allocated to grassland and arable land. The

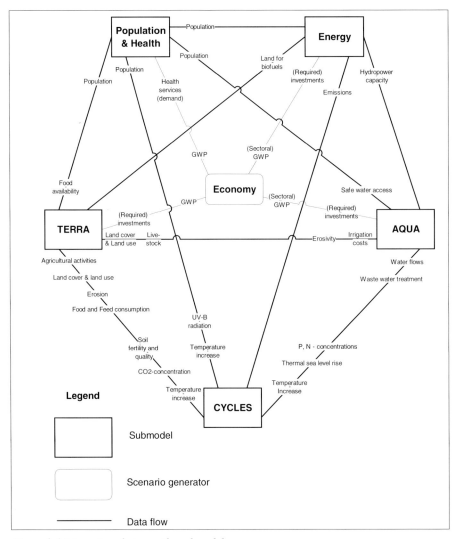

Figure 3.4 Interactions between the submodels.

expansion of hydropower is exported to the AQUA submodel and determines the demand for new water reservoirs in AQUA.

Access to safe drinking water and proper sanitation, which are exported from the AQUA submodel, affect the demographic and health dynamics in the Population and Health submodel. The hydrological cycle influences the element cycles in the CYCLES submodel: the outflows of elements from groundwater and surface water are driven by the water outflows from these water reservoirs. Also, AQUA simulates domestic and industrial wastewater treatment which determines the fate of nutrients. One of the factors determining erosion, the rain erosivity factor, is calculated in AQUA on the basis of the rainfall distribution throughout the year and exported to the TERRA submodel. Finally, the costs of irrigation are influenced by water availability and quality.

The available food per capita, exported by the TERRA submodel, affects people's nutritional status and thereby human mortality simulated in the Population and Health submodel. Land-cover changes affect the processes of evaporation, infiltration, percolation and river runoff simulated in AQUA. The area of irrigated cropland determines the irrigation water demand in AQUA, which is the main component of total water demand. Another (small) component is determined by the size of livestock in TERRA. Changes in land-cover patterns, erosion, emissions as a result of agricultural activities (fertiliser use, biomass burning and domestic animals) and food and feed consumption are exported to the CYCLES submodel. They affect the global flows of basic elements and related compounds within and between the major reservoirs in CYCLES. A global temperature increase simulated in the CYCLES submodel has an effect on malaria risk, schistosomiases and cardio-vascular diseases in the Population and Health submodel. Changes in the level of UV-B radiation affects the risk of skin cancer. In TERRA, higher CO_2 concentrations can affect the potential yield of arable land positively (CO_2 fertilisation) and temperature increase negatively (heat stress). The flow of organic matter and inorganic compounds has an effect on soil fertility and quality and thus on food production in the TERRA submodel. The concentrations of various substances in fresh groundwater and surface water, calculated in CYCLES, are exported to AQUA to determine fresh-water quality classes. Another link concerns the effects of a temperature change on the hydrological processes in AQUA. Finally, sea-level rise due to thermal expansion is exported, which is an important component of total sea-level rise simulated in AQUA.

3.6 Future work

As has been said before, we view the TARGETS model primarily as a toolbox which allows for experimentation with new concepts, methods and techniques. A model version is then a material manifestation of successful ideas, which invites testing and critical review. The model should not place scientists in a straitjacket, but rather function as a means of stimulating creative thinking. Often when a model is launched, a great deal of effort is devoted to refining the model by including more details. Such a strategy may actually not lead to a model which is more useful or scientific. On the contrary, it may result in a less transparent and rather unmanageable model, thereby losing its role as an exploratory and instructive tool. Moreover, the additional detail requires more data which are often unreliable or only partly available. Hence, future work on the TARGETS model will take another direction.

Some submodels are presently being implemented for regions, in connection with the IMAGE2.0 model. This broadens the experience with the model and refines the genericity. It also necessitates the inclusion of regional interactions, as for instance with fuel trade. A second step is to improve the dynamic representation of human actions, among them consumer behaviour (Jager *et al.*, 1997) and farming. A third step is to explore the role of the TARGETS framework as a tool to communicate issues of global change and sustainable development to a larger audience of policy makers and analysts and of scientists and interested lay people. The TARGETS1.0 model with its interactive visualisation shell is already being used in the context of the ULYSSES project (Jaeger, 1995) it will also be made available on CD-ROM. Some exploratory steps for the design of a policy exercise have been formulated (de Vries *et al.*, 1993).

From a modelling perspective, we intend to continue the application of novel approaches in the emerging field of complex systems modelling. Some case studies have been worked out to illustrate the potential benefits of an evolutionary modelling approach, among them the application of genetic algorithms to search for optimal and suboptimal trajectories (Janssen, 1996).

The Population and Health submodel

"There is no known biological reason why every population should not be as healthy as the best."
G. Rose, The strategy of preventive medicine (1993)

4 THE POPULATION AND HEALTH SUBMODEL

Louis W. Niessen and Henk B.M. Hilderink

This is the first of five chapters which focus on submodels within TARGETS. The framework of the Population and Health submodel includes socio-economic and environmental pressures, simulations of fertility, disease-specific mortality and morbidity, and their impact on population size, structure and health levels. The response subsystem comprises policies in the field of fertility and health. Whereas there are a number of separate models of fertility and population, the innovative aspect of the approach adopted here is that it is highly integrative, incorporating both population and health dynamics.

4.1 Introduction

During the past century, most populations of the world have experienced an increase in their levels of social welfare and economic development. These changes have shown a concomitant increase in the average life expectancy at birth and a decrease, although slower, in fertility levels (UNFPA, 1996; World Bank, 1993). The result has been an increase in world population size and a demand for resources unprecedented in history (UN, 1992; WCED, 1987). Reduction of health risks and the increased access to health services have resulted in a world-wide average life expectancy of more than 65 years during the past decades (WHO, 1996). Even though fertility rates are dropping, for some countries even rapidly, the world population is still growing at 1.5% per year. Presently, world population size in the year 2050 is estimated to be determined for about 50% merely by the *present* size of the fertile female population. The remaining 50% is thought to be determined for one third by the continuing increase in life expectancy and for two thirds by fertility levels above replacement level (Bos, 1995). In addition ageing societies all over the world are confronted with an excess demand for health resources due to the ageing process itself. Investments in health provision in a later stage of life need to be greater as they show diminishing returns (Kane *et al.*, 1990). In contrast, in the poorer regions of the world, population increases have led to a pressing need for continuing investments just to maintain current health standards (World Bank, 1993). The awareness has grown that in these regions the environmental assets relevant to human health, especially food and water, are scarce and diminishing (WCED, 1987).

This chapter describes the population and health modelling approach within the TARGETS context. It is based on reports by Niessen and Hilderink (1997) and by van Vianen *et al.* (1994) in which the Population and Health submodel has been described in full detail. The objective of the Population and Health submodel is to describe long-term changes in the size and structure of populations as well as the associated changes in health under varying socio-economic and environmental conditions in the past and in the future. This includes populations living in pre-industrial societies as well as those living at the highest known health levels. In our view such a long-term public health approach should consider the main input-output relationships between population, fertility and health in terms of both environmental and societal resources. The modelling framework can be used to explore and test general policy questions as well as specific hypotheses in the field. Policy needs may lead to comparative analyses of important general questions like: "What are effective ways to improve reproductive health?", and: "What are effective and efficient programme interventions to reduce disease and mortality at the population level?". The model has also been applied also at the national level: for China (Zeng Yi *et al.*, 1997), for India (Hutter *et al.*, 1996), and for the Netherlands and Mexico (Niessen and Hilderink, 1997).

4.2 Health transitions

The theory that addresses changes in population and health in one general frame of reference is the theory of the health transition (Caldwell, 1993; Frenk *et al.*, 1993; Ness *et al.*, 1993). It describes how populations can go through typical health and fertility stages as they change from living in pre-industrial agriculture-based societies to industrialised societies (*Figure 4.1*). The health transition has two components: epidemiological and a fertility one. The health transition includes the changes in these two areas as well as the concomitant changes in the organisation of social and health-related services. The resulting combined effects on population size and structure is labelled the demographic transition. The epidemiological transition shows a shift from a situation in which the infectious diseases among children are dominant to a situation in which diseases occur in the last years of life. The resulting average life expectancy at birth can rise from around 30 years to over 80. In the last stage the chronic diseases are dominant. The fertility transition describes the decline of the fertility level. The change in fertility behaviour is generally seen to be caused by the process of modernisation within societies. The fertility level, represented by the total fertility rate, can drop from seven children per woman in the early stages of the transition to, or even far below, a replacement level of 2.1 children per woman. The health transition can be categorised by the three stages in combinations of the epidemiological and fertility transition as shown in *Figure 4.1*.:

- *Stage 1*: Reproductive health needs and deficits in early stages of the health transition;

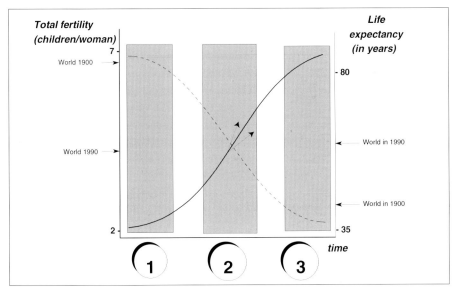

Figure 4.1 The three stages within the health transition. The right y-axis represents the level of life expectancy (——) and the left y-axis the level of fertility (-----). (arrows: see text under Stage 2).

- *Stage 2*: In the midst of the transition continuing reproductive needs and health deficits in populations with a growth in population size and competition for resources;
- *Stage 3*: Health needs of ageing populations that have reached a presumably new steady state with a zero population growth where fertility is controlled and survival improved.

Stage 1: Reproductive needs, amenable disease and premature mortality

The first stage is characterised by low life expectancy levels related to an epidemiological pattern of infectious diseases and by high fertility figures. Examples of rural and urban populations within this stage can be found all over Africa, in Latin America and in Asian regions. However, in all world regions and also in most countries fertility rates are dropping, often fast (UNFPA, 1994; UNFPA, 1995; World Bank, 1993). Still, there are huge unmet needs for broad and more specific reproductive health measures, as concluded at the conferences in Cairo and Beijing (UNDP, 1995). Accumulated mortality risks for women throughout their fertile period still might be ranging from 0.025 in South Asia to 0.05 in Africa (Graham, 1991) and their contribution to the total health burden is considerable (*Figure 4.2*). Pathological conditions related to reproduction are usually preventable and/or curable. Policies that are proposed vary from broader socio-economic programmes investing in education and employment to more health-service orientated investments related to maternal health and family planning (UNFPA, 1995; Hutter *et al.*, 1996).

Countries in this stage show many disparities in general health (Mosley and Cowley, 1991; World Bank, 1993). Agreement seems to be growing that both socio-economic inputs and essential preventive and curative health interventions can contribute to health levels. The local effects of environmental change like air pollution and global warming may mean additional stress upon the vulnerable groups in this stage and may contribute to the existing damaging health effects (Doll, 1992; Martens *et al.*, 1995b). Examples of areas in which health benefits can be gained in this stage are: infectious/childhood diseases, and maternity and perinatal care. These preventable and curable diseases still form a major part of the global total burden of diseases and amount within this stage to about 25 million deaths annually (*Figure 4.2*).

Stage 2: Population growth, economic development and environment

Interactions between population growth, economic development and the environment have been recognised as being vital for populations since the early seventies (Kiessling and Landberg, 1994; Meadows *et al.*, 1974). Changes in population are important in themselves and in relation to resource availability. In a long-term analysis, population growth is seen as being related to both fertility and mortality (Bongaarts, 1994, 1996; Demeny, 1990). In this stage death rates fall but birth rates do not (yet); hence, population size increases sharply. These changes depend in an interrelated, but different, way on socio-economic developments and

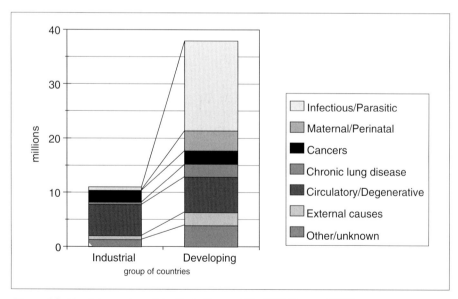

Figure 4.2 Absolute number of deaths in the world in 1985 (Lopez, 1990).

on access to services. Population growth is an important pressure on such societal resources, as the basic environmental sources of food, water, air, and space (WRI, 1994). As populations increase, the net outcome in terms of resource availability per capita and resource distribution will determine whether and how quickly they will pass through the transition. When there is a surplus of available resources, the transition might be accelerated but when resources are insufficient the transition might be delayed or even stagnate (as indicated by the arrows in *Figure 4.1*). An example is the food situation in countries in Sub-Saharan Africa. Food production in absolute figures has risen on this continent during the period of population increase of the last decades (Rosegrant *et al.*, 1995), but food per capita has declined with a consequential absolute increase in the numbers of undernourished people. Causes of death where health benefits can be gained are external, i.e. accidents and violence, but also circulatory diseases and cancer (see *Figure 4.2* and *4.3*).

Stage 3: Ageing and the demand for health care

In the last recognised stage low levels of fertility are dominant and sometimes even below replacement level (van Vianen *et al.*, 1994), while life expectancy is high and causes of death are changing (Uemura and Pisa, 1988; Ueshima, 1987; World Bank, 1993). Historical data (Alter and Riley, 1993), four modern long follow-up series (Riley, 1990) and cross-national data (Ruwaard *et al.*, 1994) suggest a general pattern

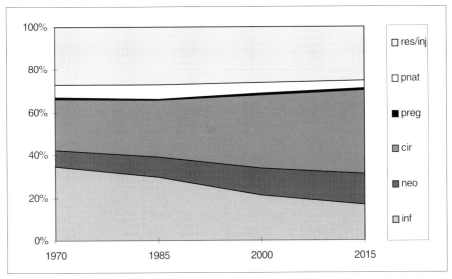

Figure 4.3 The global epidemiological transition 1970-2025: the relative distribution of causes of death (Bulatao, 1993). res/inj = residual diseases and injuries; pnat = perinatal diseases; preg = pregnancy-related (or maternal) causes; cir = circulatory diseases (including heart disease and stroke); neo = neoplasms (or cancers); inf = infections.

that the years one has to live with disease, increase with socio-economic development, as also the frail are surviving.

In this stage one can expect the total disease burden is expected to determine the demand for health care. One can observe a decreasing demand for health care among the younger age groups as their health improves, and in an increasing demand among the oldest age groups. As population size increases, however, there will be a necessity to maintain social and health-service investment levels for all these groups at a sufficient level (Feacham *et al.*, 1991; World Bank, 1993). As the young grow older, the demand for health care will rise more because of diseases related to old age and because of the increase in disease survival that might lead to increase disability (Bonneux *et al.*, 1994; Niessen and Rotmans, 1993). The final result may be that, with worldwide ageing of populations, including those in developing countries, disease levels among the oldest people will prove to be the most important factor in the total demand for social and health services (*Figure 4.3*) (Kane *et al.*, 1990).

4.3 Modelling population and health

Existing approaches

Fertility modelling approaches are widely accepted and, only recently, have existing mathematical techniques been introduced in the health area (Weinstein *et al.*, 1987; WHO, 1994). An integrated approach which incorporates both fertility and health dynamics is lacking. The Population and Health submodel of the TARGETS modelling framework will be discussed in this section . We have built a model based on current conceptual models and mathematical approaches to fertility, population and health. Before we introduce the integrated model, we will give a brief overview of existing approaches.

We have incorporated a fertility model in our approach which has been widely applied and which uses indices for the proximate fertility determinants (Bongaarts *et al.*, 1984). It uses four fertility indices which are related to the reproductive status of women: age of marriage, contraceptive use, abortion rate and infecundity. A more elaborated and data-intensive approach is the application of multi-state modelling which includes the modelling of birth cohorts of women (Bonsel and van der Maas, 1994; Zeng Yi, 1991).

In most health approaches, broad health determinants are recognised, like literacy and income status. The classical risk factors and nutritional status are recognised as proximate determinants (Frenk *et al.*, 1993; Hurowitz, 1993; Ruwaard *et al.*, 1994; Vallin, 1992). There is a need for further development of these approaches to make them more suitable for health planning and mathematical implementation. Regression analysis shows that the combined influences of five areas of society (i.e. the economy and income status distribution, nutrition, water supply and sanitation, education and medical services) can explain roughly 80% of the variation in life expectancy (Cumper, 1984; Gross, 1980). It may be clear that these regression techniques can only give suggestive evidence on the causes of population and health changes (Millard, 1994; Preston, 1975, 1976, 1980). Integrated approaches take account of the simultaneous occurrence of multiple-risk factors and diseases as well as cause-effect relationships. In a recent evaluation it is concluded that they cannot be used in the clinical area on an individual basis but are appropriate at the population level (Feinstein, 1994).

There are a number of different statistical techniques to extrapolate trends in total fertility and life expectancy to compute population projections. United Nations' population projections go up to 2200 (Bulatao, 1992). Assumptions regarding continuing fertility and mortality trends are based on the extrapolation of historical time-series, and on assumptions on when the last stage of the demographic transition is reached. Using these assumptions, one can map out a trajectory of population increase. The UN projections are based on assumptions about the period when replacement levels are reached (2.1 children/woman). The most widely used UN medium variant presupposes a steady-state level for the year 2050 with a total fertility level of 2.1 and a life expectancy increase of 15 years at world level. Another example is given by IIASA (Lutz, 1991, 1994) which assumes much lower future fertility levels for various world regions (down to 1.4 children/women).

A PSIR approach to fertility and health

The main structure of the submodel, represented in *Figure 4.4*, is based on the Pressure-State-Impact-Response (PSIR) framework as described in Chapter 2. This framework consists of:

- a *pressure module* describing the health determinants which are divided into socio-economic factors including income and literacy status, and environmental factors, among them food and water availability;
- a *state module* simulating fertility behaviour and population dynamics concerning disease and disease-specific mortality, both having their inputs to a population module distinguishing sex and age groups;
- an *impact module* that describes the quantitative and qualitative aspects of the state module, like the burden of disease and life expectancy as well as the size and structure of the population;
- a *response module* consisting of population policies influencing the fertility behaviour and health policies influencing the disease processes.

Position within the TARGETS model

Figure 4.5 shows the position of the Population and Health submodel within the overall TARGETS framework. The input from the other submodels are per capita food supply from TERRA, the fraction of the population without access to safe drinking water from AQUA and other environmental factors (temperature increase

4 THE POPULATION AND HEALTH SUBMODEL

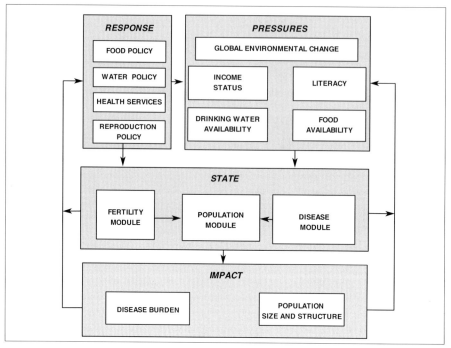

Figure 4.4 A Pressure-State-Impact-Response (PSIR) representation of the Population and Health submodel.

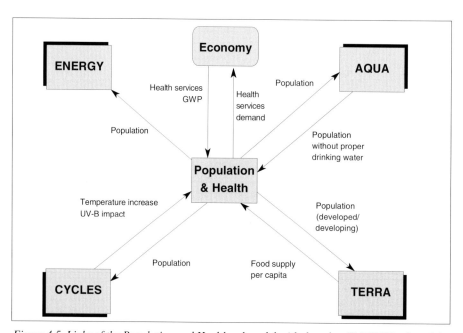

Figure 4.5 Links of the Population and Health submodel with the other TARGETS submodels.

and UV-B radiation) from CYCLES. The Gross World Product (GWP) and the availability of health services are imported from the economic scenario generator described in Chapter 3. The resulting actual total health service expenditures are calculated in the Population and Health submodel and are part of the service output.

4.4 Model description

4.4.1 Introduction

The population and health pressures represent those factors that influence the modelled proximate fertility and health determinants. The selection of determinant categories is based on the evidence regarding their supposed quantitative importance throughout the health transition as reported in the literature. They are categorised in two groups: socio-economic and environmental factors. The two variables which describe the socio-economic pressures are:

- Gross World Product (GWP) expressed in 1990 US dollars. This variable determines the available income per capita and the resources available for health services. Separate projections for the low-income countries are used in a distribution function to estimate the number of people below the absolute poverty line (World Bank, 1993);
- the female literacy level expressed as the fraction of the literate adult female population. This variable is computed as a delayed function of GWP and the Human Development Index[1] (HDI).

The variables which describe the environmental pressures are:

- food supply expressed in kilocalories daily intake per person (from the TERRA submodel). From these projections the fraction of the population suffering from malnutrition is calculated for the sub-populations that fall under the low socio-economic status categories (see below). These are based on an empirical distribution functions as reported by the FAO;
- drinking water and sanitation defined as the fraction of the population with proper access to safe drinking water and having sanitation (from the AQUA submodel) that fall under the low socio-economic status categories. In case of large discrepancies

---

[1] This index is constructed out of three components according to the UNDP, i.e. years of schooling, life expectancy and per capita GWP. Since the years of schooling are not modelled, female literacy is taken as a proximate for the second component. The values of these components are scaled with a defined minimum and maximum value and are equally weighed to obtain the HDI. Life expectancy varies between 25 and 85 years, literacy between 0 % and 100 %, and per capita GWP between $ 200 and $ 40,000 (UNDP, 1994).

between the two, the safe drinking water coverage is chosen because this factor is the most relevant of the two (Esrey *et al.*, 1985, 1991);
* temperature increase (from the CYCLES submodel). It influences the number of people exposed to malaria risks, the spreading of schistosomiasis and, by heat stress, affecting cardio-vascular disease mortality.

The state module consists of three components: the fertility module, the health module and the population module wich are discribed in the next sections.

4.4.2 The fertility module

The modelling of fertility change is based on Bongaarts and Potter (1983). We adapted this rather static approach into a more dynamic model by using the PSIR framework. In *Figure 4.6* the simplified PSIR diagram of the fertility module is represented. The main outcome of the fertility module is births by sex. The calculations of births are based on the Bongaarts model, which assumes that a biological maximum total fertility is reduced by the four proximate determinants

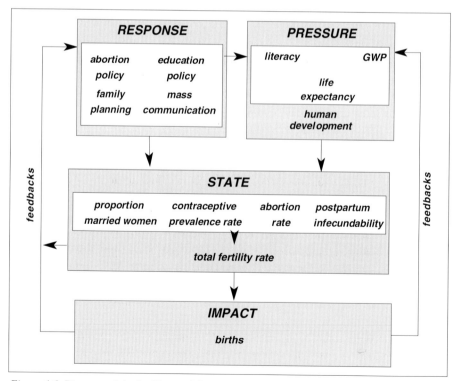

Figure 4.6 Diagram of the fertility module.

(Bongaarts and Potter, 1983). The maximum total fertility ($FERT_{max}$) is kept constant at 15.3 and is a theoretical biological maximum average that has been derived from assumptions about coital frequency, probability of conception and length of post-partum amenorrhea (van Vianen et al., 1994). The two main characteristics we added to the Bongaarts approach are the linkage of the fertility determinants to the level of socio-economic development and the modelling of the use of contraceptives as a diffusion process (Montgomery and Casterline, 1993; Rosero-Bixby and Casterline, 1993). The main driving forces representing socio-economic development are GWP, female literacy and life expectancy. These are combined in the HDI. Data have been extracted from reports by the UNDP, and from UNICEF, UNFPA and the World Fertility Survey.

The four proximate determinants modelled in the fertility module all range between one and zero. They are:

- the index of marriage (C_m). This index is based on the average age of marriage which determines the fraction of the reproductive lifespan spent in stable sexual union. This average age can range from a minimum of 16 years to a maximum of 31 years depending on the level of the HDI. In the C_m this average age is corrected for the fraction of women ultimately married;
- index of use and effects of contraceptives (C_c). This index represents the effects of losing the reproductive period due to deliberate fertility control via contraceptives. The use and effectiveness of the different contraceptive methods are influenced by the HDI but also depend strongly on the assumed diffusion rate which varies between 5 to 10% per year. The use of contraceptives is bound by an upper limit;
- the index of postpartum infecundity (C_i). This index is defined as the fraction of the fertile lifespan which is lost for reproduction purposes due to breast feeding and culturally motivated abstinence. The value of this period in the model may vary from 3 to 18.6 months;
- the index of abortion (C_a) as a function of the number of induced abortions combined with the fraction of reproductive lifespan loss due to this abortion. The number of abortions is related to the level of the contraceptive rate.

The module calculates the resulting total fertility rate (*TFR*) which represents the number of children a woman has received at the end of her fertile period. In the fertility module this period is assumed to be from age 15 to 45. In equation:

$$TFR = C_m \times C_c \times C_a \times C_i \times FERT_{max} \qquad \text{[children/woman]} \qquad (4.1)$$

To obtain the yearly number of births, the *TFR* is multiplied by the total number of fertile women as calculated in the population and health module and corrected for the length of their fertile period (30 years):

$$\text{total births} = TFR \times \text{fertile women} / \text{fertile period} \qquad \text{[children/year]} \quad (4.2)$$

To calculate the births by sex, the total births are multiplied with a constant sex ratio, defined as the ratio between the number of boys and the number of girls born.

4.4.3 The health module

The health module simulates the number of persons suffering from diseases and the number of deaths related to these diseases. The disease figures are used to estimate the health status of the population while the death figures determine the overall age- and sex-specific mortality rates. We defined a number of health-related population reservoirs by age group and by sex. The health-state reservoirs are treated in a similar way as the overall population within the population and health module: inflows and outflows of the reservoirs combined with an initial value determine the contents of the reservoir. The flows concerning the disease processes are shown in *Figure 4.7* and are described here. The process of ageing in between the selected age groups and deaths from residual mortality causes are described in the next section. Age groups will not be included in the model equations in this section.

Risk exposures

The health risks associated with the exposed population (represented as *pressure* in *Figure 4.7*) are based on the broad and proximate health determinants of the health transition (Frenk *et al.*, 1993). Health risks consist partly of the pressures distinguished in section 4.4.1. The major health determinant is *socio-economic status* (SES) (Marmot and Elliot, 1994) which is defined as a function of income status modified by the fraction of the literate population:

$$\text{high SES}_{\text{fraction}}(t) = \text{fraction}_{\text{high/mid income}} + \text{fraction}_{\text{literate}} \times (\text{fraction}_{\text{low income}} - \text{fraction}_{\text{poor}}) \qquad (4.3)$$

The fraction of low income status is defined as having an income < \$ 700 per capita (World Bank, 1993); the fraction 'poor' is defined as those people living below the absolute poverty line (< \$ 350 per capita per year). The global population is clustered within these two SES categories (*Figure 4.8*):

- high socio-economic status (SES) in combination with health risks related to modern life style: *high SES health determinants;*
- low socio-economic status divided into a category without the presence of environmental or other risks and a category with environmental risks including

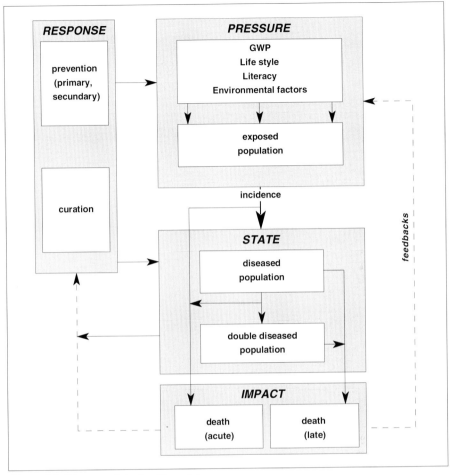

Figure 4.7 Diagram of the health module.

lack of sufficient food and/or safe water and the presence of vector-transmitted parasites: *low SES health determinants*.

The combinations of the health determinants, or multiple-stress factors, are depicted in *Figure 4.9*. Assumed is an independent occurrence of most categories except for the combination of malnutrition and no access to safe drinking water where a correlation/clustering factor of 1.1. is used. This figure proved to be the maximum still possible to explain empirical data (World Bank, 1993). In this way a clustering of health risks among the populations with low economic status can be simulated. Criteria for selecting exposure types have been the availability of empirical evidence of a quantifiable relationship with the occurrence of risk(s) in the epidemiological literature, the magnitude of the influence on disease and mortality

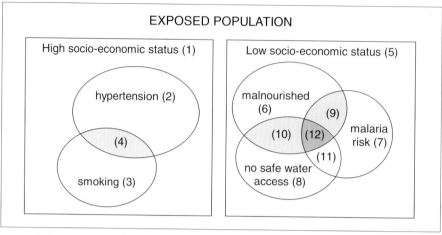

Figure 4.8 The twelve multiple exposure categories within the exposed population.

levels in societies (Schofield *et al.*, 1991; van der Walle *et al.*, 1992) and the availability of population-based statistical data. We will briefly discuss both SES categories in more detail.

The high SES population is further divided into those exposed to smoking or not and those having high blood pressure or not. These exposure categories are modelled on a sex- and age-specific basis and estimated from empirical time-series. It is assumed that only people who reach this level of social and/or material welfare will be exposed to these risks. There is a WHO world *smoking* level estimate of 52% for men and 10% for women. The input series for smoking levels are based on estimations of the world tobacco production since 1900 (WRI, 1994). Although the occurrence of related health problems has been rising, first in the sixties and seventies in Europe, the mortality figures are now rising worldwide, although much more among men than among women (Peto *et al.*, 1992). The relationship between *high blood pressure* and vascular conditions (heart disease and stroke) has been recognised and quantified. There are also estimates of the potential health gain by treatment or prevention (Gunning-Schepers, 1989). Input data are based on existing survey data and reviews, or existing models (Niessen and Rotmans, 1993; Ruwaard *et al.*, 1993; Tunstall *et al.*, 1994).

The low SES population, i.e. illiterates having low-income status combined with those below the absolute poverty line, are subdivided into eight exposure categories consisting of combinations of nutritional status, drinking-water supply and malaria risk but *not* high blood pressure and smoking. *Malnutrition* or low nutritional status is parameterised as a function of per capita daily food intake based on the empirically fitted relationship given by the FAO in their fifth World Food Survey (1987). Food intake is equated to food supply (see Chapter 7). *Access to safe drinking water* depends on the total public water supply investments. The population without public water supply and relying on private water supply only is regarded as having no access to safe drinking water (see Chapter 6). The third low SES-related determinant included is the proximity of people to areas with malaria mosquitoes carrying *malaria parasites*. The fraction of the population exposed to such a malaria risk is based on international estimations (0.44; Martens *et al.*, 1995b). This fraction is influenced by temperature but can also be influenced by investments in vector control programmes as a fraction of the total health service expenditure.

The presence of disease
The second part of the health module contains the reservoirs which represent the number of people afflicted by the occurrence of a disease. The disease groups are selected based on their empirically estimated contribution to mortality and disease levels in societies as is known from the international registries (*Figure 4.2*) (Ghana Health Assessment Project Team, 1981; Mosley and Cowley, 1991; Parkin, 1994; Thom and Epstein, 1994). Some of the more important causes of disease and mortality (like tuberculosis, accidents) have not been modelled explicitly as their occurrence has not been documented in a way which allows for relating them to particular health determinants and/or interventions, except to the general category 'socio-economic change'. Other diseases seem to occur at a more or less constant rate in populations like some neuro-psychiatric disorders (Cowley and Wyatt, 1993). Both categories of diseases are included in the 'residual mortality' category. This selection resulted in modelling the specific diseases as listed in *Table 4.1*.

The subdivisions of the exposed population into the twelve exposure categories determines the absolute risk for these diseases. The absolute level of the annual disease risk is the outflow from the exposed population towards the disease reservoirs. It is obtained by a multiplication of a constant base-desease risk for the

Category	Specific diseases
Infectious/parasitic diseases	Gastro-enteritis
	Acute respiratory infections
	Measles
	Malaria
	AIDS
	Schistosomiasis
Maternal/perinatal diseases	Prenatal mortality
	Maternal mortality
Cancers	Lung cancers
	All other cancers
Pulmonary disease	Chronic obstructive pulmonary disease
Circulatory and degenerative diseases	Acute infarctions
	Chronic heart failure
	Cerebro-vascular accidents (stroke)
Residual diseases	

Table 4.1 Disease categories within the health module.

unexposed populations with an excess relative risk (*RR*) which depends on the exposure category and the disease type. The base-disease risk is calculated from the population-attributive risk fractions (Walter, 1976). The *RR* represents the excess chance of getting a disease because of a health risk exposure as compared to a disease-specific base risk. The values of these *RR* are derived from epidemiological literature. The disease risks of the infectious diseases malaria and schistosomiasis are based on the outcomes of a more detailed epidemiological module described in Martens *et al.* (1995a). The equation for the disease risks for all the exposure and disease categories is as follows:

$$\begin{aligned} \text{exposed out}_{sex,age}(t) &= \text{disease risk}_{sex,age,exposure,disease}(t) \\ &= \text{base disease risk}_{sex,age,disease} \times RR_{sex,age,exposure,disease} \\ &\quad \times \text{exposed population}_{sex,age,exposure}(t) \quad \text{[person/year]} \end{aligned} \quad (4.4)$$

This flow from the exposed to the diseased reservoirs is divided into three events (*Figure 4.7*). The first, the event directly related to the initial disease risk, is the case-fatality rate (*CF*), defined as the probability of dying during the acute episode of the disease. The levels of the case fatality are age, sex and disease-specific. They are determined by the level of curative health service expenditures per capita. For each disease a minimum and maximum value for the case fatality are defined. These correspond to the lowest, highest known levels, respectively reported in the literature. A cure-effectiveness function determines the actual *CF* between the extreme values:

$$CF\ fraction_{sex,age,disease}(t) = CF^{min}_{sex,age,disease} - \text{cure effectiveness}(t) \times (CF^{max}_{sex,age,disease} - CF^{min}_{sex,age,disease}) \quad (4.5)$$

The second possible event is being cured within a year after becoming incident. Similar to the case fatality, the level of cure is determined by a cure-effectiveness function modified by the level of curative health-service expenditure. The third event is that new cases from the modelled infectious and chronic diseases, i.e. those who neither die nor recover, flow into the diseased reservoir. They become chronically ill from the disease involved, an event which is assumed to be irreversible.

Next, there are two ways to leave the diseased reservoirs (*Figure 4.7*). The first is the chance of dying due to the overall base mortality increased with a disease-specific delayed mortality risk. This higher mortality chance is defined as late mortality fraction. This fraction is also based on a minimum and maximum value and influenced by the effectiveness of curative health service expenditure. The second way to leave the diseased reservoir is by getting another disease. Especially among the elderly but also among those with a respiratory disease this is frequent. 'Double-diseased' reservoirs have been defined to include the eight most frequent combinations of chronic diseases. The events related to this double diseased reservoir are similar to getting a first disease: being cured and dying are treated similarly, using the same functions and the same absolute values as for single diseased. The mortality rates by sex and age related to these fourteen disease groups are obtained from the figures on the single and double-diseased reservoirs and are used within the population module.

4.4.4 The population module

Within the population module the dynamics of births (from the fertility module) and disease-specific deaths (from the health module) are taken into account, determining the size and changes in the population age and sex groups. In addition to a distinction between males and females the population is divided into five age groups: 0-14, 15-44, 45-64, 65-74 and 75 and older. This selection of age groups is mainly based on the epidemiology of diseases during the observed health transitions: childhood diseases, maternity-related diseases, degenerative diseases and the possible evolution of old age diseases. The changes in these population subgroups equal:

$$\frac{dpop_{sex,age}(t)}{dt} = pop\ in(t)_{sex,age} - pop\ out(t)_{sex,age} \qquad [\text{person/year}] \qquad (4.6)$$

where:

$$pop\ in_{sex,age}(t) = births_{sex}(t) \qquad \text{if age} = 1$$

$$pop\ in_{sex,age}(t) = [pop_{sex,age-1}(t) - deaths_{sex,age-1}(t)] \qquad [\text{person/year}] \qquad (4.7)$$
$$\times aging_{sex,age-1}(t) \qquad \text{if age} > 1$$

$$pop\ out_{sex,age}(t) = deaths_{sex,age}(t) +$$
$$[pop_{sex,age}(t) - deaths_{sex,age}(t)] \times aging_{sex,age}(t) \qquad [\text{person/year}] \qquad (4.8)$$

The outflow of the different subpopulations is defined as the number of people who have died added up with the ageing of the rest of the subpopulation involved. The calculation of the number of deaths is the multiplication of the fraction of dying (*total mortality fraction*) and the population in absolute figures:

$$deaths_{sex,age}(t) = pop_{sex,age}(t) \times total\ mortality\ fraction_{sex,age}(t) \qquad [\text{person/year}] \qquad (4.9)$$

The yearly ageing of the various subpopulations depends on the size of the age group and is roughly the inverse of that size in years (for example, the ageing out of age group 0-14 is 1/15). Because of the non-uniform mortality distribution within the age groups, it is adapted to the distribution of people over the age group. This distribution is represented by the fraction of people surviving up to the last year of the age group (in the example up to 14 years) divided by the length of the age period. Thus:

$$aging_{sex,age}(t) = \frac{(1 - total\ mortality\ fraction_{sex,age})^{lengthgroup_{age}}}{\sum_{i=1}^{lengthgroup_{age}}(1 - total\ mortality\ fraction_{sex,age})^i} \qquad (4.10)$$

The total mortality consists of three components. The disease mortality fraction is the mortality that can be explained by epidemiology. It is determined by the exposure of people to various broad and specific health risks and the subsequent occurrence of disease and death. The maternal mortality fraction is calculated by multiplication of the number of women of the ages between 15 and 45 years with the number of births and a maternal mortality risk based on the level of curative health care services. The third component is 'residual mortality'. It is a presumably biological baseline level of mortality, sex and is age-specific, defined as the yearly mortality that cannot be related to a particular cause. This mortality fraction is quantified using a golden standard life table.

Population and health impact components

The impact component of the population and health system represents the quantitative and qualitative aspects of the population change: population size and structure, and the disease burden and its socio-economic consequences. Population size is calculated each time-step by taking the sum of the sex and age-group reservoirs in the population state module. Besides these impacts the Population and Health submodel provides information about the levels of fertility and mortality represented in the crude birth rate (CBR) and the crude death rate (CDR), defined as the respective number of births and deaths per 1000 persons per year. The structure of the population is a direct outcome of the population model. From this structure, a dependency ratio can be obtained. Another general measure is the loss of disability-adjusted life years (dalies) per thousand people (World Bank, 1993). In this approach the time lost due to disease, acute or chronic, are added to the time lost due to premature death as compared with the standard life table North (Murray and Lopez, 1994).

We adapted this approach to calculate the loss of disability-adjusted life expectancy (*DALE*) in a standard-demographic manner based on life table analysis (Niessen and Hilderink, 1997). This methodology accounts for both the loss of health through premature death as well as through disease. To calculate the number of life years lost, they are compared to an assumed estimated upper limit of 82 years for men and 88 years for women (Murray and Lopez, 1994). This origin of the losses of health by category can be obtained from the health model as the model in each computation records the relative contributions of exposure categories to the occurrence of disease and death. In this way the years lived and the average loss of health can be clustered into four categories measured in *DALE* years (Section 9.5.1). These are:

- the net healthy years lived and the loss of life years by three groupings of exposure categories;
- loss of years due to high SES and modern life style-related health risks;
- loss of years due to low SES only;
- loss of years due to environmental and other factors in combination with low SES.

The occurrence of diseases is only assigned to the last three categories while the category 'mortality from residual diseases' is not included. The contribution of the different diseases to a loss of health is distinguished for acute episodes and chronic cases. The duration of an acute period depends on the disease: in general two weeks for infectious diseases and one month for chronic diseases. Chronic cases are counted as lasting a whole year. The time spent with disease is weighed by degree of related disability (Murray and Lopez, 1994).

4.4.5 Policy response

The response module accounts for policy options related to (i) population policy focusing on fertility behaviour and (ii) health policy in the form of investments in health care. Population policy measures are taken on the basis of a relationship between the rate of population growth and the rate of economic growth. The priorities for the four policy options: abortion, family planning, education and mass-communication are allocated exogenously (van Vianen *et al.*, 1994).

Health policy measures are simulated by linking the lack of population health status to a resource allocation mechanism which assumes a demand/supply feedback. Health service expenditures are exogenously allocated among the following three categories:

- primary prevention: health services which, in the WHO definition, aim at reducing the exposure fractions by preventing people of getting exposed or taking away the exposure (food policies, vector control, anti-smoking campaigns);
- secondary prevention: health services, in the WHO definition, preventing people becoming ill, e.g. immunisation campaigns (Basch, 1994).
- curative services: health services reducing the mortality risks (case fatality and late mortality) and increasing the chances of getting cured. In the aggregated model, the effects of curative interventions on survival are simulated with one general function while the minimum and maximum values of the effects are specific for the disease, the disease stage, age and sex.

The required increase in health services per capita is initially determined by the calculated burden. The new demand for health services is calculated from this. The resulting public health expenditures are assumed to be available as part of the economic service output in the next year, unless it exceeds a certain fraction of GWP, derived from statistical cross-country correlations. The full loss of *DALE* years forms the upper limit to the desired health gain.

4.5 Calibration and validation

Data sources and use
Empirical data have been used for historical calibration and validation for the period 1900-1990. We extracted these data from four types of sources:

- international data registrations as reported or made available by different levels of aggregation (national, regional or global);
- population-based epidemiological surveys as reported in the literature or as used by other models by different levels of aggregation (national, regional or global);
- clinic-based research as reported in the literature;
- other surveys at the population level in the field of the social sciences.

We used the data mostly as input for the Population and Health submodel without further calculation, interpolation or smoothing. As can be expected, one finds many different parameter values in the literature. Our selection is based on the criterium that the study population involved should be similar to the defined (sub-)population in the model, e.g. belonging to the high or low SES categories for increased risks of diseases. We refer to the appendices of the background report (Niessen and Hilderink, 1997) for the chosen values for disease risks, relative risks, case fatality and late mortality by age group, sex, exposure category and disease group as well as for the literature sources.

Calibration
The modelling framework uses relatively basic mathematical functions. Also in the definition of parameters it attempts to remain as close as possible to empirical data and already established demographic and public health concepts. Still, in face of the complexity of the model system, it is necessary to adapt some input variables to be able to reproduce time-series of existing data on fertility, population and health. This process we define as empirical calibration. The variables we used for calibrating the model outcomes are shown in *Figure 4.9*.

It shows that there is an overestimation of the simulated global total fertility rate as compared to the empirical figures. The latter are most likely underestimations as there is probably an underreporting of births and deaths in the perinatal period and in

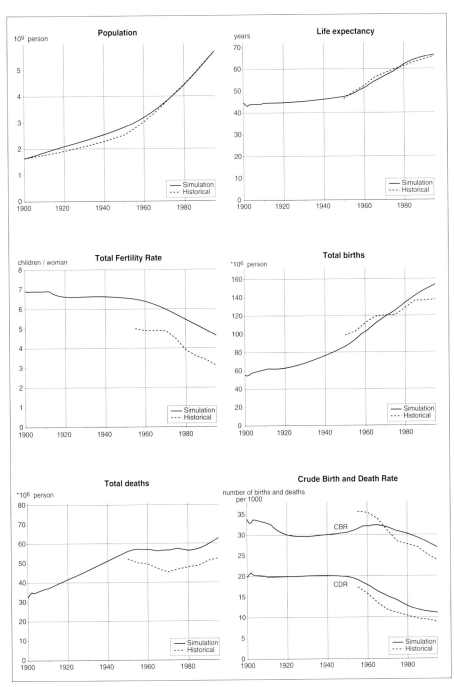

Figure 4.9 Six calibrated model variables compared with the empirical time-series for the period 1900-1990. The data are extracted from UN statistical summaries.

early infancy. Within the population module, the number of perinatal deaths is increased by including the new-born in a malnourished exposure category as in the case of the first age group. Hence, the simulated number of deaths exceeds the historical estimates. During model runs this malnourished fraction disappears later in the health transition as food availability improves and, consequently, perinatal mortality is lowered. The acceleration in the population increase starting around the 1950s cannot be simulated more accurately due to the large sizes we use for age groups. The steady drop in perinatal and infant mortality that has been observed shows up as only a slow increase in population because it is 'diluted' over a period of 15 years (the size of the first age group).

Structural validation
Once calibrated, the model is run and tested to compare its output to *other* existing data from empirical surveys. We define this as structural validation. Two fluxes related to the disease reservoirs, the disease incidence and the disease-specific mortality, are relatively difficult to quantify, especially at the more aggregated population level. We tested the consistency of the model values by eliminating the disease-specific reservoirs and directly calculating the excess total age and sex-specific mortality by exposure category for the year 1990. When dividing these by the high SES and low SES age and sex-specific mortality, one gets simulated exposure-attributable figures on the excess risk of total mortality in the age and sex category.

Table 4.2 shows that the model produces relative risks in the same order of magnitude as reported in empirical surveys. One can also observe an appropriate age gradient in the low SES categories. The empirical relative risks for malnutrition are based on a large number of study populations throughout the world (Pelletier *et al.*, 1995) and is in between the simulated values. This can be expected as malnutrition clusters with a number of other health risks. The reported relative risks for lack of safe drinking water shows a large range, although we selected the values from the most rigid surveys.

4.6   Conclusions

This section reports the conclusions on the results of some simulation experiments on the net effects of single population pressures and the overall conclusion on the submodel's validity as a whole. In *Figure 4.10* results of the model are shown from single factor analysis using a stand-alone version of the Population and Health submodel. Zero growth paths (constant 1900 values) have been introduced for all health services collectively as well as for food supply per capita and safe water supply over the period 1900-1990. All other parameter values have been kept the same as in the historical calibration. These changes influence the disease and

4 THE POPULATION AND HEALTH SUBMODEL

Exposure (A)		BP	NIC	NIC + BP	Low SES	NUT	MAL	WAT	NUT+ MAL	NUT+ WAT	MAL+ WAT	NUT +WAT +MAL
Men	1-14				1	1.6	1.9	1.9	2.5	2.2	2.8	3.1
	15-44	1.0	2.4	2.6	1	1.2	2.2	1.3	2.5	1.5	2.6	2.8
	45-64	1.3	2.3	2.5	1	1.1	1.6	1.1	1.7	1.2	1.7	1.8
	65-75	1.3	2.1	2.3	1	1.0	1.3	1.1	1.3	1.1	1.3	1.4
	75>	1.4	2.5	2.8	1	1.1	1.2	1.1	1.3	1.2	1.3	1.4
Women	1-14				1	1.6	1.9	1.9	2.5	2.2	2.8	3.1
	15-44	1.0	2.6	2.6	1	1.2	2.0	1.2	2.2	1.4	2.3	2.5
	45-64	1.1	1.9	2.0	1	1.1	1.5	1.1	1.6	1.2	1.6	1.7
	65-75	1.3	1.8	2.1	1	1.0	1.3	1.0	1.4	1.1	1.4	1.4
	75>	1.6	2.1	2.5	1	1.1	1.4	1.1	1.5	1.2	1.5	1.6

Exposure (B)		BP	NIC	NIC + BP	NUT	MAL	WAT
Men	1-14				2.5	1.5-1.6	1.3-5.0
	15-44	1.2	3.05	2.61			
	45-64	1.3	2.31	2.61			
	65-75	1.3*	2.09	2.61			
	75>	1.3*	1.54	na			
Women	1-14				2.5	1.5-1.6	1.3-5.0
	15-44	NA	2.69	na			
	45-64	NA	2.52	na			
	65-75	1.4*	2.00	na			
	75>	1.4*	1.44	na			

SES: Socio-economic status; BP = High blood pressure; NIC = smoking; NUT = malnutrition; MAL= Malaria exposure; WAT = no safe drinking water; na = not available.
Sources: High blood pressure: MRFIT-study in Marmot and Elliot, 1994; * Chicago Heart Association Detection Project in Marmot and Elliot, 1994; Smoking: Peto et al., 1992; High blood pressure + smoking (for CHD only): MRFIT-study; Malnutrition: Pelletier, 1995; Water risks: Esrey, 1985; Malaria risk: Kisumu and Garki project, Najera et al., 1993.

Table 4.2: Simulated (A) and empirical (B) age- and sex-specific total relative mortality risks.

mortality levels directly and the fertility levels indirectly through the effects on life expectancy. Next, the combined effects of the two scenarios are shown. Some important conclusions are:

- the net effects of the two zero scenarios on the increase in total fertility, on the reduction in population size and on life expectancy are considerable and are of the same order of magnitude;
- with an almost zero growth in life expectancy, there is a considerable population growth in this century due to a high fertility ratio.

- the fertility rates are declining less due to the lowering effect of life expectancy on the human development index. This is an important compensating mechanism in the growth of the population: in spite of a lower number of fertile women alive, there are more children born per woman.
- at the same time the increase in life expectancy causes a much higher population growth rate, in spite of the concomitantly declining birth rates and the additional number of women alive. This is a reverse buffering mechanism.

General conclusions

The Population and Health submodel describes the changes in the global population divided into age groups, sex and health-risk status. The model calibration overestimates the number of births and deaths. This is both due to the model age-group structure and the fact that reporting at the global level is not well developed. We have chosen to report the inconsistency of the data and the model instead of further adapting it. The structural validation of the model shows a striking consistency with the empirical data. We conclude that the model describes the changes in fertility, population and life expectancy since 1900, as well as the intrinsic dynamics of these changes. It also describes the relationship of the changes to those in population pressures throughout the global population transition.

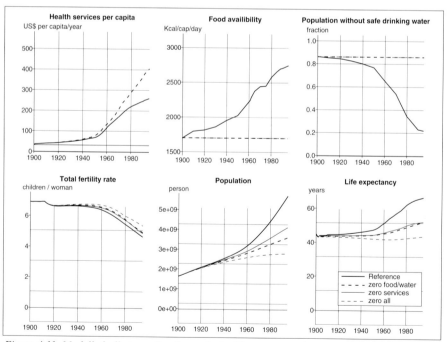

Figure 4.11 Modelled effects on the change in fertility rate, population size and life expectancy from 1900 to 1990 by major clusters of health determinants. Zero growth scenarios have been used for health services as well as for food availability and acccess to safe drinking water.

Validation by applying the model at different levels of aggregation

The modelling framework has been used to simulate health transitions in different populations and at different scales. This contributes to the validity of model dynamics. We applied the model to three countries: India, Mexico and the Netherlands (Bakkes,1997; UNEP, 1997). We selected the case studies by three criteria: (1) each population should presently be living in a different stage of the health transition, (2) the availability of population data, and (3) different population sizes.

India shows the demographic pattern of a late first transitional stage with still a high level of fertility and mortality. Mexico finds itself in the second stage. Death rates have been declining for many decades but the changes in fertility are lagging behind for a rather long time, some 30 years. The Netherlands, in the next stage, have experienced an early decrease in mortality levels and a relatively slow decline in births rates to below replacement level. The crude death rate rises as the Dutch population started ageing.

A number of model adaptations have been made for a similar historical calibration. These adaptations are:

- The start values related to fertility, population, exposure and disease reservoirs. For each of the applications these have been adapted based on the historical input scenarios and available additional data. If these were not available they were estimated by doing an initialisation run for six hundred years to calculate the values for a supposedly typical pre-industrial society in an early phase of the transition.
- The adaptations within the fertility module have been most fundamental especially on:
  - the timing of technology: in casu the start of the contraceptive use diffusion process;
  - the speed of the diffusion process (world version 7.5%); the desired family size and the relation with human development;
  - ultimate percentage of women married (world version 92.5%);
  - the inclusion of the effect of son preference (in the Indian version);
- The adaptations made in the disease model have been less extensive:
  - the timing of technology; in casu the onset of the effect of medical technology by changing the effectiveness function.

The adaptations are basically value adaptations. They do not change our postulate on validity the of generic nature of the health transition assumed in the model.

India[a]	1900	1950	1970	1990
Crude Death Rate	42.6	22.8	15.4	10.3
Crude Birth Rate	49.2	40.9	37.8	30.6
Life expectancy	23.0	41.0	50.0	58.0
Children/women	6.3	5.9	5.8	4.1
Mexico[b]	**1900**	**1950**	**1970**	**1990**
Crude Death Rate	45.0	16.1	9.8	5.2
Crude Birth Rate	49.0	46.7	43.5	28.9
Life expectancy	33.0	50.0	61.0	70.0
Children/women	6.3	6.8	6.4	4.2
the Netherlands[c]	**1860**	**1900**	**1950**	**1990**
Crude Death Rate	25.6	18.0	7.6	8.6
Crude Birth Rate	35.9	31.7	22.9	13.1
Life expectancy	38.0	48.0	67.0	78.0
Children/women	6.1	5.4	3.4	1.8

[a] for Mexico see van Vianen et al. (1994) and Bobadilla, (1993).
[b] for India: see Hutter et al. (1996) and Arnold (1993).
[c] for the Netherlands: see CBS (1990); Woytinski (1953); Houwaard (1991) and Mackenbach (1991).

Table 4.3 Historical population figures for the three national case studies.

The energy submodel: TIME

*"Oil has helped to make possible the mastery over the physical world...
Much blood has been spilled in its name."*
D. Yergin, The prize (1991)

5 THE ENERGY SUBMODEL: TIME

Bert J.M. de Vries and Marco A. Janssen

This submodel simulates the supply and demand for fuels and electricity, given a certain level of economic activity. It is linked to other submodels, for example through investment flows, population sizes and emissions. The energy model consists of five modules: Energy Demand, Electric Power Generation, and Solid, Liquid and Gaseous Fuel supply. Effects such as those of depletion, conservation, fuel substitution, technological innovation, and energy efficiency are incorporated in an integrated way, with prices as important signals. Renewable sources are included as a non-thermal electricity option and as commercial biofuels.

5.1 Introduction

Modern societies as they have developed over the last two centuries require a continuous flow of processed fuels and materials. Until some 200 years ago energy needs were largely met by renewable fluxes such as water and biomass. Since then energy has increasingly been derived from the fossil fuels coal, oil and gas. To be useful these fuels have to be extracted, processed and converted to heat and chemicals. For all these steps the production factors labour, land, capital, and energy and material inputs, are required. All three steps are also accompanied by waste flows, the largest being the emission of carbon dioxide (CO_2) during combustion. *Figure 5.1* shows the use of fossil fuels in million tonnes of oil equivalents over the period 1800-1990. The graph shows an increase in the use of coal, followed by the penetration of oil and later natural gas. Superimposed on this are the flows of hydropower and nuclear energy, both in the form of electricity. Traditional biomass (not shown) is also an important energy source; its share is estimated in the order of 55 EJ/yr, i.e. about 13% of total world energy use (Hall and House, 1994).

Coal is a relatively abundant resource in comparison with liquid and gaseous carbon fuels. It fuelled the Industrial Revolution to a large extent and as late as 1930 it was still the dominant commercial fuel. In the 1950s the coal industry was still one of the major industries in the world, employing 1.6 million people; almost two-thirds of world output was concentrated in Great Britain, Germany and the USA (Gordon, 1970; Woytinski and Woytinski, 1953). Since then the contribution made by coal in the commercial energy market has been declining and China, with 26% of world output, has become the largest producer, followed by the USA, with 24% (Anderson, 1995a). The main reason for this is the growing availability of cheap and convenient oil and gas.

Crude oil and a variety of fuels derived from oil have provided an increasing

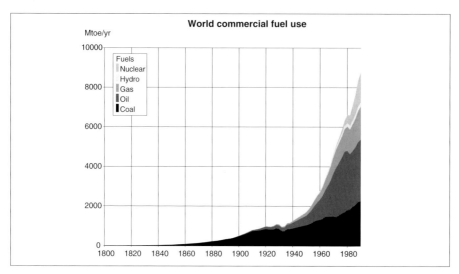

Figure 5.1 Use of fossil fuels and hydro and nuclear power in the world, 1800-1990 (Klein Goldewijk and Battjes, 1995).

proportion of the world's energy needs. Oil products dominate the transport sector: in 1990 transport in the OECD used 37.5 EJ, 99.2% of which was in the form of oil and oil products (Statoil and Energy Studies Programme, 1995). Oil and oil products are among the most widely traded commodities in the world with 80% of all oil produced being traded internationally (Subroto, 1993). Oil exploration, production and processing are to a large extent controlled by multinational oil companies. The involvement of national governments is important because in several countries oil is a significant and in some cases dominant source of government and export income[1]. Since the 1930s natural gas has become a major commercial fuel, first in the USA and later in Europe and Russia. Convenience of use gives it a clear premium value, but transport costs per unit of energy are still much higher than for coal and oil. Flaring of natural gas is becoming less common but still accounts for an estimated 10% of world production. Electric power generation is an important and growing part of the energy-supply system. In the industrialised countries the share of electricity in total final energy use rose from less than 7% around 1950 to more than 17% around 1990 (Nakicenovic, 1989). Construction of power plants and transmission and distribution networks absorb a sizeable proportion of national investments, especially in the early stages of establishing power supplies[2]. Thermal electric power plants require large amounts of fossil fuel, causing major emissions of oxides of carbon, sulphur and nitrogen.

---

1   This is not only true for OPEC countries like the Arab states, Venezuela, Nigeria, Mexico and Indonesia, but also for oil and gas producers such as Norway, Great Britain and the Netherlands.

2   Annual investments in the electric power sector in the 1990s in the developing countries are estimated at $US 10^{10}, equivalent to 12% of total domestic investments (Nakicenovic and Rogner, 1995).

In recent decades numerous analyses have been published on the future of the global energy system, but the emphasis has shifted from the issue of depletion of accessible oil and gas reserves to analyses of the costs and potential of nuclear and renewable energy options. With the fall in the oil price in the 1980s, the focus shifted to the environmental impacts of continued fossil fuel use, which has revived the need for research and development programmes for renewable energy and safe nuclear reactors. With rising oil imports in some OECD countries, strategic issues are becoming prominent again.

In this chapter we first discuss the major issues in energy policy. Next, the energy submodel is described, first as part of the integrated TARGETS framework and then in terms of the separate modules. The focus is on the links between demand and supply, with prices as an important signal for investment and fuel use decisions. Finally, we discuss the calibration of the model for the world at large and the calibration results for the period 1900-1990.

5.2 Energy issues

The major issues for long-term energy policy are how energy use per unit of activity will develop, the extent to which fossil fuels will be available at what costs, whether fossil fuel combustion will have to be constrained because of environmental impacts from emissions, and if so what alternatives will be available and at what costs. Underlying these issues are questions of technology development and transfer, energy prices, and industrial restructuring and consumption patterns. What has been called the energy transition (Naill, 1977) is primarily seen as the shift from fossil fuels to biomass and other solar-based forms of energy. Here, we briefly indicate four major themes; the controversies surrounding them are discussed in more detail in Chapter 13.

Declining energy intensity
In the last few decades energy – and material – intensities have been declining in the industrialised regions. The major reason is a change in activities, products and processes, in combination with new technologies and materials (Grübler and Nowotny, 1990)[3]. It is as yet unclear whether this trend will persist. On the one hand, it is counteracted by trends which go with rising income, e.g. an increasing number of luxury cars and decreasing household size. It may also be reinforced, for instance, through saturation tendencies, less emphasis on material goods and increasing support for 'green' technologies and investments.

---

[3] The change has been variously described as a transition to a service economy, the information age, the prosumer society and the like. It is also denoted by such concepts as dematerialisation and ecological restructuring.

The less industrialised countries are experiencing an industrialisation process which in some respects is similar to the earlier one in Europe and North America. This has resulted in a rise in energy intensity but to levels well below the ones observed in the past for the present OECD countries. In our model changes in energy intensity due to changing activity patterns, products and processes ('structural change') are distinguished from energy conservation. A further distinction is that the latter is split into autonomous and price-induced parts. Unfortunately, even at the sectoral level it is difficult to separate the structural and the price-related changes in energy intensity from autonomous trends (Schipper and Meyers, 1992).

Depletion of fossil fuel resources
The debate about the quantity and quality of (energy) resources has a long history. In some periods, the general mood was dominated by concern about imminent depletion – as in the report to the Club of Rome, 'Limits to Growth' (Meadows *et al.*, 1972), in which depletion of natural resources may become a major cause of industrial collapse. In other periods, it was a non-issue or the general attitude was that undiscovered resources were vast. What really matters is resource quality (in terms of depth, seam thickness, composition and location). In combination with geological probability and prevailing technology and prices, resource quality determines which part of the resource base is considered to be the technically and economically recoverable reserve. There is general agreement that the coal resource base is large enough to sustain present levels of production throughout the next century without major cost increases (Edmonds and Reilly, 1985). Estimates of long-term supply cost curves for conventional crude oil and natural gas are more controversial (McLaren and Skinner, 1987). Liquefaction and gasification of coal and unconventional oil occurrences like tar sands and oil shales also play a recurrent role in the debate.

Emissions from fuel combustion
Fossil fuel (product) combustion is by far the largest source of anthropogenic emissions of carbon dioxide (CO_2), sulphur dioxide (SO_2), nitrogen oxides (NO_x, N_2O), methane (CH_4) and carbon monoxide (CO). Coal is the major culprit, having a specific CO_2 emission coefficient twice that of natural gas. Emissions of SO_2 and NO_x, which are causing serious air pollution, can technically be reduced but the necessary measures are costly (RIVM, 1991). Reduction of CO_2 emissions is possible by increasing energy efficiency, reducing activity levels or switching to non-carbon fuels; the option of CO_2 removal may become feasible in the future for large-scale combustion processes. Assuming relatively scarce low-cost oil and gas resources, many official forward projections indicate an increase in coal use and in CO_2 emissions (IIASA/WEC, 1995; Leggett *et al.*, 1992). Of course, this hinges to a large extent on the assumptions about energy demand growth and on the role of non-carbon energy sources, as is discussed in more detail in Chapter 13.

Alternatives to fossil fuel
There is a long-term trend in the global energy system towards fuels with a lower carbon to hydrogen ratio: away from coal and towards methane. Major options for a further decarbonisation are nuclear energy and electricity from renewable sources. Whereas expansion of hydropower is less than expected due to the increasing awareness of side-effects of large dams, the prospects for electricity from solar photovoltaic cells and wind turbines are improving as costs are declining. Another option to reduce net anthropogenic CO_2 emissions is the production of liquid and gaseous fuels from biomass. Apart from food and fibre biomass is an important source of both energy and materials. After upgrading, biomass can become a substitute for gasoline as is the case in Brazil and the USA, or can be used in electric power generation. There are still major uncertainties about the rate at which biomass fuels can penetrate the market (Johansson *et al.*, 1993).

Other issues with regard to the energy system are strategic dependence and capital requirements. OECD countries are again becoming more dependent on oil from the Middle-East oil; for the fast growing economies of East Asia oil may also soon become a security issue (Calder, 1996). Expansion of the energy system will require enormous investments, an increasing share of which will be needed in the presently less developed regions (Dunkerley, 1995). Capital shortage and the resulting electricity shortages are already thwarting economic growth aspirations in several countries.

5.3 Position within TARGETS

The Energy submodel has been developed as part of the TARGETS and IMAGE models, hence its acronym TIME (Targets IMage Energy model). It simulates the demand for commercial fuels and electricity, given economic activity levels, and calculates the required investments and land to supply these fuels as well as the costs – which then affect demand. The energy model consists of five modules: Energy Demand (ED), Electric Power Generation (EPG), Solid Fuel (SF), Liquid Fuel (LF) and Gaseous Fuel (GF) supply. Energy demand is calculated from sectoral activity levels, which are calculated in the economic scenario generator. This demand is converted to demand for solid, liquid and gaseous fuels, and for electricity, taking into account autonomous and price-induced changes in the energy intensity and price-induced substitution between fuels. Demand for electricity is supplied from either thermal or non-thermal power plants. Demand for secondary fuels, including that for the generation of electricity, is met by primary energy from the three supply sectors. *Figure 5.2* overviews the five modules. A more detailed model is given in de Vries and van den Wijngaart (1995).

The Energy Demand module is the pressure module within the PSIR framework set forth in Chapter 2. It results in capital stocks which exploit and process fossil

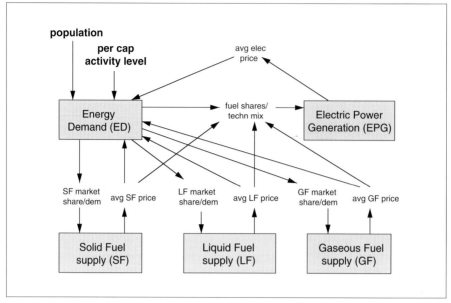

Figure 5.2 The five modules in the Energy submodel.

fuels, generate electricity and increase energy efficiency. Along with the remaining fossil fuel resources, these capital stocks represent the state of the system. Emissions from fossil fuel combustion and use of land for biomass are among the impacts. The response system is endogenous insofar as energy conservation, fuel demand and investment decisions in fossil fuel supply and electricity generation are determined through costs and prices as intermediary variables. There are also exogenous response variables, the most important of which is the levying of fuel taxes and the implementation of demonstration programmes for non-thermal electricity technologies and commercial biofuels.

There are a number of links between the Energy submodel and the other submodels within TARGETS. First, there is interaction between the Energy submodel and the Population and Health submodel: energy demand depends on the exogenous levels of economic activity in absolute terms but also in per capita terms. A second, important link is the one between the Energy model and the CYCLES model. The combustion of fossil fuels generates emissions of CO_2, SO_2, NO_x, N_2O and CH_4, which serve as input for the CYCLES submodel. The land requirements for biofuels are supplied to the TERRA submodel and allocated to grassland and arable land. The expansion of hydropower is linked to the AQUA submodel. The required investments for the energy system are used in the economic scenario generator. *Figure 5.3* indicates the Energy submodel and its interactions with other TARGETS submodels.

During the construction of the various modules we were guided by a few explicit objectives. First, the modules should adequately reproduce the 1900-1990 data on sectoral secondary fuel use, exploration and exploitation in the fuel supply sectors and electricity generation for the world at large. The issue of calibration is dealt with in section 5.7. Secondly, depletion in the form of rising average production costs and technological progress in the form of learning-by-doing have to be incorporated. Thirdly, fuel prices are calculated from capital and labour costs, and should function as signals to direct investment behaviour. Finally, the modules have to allow for at least two non-carbon alternatives, one in the heat market and one in the electricity market. Most modules have also been implemented for the USA and India for 1950-1990 as a basis for validation (van den Berg, 1994). The fossil fuel submodels build on previous energy models, like the Fossil-2 model (AES, 1990; Naill, 1977) and a system dynamics model of the US petroleum sector (Davidsen, 1988). The Energy Demand and Electric Power Generation module build on work from, for example, Baughman (1972), de Vries *et al.* (1991) and Schipper and Meyers (1992).

Evidently, there are deficiencies and omissions in the Energy model, partly a consequence of the very attempt to construct a generic model from regional/local-scale observations and descriptions to be applied at a global scale. Some of these are less relevant because they hardly affect the overall long-term system behaviour. For example, there is the aggregation of various solid fuels into a single one with fixed characteristics. Others may be relevant but more simulation experiments are needed before their consequences can be assessed. For example, the price-driven investment

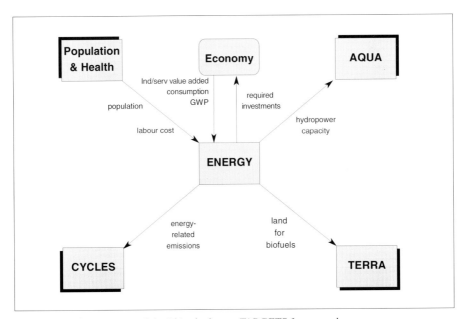

Figure 5.3 The Energy model within the larger TARGETS framework.

behaviour within the oil and coal sectors may be adequate for the USA but fail to capture crucial dynamic factors in regions like China and India. Some elements are not included, which restricts the domain of applicability. Among these are the traditional fuels; the use of fossil fuels and biofuels as feedstocks are not considered either[4]. The options of coal liquefaction/gasification and of Combined Heat and Power are not implemented; capital, labour and land markets are absent; prices and revenues are related to investment decisions but without tracing the corresponding money flows and their macro-economic consequences.

5.4 The energy demand module

End-use energy demand before technology and prices
The main elements of this module were first developed, as part of the ESCAPE and later, as part of the IMAGE2.0 project (Alcamo, 1994; Rotmans *et al.*, 1994). The background for the Energy Demand module is based on the distinction between three determinants of energy intensity: changing activity patterns, products and processes ('structural change'), Autonomous Energy Efficiency Improvements (AEEI, 'technology') and Price-Induced Energy Efficiency Improvements (PIEEI, 'prices'). In the model we first calculate the end-use energy demand which would result without any changes in technology or prices. This, it should be noted, is a non-observable quantity. It is calculated for five different sectors: residential (or consumption), industrial, commercial (or services), transport and others (Toet *et al.*, 1994). Two forms of sectoral end-use energy forms, heat and electricity, are distinguished[5]. Aggregate electricity end-use demand and sectoral heat end-use demand is driven by the product of population and the so-called structural change multiplier, *SCM*. This multiplier, a function of a per capita activity indicator, is calculated for each sector. The multipliers capture the effects of structural change, (i.e. the change in composition of the economic activity) on end-use energy demand. In equation form:

$$SCM_t = \varepsilon_t / \varepsilon_{1900} = [\,\varepsilon_{A \to \infty} + (\beta_1 + \beta_2 A_t)\, e^{-\beta_3 A_t}\,] / \varepsilon_{1900} \qquad (5.1)$$

where ε_t is the energy intensity (in GJ per $) and A_t the sectoral per capita activity indicator in year t. Equation 5.1 shows that the SCM multiplier is normalised to the energy intensity in 1900 and that the energy intensity will drop to some lower limit $\varepsilon_{A \to \infty}$, when the per capita activity reaches very high levels. Depending on the choice

---

[4] Due to their variety of supply and use (crop residues, animal dung, charcoal, fuelwood) and their low status as the 'poor man's fuel', traditional fuels are rarely found in the official statistics and are inadequately measured.

[5] Heat is a shorthand way of referring to all non-electric end-use applications of energy for which commercial secondary fuels are used.

of the parameters, the SCM multiplier may decline from 1900 onwards or first rise and then decline.

Autonomous and Price-Induced Energy Efficiency Improvement

The calculated end-use energy demand is multiplied by the Autonomous Energy Efficiency Increase (AEEI) multiplier to account for the historical fact that even with falling energy prices energy intensity has dropped in many sectors. Formalisation of the underlying technology dynamics is beyond the scope of the present submodel. Hence, we have introduced this autonomous increase for each sector as an exogenous factor which declines exponentially to some lower limit and is linked through a delay to the turnover rate of sectoral capital stocks. The expression for the AEEI factor is:

$$AEEI = \varepsilon^*_{limit} + (1 - \varepsilon^*_{limit}) \, e^{-c \times (t - 1900)} \qquad (5.2)$$

where c is the time-dependent, exogenous annual rate of efficiency increase and ε^*_{limit} the lower limit on the reduction that can be achieved through AEEI. Although the value of ε^*_{limit} is related to the second law of thermodynamics, it is hard or even impossible to base it on physical considerations if output is measured in monetary units.

To incorporate the effect of rising energy costs to consumers, we have opted for an intermediate approach between the bottom-up engineering analyses and the top-down macro-economic approach. It is based on an energy conservation supply cost curve which represents the costs and effectiveness of energy conservation options. This curve is assumed to shift over time; its shape determines, in combination with a

Energy conservation and prices

The sectoral PIEEI multiplier is given by:

$$PIEEI = B_{max} - 1 / [\sqrt{(B_{max}^{-2} + UECost \times PBT \times (1 + d)^{t-1975} / \alpha)}] \qquad (5.3)$$

where B_{max} is the ultimate reduction achievable, UECost the average end-use energy cost and PBT the assumed payback time, which energy users apply within the sector. The time-dependent parameter d reflects the autonomous rate at which energy conservation investments become cheaper. It starts in1975 on the assumption that before 1975 no price-induced changes have occured. The parameter α is a scaling constant which allows gauging the curve to empirical estimates. For example, for $B_{max} = 0.9$, the choice of α / B_{max} indicates the level of the average investment costs per GJ conserved, at which a total reduction in energy intensity of 62% is realised.

The UECost is calculated by dividing the fuel costs by an average (fuel-dependent) conversion efficiency and adding a (fuel-dependent) fixed capital cost component. It should be noted that this formulation implies the use of a price elasticity which depends on the degree of conservation, the energy cost and on time. The price elasticity tends to go down when energy prices go up, reflecting the phenomenon that price changes induce fewer conservation investments once the cheapest options are introduced. The empirical basis for equation 5.3 is given with the energy conservation curve, which represents the cumulative investments as a function of the PIEEI factor, i.e. the price-induced reduction in energy intensity. A variety of such curves has been published in the literature over the past 5-10 years (Blok et al., 1993; Bollen et al., 1996). Reliable estimates are only available for a few countries.

return-on-investment criterion, how many energy efficiency investments are made. Energy demand after AEEI is multiplied by a factor, 1 – PIEEI. The value of the Price Induced Enery Efficiency Improvement (PIEEI) multiplier is determined by end-use energy costs which in turn depend on prices and market shares of secondary fuels. Key parameters are the gradient of the sectoral conservation investment cost curve and the desired payback time which consumers use in deciding to invest in energy conservation. This mechanism, applied irreversibly and with a delay in the sense that action is only taken if energy end-use costs go up, is extended with another factor which lowers the cost curve over time according to an exogenously set rate. This is a simple way to account for the fact that regulation and mass production will tend to make many energy-efficiency measures cheaper over time.

Secondary fuel demand
Electricity demand after AEEI and PIEEI is met by electric power generation as described in the EPG module (Section 5.6). Heat demand after AEEI and PIEEI is satisfied by a price-determined mixture of solid, liquid and gaseous fuels. The next step is to convert this into a demand for secondary fuels. We distinguish four commercial fuel types in the TIME submodel: solid, liquid and gaseous fuels, with the liquid fuels split into light (LLF: gasoline, kerosene etc.) and heavy (HLF: fuel oil and distillates). Fuel wood in the residential sector and all kinds of agricultural and industrial waste flows used for energy functions are not (yet) included. The market shares of these four commercial fuels are calculated for each sector from their relative prices through a multinomial logit function (Bollen *et al.*, 1995, 1996). In the model, actual market shares follow, with a delay, these economically indicated market shares. The change in market shares affects the end-use costs, which in turn determine the degree to which energy conservation actions are taken in year t+1.

There are two additions to this statement. First, the consumers in the five sectors are faced with different prices because transport and storage costs, and taxes and/or subsidies, differ. Moreover, non-price factors influence the decision to use certain fuels, e.g. strategic and environmental. We have therefore introduced a so-called premium factor to incorporate price components which are not included and to account for differences between perceived and actual market prices. These premium factors have also been used to calibrate secondary fuel use. Secondly, the available user technologies and distribution networks did not always allow an unconstrained choice of one of the three secondary fuels. In some cases it turned out to be logical and necessary to constrain the substitutable part of useful energy demand[6]. This we considered a conceptually more plausible approach than adjusting the premium factor to unrealistically high values.

---

6   For example, road transport was not an alternative for rail transport at the beginning of the century so we confined the market share of the transport sector, for which coal was a possible substitute, to 90% around 1900 to 10% around 1990.

5.5 The fuel supply modules

The three fuel supply modules (solid, liquid and gaseous) have a few aspects in common which will be discussed briefly. First, for all three resource bases (coal, crude oil and natural gas) the exploitation dynamics are governed by a depletion multiplier and a learning parameter. The former reflects the rising cost of discovering and exploitation of occurrences when cumulated production increases. The latter works to the contrary by assuming that the capital-output ratio will decline with increasing cumulated production due to learning-by-doing in the form of technical progress. These effects are taken into account by multiplying the respective capital-output ratios of coal, oil and gas with a depletion multiplier and a technology multiplier (de Vries and van den Wijngaart, 1995). Conceptually, we follow here the often used assumption that the cheapest resource deposits are exploited first. In the past, this has obviously not been the case at the world level. For example, an obvious violation was the discovery of the giant low-cost oil fields in the Middle East (Yergin, 1991). We have therefore inserted these discoveries as exogenous, zero-cost exploration successes. However, the hypothesis may be increasingly seen to be correct because of trade liberalisation and the downward trend in transport costs. For oil there is already effectively one world market; a world coal market is in rapid development (Ellerman, 1995). For natural gas, this is not yet the case due to high transportation costs. Transporting gas in an onshore pipeline might cost seven times as much as oil; to move gas 5000 miles in a tanker may cost nearly 20 times as much (Jensen, 1994).

A second important element in the liquid and gaseous fuel module is the possibility of a non-carbon based alternative fuel penetrating the market. This alternative is confined at present to a biomass-derived liquid/gaseous fuel alternative, for which land will be an important input. Labour may be an important input, especially in low-labour productivity regions. In fact, biofuels may initially only have a competitive advantage – apart from strategic considerations – because large amounts of cheap labour can be absorbed. More specific conversion routes, e.g. hydrogen from biomass, solar heat or electricity, have not explicitly been modelled in the current version. We will now discuss each of the three modules in more detail.

The Solid Fuel (SF) module

The SF module is represented in *Figure 5.4*. The most important short-term loop is the demand–investment–production–price loop. Given a demand for solid fuels from the ED module, the anticipated demand generates investments into new production capacity. These investments form a fraction of the revenues, depending on the price-to-cost ratio, and are distributed among underground and surface coal mining operations on the basis of the production cost ratio. For underground coal the capital-labour ratio rises according to an exogenous time-path. An important longer term loop is the solid fuel price changing in response to depletion and learning, which in turn affects coal demand calculated in the ED module. Learning is incorporated by

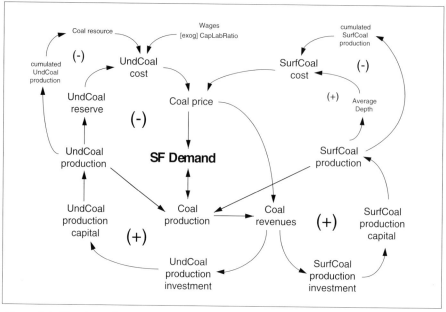

Figure 5.4 The demand-investment-production-price loop in the Solid Fuel (SF) module. The left-hand side represents the demand-driven exploitation loop for Underground Coal (UndCoal) with depletion and capital-labour substitution. The right-hand side is the same for Surface Coal (SurfCoal), along with depletion and learning (COR Capital Output Ratio; CapLabRatio Capital Labour Ratio).

multiplying the capital-output ratio for surface-mined coal by a factor which declines as a linear function of the logarithm of cumulated production. This multiplier is set at one for 1980. The coal price is calculated by adding the capital costs for upgrading and transport. We consider only one generic type of coal, at 29 GJ/tonne, also referred to as 'solid fuel'. The use of coal as feedstock is not accounted for, except in the case of coking coal for pig iron production, where it is part of industrial fuel use.

The life cycle of coal is based on the distinction between the resource base, identified reserves and cumulated production. The resource base is explored and discovered, i.e. converted into identified reserves. An exogenous discovery rate is entered to match trends in past reserve estimates. Coal companies decide to invest in coal producing capacity on the basis of anticipated demand for solid fuel. This anticipated demand represents a trend extrapolation over a time horizon of T years of the form $(1+r)^T$ with r being the annual growth rate in the past 5-10 years. How much is invested in coal production is based on the return on investment value: the larger it is, the higher the fraction of coal revenues re-invested in the industry. The share of this investment flow that goes into underground mining depends on the cost ratio between underground and surface-mined coal in accordance with a multinomial logit function.

The investments add to the coal-producing capital stocks, the output of which is determined by the capital-output ratios, γ_{prod}. These are assumed to depend on three trends which have been observed in the past in various degrees and combinations:

- as exploration proceeds, newly discovered deposits tend to be of lower quality, i.e. deeper, narrower and more distant. This is modelled by dividing γ_{prod} by a depletion-cost multiplier (<1);
- in the labour-intensive underground coal mining, labour productivity increases over time as more capital per labourer is used (Cobb-Douglas form of production function); the capital-labour ratio is exogenous input;
- over time, capital costs to find and produce one unit of coal tend to decline due to technical progress of all forms. For underground mining this is implicit in the capital labour substitution. For surface mining it is modelled by multiplying γ_{prod} by a technology factor (<1), which is a function of cumulated production.

After the calculation of capital stocks, actual coal production equals coal production capacity unless the ratio between coal demand and coal production capacity exceeds 0.9, in which case the coal capital utilisation rate increases to 1.0 for a capacity shortage of 20%.

An important input for the ED module is the coal price. The capital costs of coal are calculated as an annuity factor times the production capital stock, divided by the annual production. For underground mining the labour costs are also included. For surface-coal mining, labour costs are taken to be a fixed and small fraction of the capital costs. The wage rate is assumed to be a time-dependent fraction of average consumption per capita. The average coal cost c_{SF} is a weighed average of the cost of underground and surface coal. The coal price is also influenced by the demand-supply (im)balance through the Supply Demand Multiplier SDM. The average mine-mouth price is now given by:

$$p_{SF} = SDM \times a \times c_{SF} / P_{SF} \qquad [\$/GJ] \quad (5.4)$$

with a the annuity factor[10] and P_{SF} the annual Solid Fuel or coal production. If the price changes in response to an excess or shortage of capacity, this decreases or increases revenues, which in turn generates lower and higher investments, respectively with a delay. The last step is to incorporate the capital requirements and resulting add-on costs for transport and upgrading of coal. This is done with a constant factor which also accounts for conversion. It is assumed that 90% of these additional costs are in the form of annuity payments for investments. Energy, mostly Heavy Liquid Fuel, for coal transport is not explicitly included.

---

10 The annuity factor $a = r/[1-(1+r)^{-EL}]$ with r being the interest rate and EL the economic lifetime of the investment.

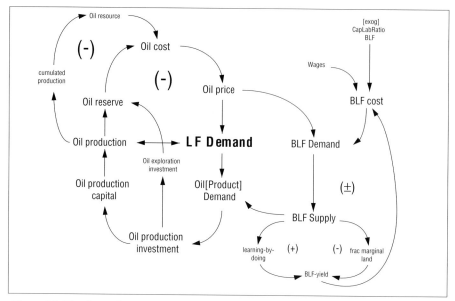

Figure 5.5 The demand-investment-production-price loop in the Liquid Fuel (LF) module. The left-hand side represents the demand-driven oil exploitation loop with price-induced exploration, depletion and learning; the right-hand side represents the penetration, depletion and learning dynamics of BioLiquidFuels (BLF) (COR Capital Output Ratio; CapLabRatio Capital Labour Ratio).

The Liquid Fuel (LF) and Gaseous Fuel (GF) modules

Figure 5.5 shows the LF module as a causal loop diagram. The GF module has an almost identical structure and therefore will be discussed only where it differs from the LF module. Some model elements are similar to the SF module. The most important short-term loop is the demand – investment – production – price loop. The anticipated demand for liquid fuels generates investments into new production capacity. Exploration investments are determined by the desire to keep the reserve-production ratio (RPR) sufficiently high. A certain profit level in terms of revenues over costs is required to sustain exploration investments. In a second loop, the liquid fuel price changes in response to depletion and learning dynamics due to oil production, which in turn affects liquid fuel demand. The underlying mechanisms are a capital-output ratio which rises with declining marginal resource quality and which decreases due to learning-by-doing. Another long-term element is the penetration of biofuels: if biofuels can be produced at competing cost levels, investment will be made in biomass plantations. Costs depend on technology through learning-by-doing and on depletion due to the use of less productive land. There is an impact on food production in the TERRA submodel through land use for plantations.

As with coal, the life cycle of oil and gas is based on the distinction between the resource base, identified reserves and cumulated production. Oil is discovered, i.e. converted from resource into identified reserves, produced and combusted. The ultimately recoverable oil at the technology and price levels throughout the simulation period is itself a function of learning with expert bias (Sterman and Richardson, 1983). The ED module simulates the demand for liquid fuels in two forms: Heavy Liquid Fuels (HLF) and Light Liquid Fuels (LLF)[11]. The fraction of total demand, which is in the form of LLF, is given exogenously. Only a fraction of LLF demand is satisfied by oil products. The remaining market share (μ_{BLF}) is supplied by commercial biofuels (BioLiquidFuel BLF) and depends on the relative cost of LLF and BLF. The required production of crude oil can now be calculated using an overhead factor covering exploitation and refinery energy use and losses. This determines investments in the oil exploitation, and in transport and refining. If the reserve-production ratio (RPR) is below a desired level (RPR$_d$) and the average market price, $p_{CO,avg}$, is sufficiently high, oil companies will also invest in crude oil exploration[12].

How much to invest in oil production capacity depends on the capital-output ratio, γ_{prod}, of the crude-oil-producing capital stock. New investments are equated to the depreciation of the existing capital stock plus the required additional capacity. The resulting equation is of the form:

$$dC/dt = C_{req} - C/EL \qquad [\$/yr] \qquad (5.5)$$

with EL the economic lifetime and C_{req} the additional required capacity, which depends on demand and identified reserves. If the market price is less than the price required to make a profit, investments will be lower than required - as with exploration investments. One assumes it will take some years before investments generate new reserves or lead to oil production. The next step is to calculate the cost of oil products and gas. As with coal, the key factor is the capital-output ratio for production, γ_{prod}, which changes over time, as has been discussed for coal. Capital costs are calculated as an annuity factor times the production capital stock plus the exploration investments divided by the annual oil production. When the ratio between required and potential production approaches or exceeds one, the crude oil price will go up and this will increase exploration and exploitation investments (Supply Demand Multiplier SDM, see equation 5.4). Capital costs for transport and downstream operations (refining) are linked to production capacity and to the LLF fraction to account for be additional cost of 'whitening the barrel'. From this, the costs to deliver HLF and LLF are calculated.

Biofuel penetration is simulated using a production function with capital, labour and land as production factors. A fixed capital-output ratio, g_{BLF}, and an exogenously increasing capital-labour ratio, CLR_{BLF}, reflect the transition towards less labour-intensive techniques. Land requirements are derived from a land-output ratio, β,

---

[11] In the GF model no distinction is made between various types or grades.

[12] The dynamics of recovery technology, which allows a larger fraction of the oil-in-place to be produced, is not explicitly taken into account (Davidsen, 1988).

which increases due to technology and decreases when the exogenously set supply potential is reached. The latter represents the assumption that increasingly less productive land is used for biomass plantations. Given some initial estimate of the cost of BLF, the penetration dynamics rests on the assumption that the market share for commercial biofuels is a function of its cost relative to the LLF price. The economically indicated market share as determined from a multinomial logit formula induces either private or public firms to invest into plantations producing biofuels. Calculating the required amount of labour L_{BLF} from the exogenous capital-labour ratio, the cost of biofuel can be expressed as:

$$c_{BLF} = [a\,(C_{BLF} + P_{BLF}/\beta) + (\mu_{BLF} \times D_{LLF} \times g_{BLF}/CLR_{BLF}) \times p_L\,]/P_{BLF} \quad [\$/GJ] \quad (5.6)$$

with P_{BLF} being the actual BLF supply, a the annuity factor and p_L the price of labour. The BLF price – equated to BLF costs plus a fixed profit margin – in relation to the LLF price will determine its future market share.

5.6 The electric power generation module

Figure 5.6 contains causal loop diagram of the major elements in the EPG module. Most important is the demand–investment–price loop. It simulates the planning process in which a future demand is anticipated on the basis of which new capacity is

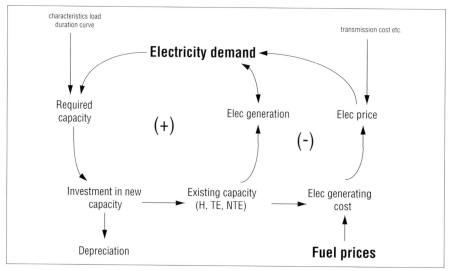

Figure 5.6 The demand – investment – price loop in the Electric Power Generation (EPG) module. Electricity demand leads to investments in new capacity, which determines with the fuel costs the electricity generation costs. The learning dynamics of non-thermal electrical (NTE) power generation is not indicated.

ordered and put into operation if there are no capital or other constraints. In combination with fuel and transmission costs, this determines the electricity price which in turn affects the demand for electricity. We assume that the investor decides on new capacity (in MWe) by anticipating growth in electricity demand in combination with a preferred reserve factor. The three capital stocks represent hydropower, thermal and non-thermal electricity generation[13]. Expansion of hydropower (H) capacity is an exogenous scenario, assuming increasing marginal specific investment. Whether the remaining new capacity ordered is thermal electric (TE) or non-thermal electric (NTE) is based on the difference between the production costs. The characteristics of the three capital stocks, hydropower, thermal and non-thermal electric, change over time. For thermal power plants, conversion efficiency and specific capital costs (in dollars per MWe) are exogenous time paths. For the non-thermal power generating options, cumulated production induces learning, which shows up as decreasing specific investment costs and hence lower total costs. This in turn will accelerate the share of these options in investments. The market share of each of the three fuels (solid, liquid, gaseous) is based on relative fuel prices[14].

Operation of electric power systems is done on the basis of rather sophisticated operational rules (de Vries *et al.*, 1991). A number of simplifications have been introduced. First, the net demand for electricity from the ED module is converted into anticipated gross demand similar to fuel demand and split into two fractions: base load and peak load. The calculation of the required capacity, and hence the required investments, is then derived from the assumption that each generating option has a constant load factor, i.e. fraction of the year that it is operated. From this, the thermal capacity required for base-load operation is calculated[15]. The required peak-load capacity, E_p, is then calculated as:

$$E_p = (1 - BF) \times ED_{gr} / (PLF_{max} \times \beta) \qquad \text{[MWe]} \qquad (5.7)$$

where ED is gross electricity demand, PLF_{max} the maximum load factor for capacity operated in the peak-load periods, $PLF \leq PLF_{max}$, and β the conversion factor from GJ to MWe ($\beta = 8760 \times 3.6$). The total required installed capacity is the sum of required base-load plus peak-load capacity, including a reserve margin to guarantee a desired level of reliability in the load-factor estimates. From this the required investments are calculated.

---

13 For the world at large, this is not unrealistic; for smaller regions resources like hydro and windpower, with their seasonal variations, cannot be simulated accurately in this way.

14 There is one generic type of thermal power plant. Differences in capital and operating costs, and efficiencies, are assumed to average out and all thermal capacity is, with a delay, assumed to be multifiring.

15 With a large expansion programme for non-thermal and hydro capacity, this may become negative, in which case it is set equal to zero.

If the required production implies $PLF > PLF_{max}$, there is capacity shortage and only the fraction PLF_{max}/PLF of peak demand is produced. Such a situation can result from an unexpectedly fast increase in demand combination with long construction periods or delays, or when the economy cannot or does not sustain the required investment flows. Also the fairly low reliability of power stations and transport systems contribute to capacity shortages and unsatisfied demand, a situation which occurs in various parts of the world. The reverse, overcapacity, shows up as increasing costs which negatively affect demand. If the ratio between the actually installed and the required system's capacity drops below one, the anticipated required electric power capacity is divided by this ratio; this provides an additional signal to install new capacity. If there is no capital constraint, the investments lead to expansion of the three electricity-producing capital stocks. The capital stock for transmission is taken to be proportional to the system's installed capacity.

For thermal electric power generation an important question is which fuels are used. In the EPG module the answer to this question is based, as in the other modules, on relative prices. A premium factor is used to allow for differences between fuel costs and the prices as perceived by utilities[16]. The next step is to calculate electricity prices, since they are input for the ED module. This is done in a way similar to the cost calculations in other modules: capital costs are put on an annuity basis and fuel costs are derived from fuel use times fuel prices. The penetration dynamics of non-thermal electric power technology (e.g. nuclear, solar) is governed, as with biofuels, on the basis of the relative generation costs of the thermal and the non-thermal option. Again, a multinomial logit formulation is used. It implies that the learning coefficient is crucial for the penetration of non-thermal capacity because this largely determines the rate at which specific investment costs decline as a function of cumulated production. This is a positive, reinforcing loop. If the sum of hydropower and non-thermal capacity exceeds the required base-load capacity, non-thermal capacity will also be put into operation for peak load whenever thermal capacity is less than the required peak-load capacity (see equation 5.7)[17]. Hence, its average load factor decreases; this drives up non-thermal generating costs, which in turn will slow down its penetration rate – a negative, stabilising loop.

5.7 Calibration

Procedure and assumptions
Calibration has been done for the period 1900-1990 for the whole world. The

---

[16] For example, for electricity generation in OECD Europe, Moxnes (1989) has found that as of 1983 coal has a premium equivalent to a price discount of 29%, whereas natural gas has been discriminated against at the equivalent of a 12% price increase.

[17] It is assumed that hydropower will never exceed the required base-load capacity.

statistical data used for the calibration come from a variety of sources (International Energy Agency IEA, 1990; Klein Goldewijk and Battjes, 1995). First, historical data on commercial fuel use have been collected and used to calibrate the ED module. For this, sectoral activity levels[18] and sectoral fuel prices to drive the model are employed. Important parameters for the calibration are: the end-use energy demand as a function of activity level (Equation 5.1), the rate of autonomous energy efficiency increase (AEEI) and its lower boundary (Equation 5.2), the form of the conservation investment cost curve and its rate of change (Equation 5.3), and the fuel cross-price elasticities and premium factors.

In a second step, each of the four supply submodules has been calibrated, using historical supply and price paths. Supply has been set equal to demand. The most important variables in the calibration are: the resource base estimates; the capital-output ratios and the corresponding depletion multipliers for coal, oil and gas; the substitution coefficients for the various investment allocations; the efficiency of thermal capacity and the premium factors for fuels used for electric power generation. The learning coefficients for surface coal mining, oil and gas exploitation, biofuels and non-thermal electric power generation have also been adjusted. The costs of nuclear-electricity, the dominant NTE option, have increased because of additional safety measures and long construction delays. In the model this has been reproduced by assuming a negative learning rate for the period 1965-1990. The supply-demand multiplier relations (Equation 5.4) are adapted from Naill (1977), Davidsen (1988) and Stoffers (1990)[19]. For a more detailed discussion, see de Vries and van den Wijngaart (1995), de Vries and Janssen (1996) and van den Berg (1994).

To perform the calibration for the integrated submodel, we needed a number of iterations during which a limited set of parameters within the four supply modules had to be adjusted to correct for minor discrepancies between simulated and historical values. It should be noted that model calibration is not an unambiguous procedure. In the ED module, for example, end-use energy demand is a non-observable quantity: it is implicit in the actual observations of secondary fuel use and activity level. Hence, a multiplicity of parameter calibrations is possible. The same holds for the relative importance of technology vs. depletion in the fuel supply modules.

Results
Figures 5.7a-h show a series of simulation results for the world of 1900-1990. Simulated total secondary energy use is compared with historical primary energy use

---

18 We have chosen the indicators used in IMAGE2.0: value-added in stable (1990) US dollars for industry and commerce, consumption expenditures in stable (1990) US dollars for residential areas, and GWP in stable (1990) US dollars for transport and other (Toet et al., 1994).

19 Two exogenous events were introduced which cannot be expected to be simulated, as the underlying dynamics do not form part of the model formulation: discovery of large oil fields between 1950 and 1970 (Middle East) and a crude oil-price increase of 50%-400% between 1973 and 1987 (oil-price crises).

because there is no data on secondary energy use (*Figure 5.7a*). As can be expected, primary energy use is higher but the trends are correct. Simulated electricity use closely match with historical data. The key assumptions – and hence ambiguities – concern the structural change multiplier and the rate of autonomous efficiency improvements. We had to assume rapidly increasing energy intensity for the transport sector and electricity to represent the emergence of new transport modes and electrical applications. The substitution dynamics from traditional to commercial fuels, especially relevant for the residential and industrial sector, are implicit in the structural change parameter estimates. Only after the rise in fuel prices in the 1970s, do the price-induced energy efficiency improvements cause a slightly faster decline in energy intensity. It turned out that for all sectors premium factors different from unity are required to simulate the substitution among secondary commercial fuels. There is a variety of possible and sometimes plausible explanations, one of them being that we have kept the conversion efficiency (from secondary fuel to useful demand) constant[20] and another one being the differences in quality and convenience.

Electricity was supplied by hydropower, thermal and non-thermal electric power capacity. *Figure 5.7b* shows the emerging dominance of thermal capacity, with coal as the major fuel and the reduced growth in fossil fuel use due to the introduction of nuclear (NTE) capacity. Coal is the major fuel used to generate electricity (*Figure 5.7c*). The rise in oil prices in the 1970s shows up as a declining share of Heavy Liquid Fuel. For coal we had to apply a cost reduction which reflects the lower coal prices for large-scale utility users.

The fuel supply side is shown in *Figure 5.7d-f*. For all commercial fuels, the model generates declining prices until 1970, when exogenous price shocks were applied (*Figure 5.7g*). The jumps in the two first decades are partly caused by the

Prices and technology – their relative importance

To understand the role of energy prices in the model, we did some experiments in which the exogenous oil price crises were left out. It turns out that its impact mean a slow-down in energy use and a smaller market share for oil, as expected. Setting all premium factors to zero causes oil and especially gas to penetrate much faster than has happened historically. Evidently, the premium factors also account for the lack of infrastructure (pipelines, equipment) which significantly delayed the use of natural gas. If the exogenous technological constraints in the model are also removed, the system immediately jumps to the present market shares for oil and gas, which is a price-determined equilibrium. The longer term consequence is that oil and gas are depleted more rapidly and coal is regained earlier and stronger. These simulation experiments point to the importance of non-economic factors in explaining the energy system evolution over the past 90 years. Our simple way of introducing the complex dynamics of technical innovations in the form of exogenous constraints to market penetration turns out to be a decisive factor in calibrating the model.

---

20 The conversion efficiencies from secondary fuel to end use are the same for all sectors and constant, at 0.65 for coal, 0.75 for liquid fuels and 0.85 for gaseous fuels.

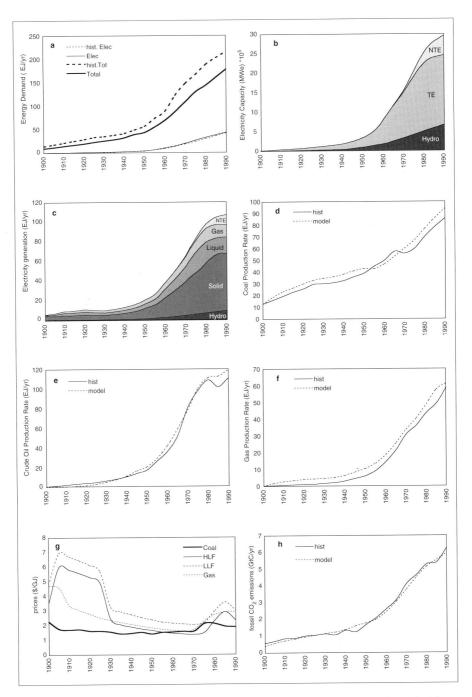

Figure 5.7a-h Simulated primary fuel use and electricity use follow historical trends (a); so do coal, crude oil and gas production (d - f) and CO_2 emissions (h). Prices decline until 1970. Then exogenous price shocks are applied (g).

model initialisation. For thermal electric power generation, costs are also declining but for nuclear power, adjustment of the learning factor causes the historical rise in costs. Rising underground-mined coal costs due to rising labour costs and depletion are mitigated by the penetration of lower-cost surface-mined coal. For oil and gas, the exploration and production costs are low and declining, because between 1900 and 1990 learning-by-doing is assumed to have compensated the increasing scarcity of the reserve base. For natural gas, the actual market price for consumers is initially much higher than what is calculated from supply-side considerations because of a high premium factor. Total primary energy use and their shares can be reproduced fairly well with the simulated demand and prices. The resulting CO_2 emissions from fossil-fuel combustion are within 5% of estimates in the literature (*Figure 5.7h*).

5.8 Conclusions

On the basis of simulation experiments carried out thus far, several conclusions can be drawn which highlight some characteristics of the Energy submodel. A first conclusion is that energy conservation in response to rising secondary fuel and electricity prices can be expected to slow down unless one assumes that standardisation and learning etc. continuously reduce the costs at which such efficiency improvements can be realised. Secondly, substitution between fuels tends to dampen price increases in any one particular fuel, unless secondary fuel prices are linked through markets or government agreements. This effect, however, is rather small and is influenced by the assumption of constant conversion efficiencies from secondary fuels to end-use. Thirdly, autonomous increase in energy efficiency can be expected to be the major determinant of sectoral fuel and electricity use, but the level of useful energy per monetary unit of activity is an equally important factor in the longer term.

With regard to fuel supply, the main conclusion is that the simulations correctly reproduce the historical time paths of reserves, production and costs until 1975. However, this requires a rather intuitive assessment of the relative importance of depletion and technology effects. The oil crisis of the 1970s caused a sequence of events which can only be reproduced by adjusting parameters in such a way that they implicitly account for mechanisms and behaviour which are absent in the model. The absence of an explicit coupling between coal, oil and gas prices is one of the causes of discrepancies between simulation results and historical data. A few topics, among them the formation of secondary fuel prices, the availability and cost of labour, and the quality characteristics of the reserves deserve closer scrutiny. Because the simulation experiments presented here are for a single global aggregate, interactions between regions are absent which, may be highly relevant in the real world. This aspect will be introduced in the next, regionalised version of the model as part of the IMAGE 2 model.

The water submodel: AQUA

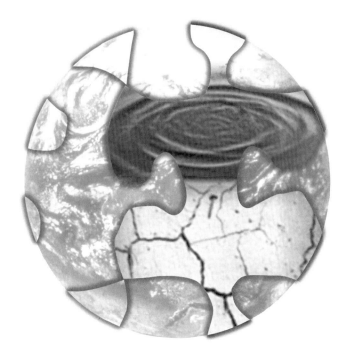

*"I can foretell the way of celestial bodies,
but I can say nothing about the movement of a small drop of water."*
Galileo Galilei

6 THE WATER SUBMODEL: AQUA

Arjen Y. Hoekstra

This chapter introduces the integrated water assessment tool AQUA. This model has been developed to assist policy analysts assessing complex global water problems. A number of major long-term water policy issues are reviewed and AQUA is calibrated and validated against values in the literature. The model has been further validated at the regional level in studies that focus on the Zambezi and Ganges-Brahmaputra river basins.

6.1 Introduction

Managing water resources has become an independent field of expertise and a separate domain of public policy. Nowadays, however, many problems of water scarcity and pollution interact with socio-economic development and environmental change to such an extent that a single discipline or sector approach can no longer provide satisfactory solutions. Water flows are purposely being regulated via reservoirs, dikes, canals, and irrigation and drainage schemes. However, runoff is unintentionally being affected by changes in land cover and soil degradation in many parts of the world; future water availability may be affected by climate change; and water pollution is often part of a disturbance of total element cycles. In developing countries, limited availability of clean, fresh water is a major cause of many diseases. The world's food supply increasingly depends on irrigated agriculture and thus again on the availability of fresh water. There is a growing recognition that studies should focus on the interlinkages and feedback mechanisms between water and (other) environmental and human systems (Young *et al.*, 1994).

This chapter discusses an integrated water assessment tool, the AQUA submodel, that has been designed to analyse complex water problems which cannot be understood without adopting a comprehensive approach towards environmental change and socio-economic development. The model describes changes in the hydrological cycle as a combined effect of human water consumption, reservoir building, land-use changes and climate change. The purpose of the AQUA submodel, an integral part of the TARGETS model, is to support analysts in developing and evaluating water policy as part of a policy for sustainable development. In this chapter, we discuss the types of issues that can be tackled with AQUA, the methods used and the structure and underlying assumptions of the tool. The chapter concludes with some calibration results and a discussion of the validity of the tool. A full description of the model is given in Hoekstra (1995, 1997), while in Chapter 14 a series of model experiments is presented.

6.2 Water policy issues

An effort is made below to identify the major water policy issues with which we are confronted today. The discussion focuses on long-term developments and aims to trace the main mechanisms underlying water-related problems.

Fresh-water availability
The availability of fresh water depends on the fresh-water recharge rate, i.e. precipitation minus evaporation. A first mechanism affecting fresh-water recharge is increased evaporation due to water use by humans and the construction of fresh-water reservoirs. Evaporation rates also change as a result of land cover changes such as deforestation and loss of wetlands. The availability of fresh water not only depends on the recharge rate, but also on the part of this recharge that forms *stable runoff*, i.e. runoff available throughout the year. The stable runoff is often increased by the construction of reservoirs, but reduced by intensified land use (deforestation, urbanisation, erosion) and river canalisations. Finally, the availability of *clean* fresh water is affected by pollution. For the longer term, global climate change as a result of increasing atmospheric concentrations of greenhouse gases, will not only result in changing temperatures but also in changes in evaporation and precipitation patterns. The global average rates of evaporation and precipitation will increase - the so-called intensification of the global hydrological cycle - but locally changes may differ considerably: some areas are expected to become drier and others to become wetter. To set priorities in water policy-making, it would be desirable to have an estimate of the relative importance of the different factors affecting future availability of clean water.

Water pollution
Human activities have increased the concentrations of various substances in both surface water and groundwater throughout the world. Total dissolved nitrogen and phosphorus concentrations in surface waters, for example, have globally increased by a factor of two and locally – in Western Europe and North America – by factors of 10 to 50 (Meybeck, 1982). This increase has become manifest in many rivers and lakes as eutrophication: richness of nutrients, excessive plant growth and deprivation of oxygen. The increased nutrient concentrations are due to domestic and industrial wastewater disposal, agricultural fertiliser use, increased erosion and increased atmospheric deposition. Other types of water-quality deterioration result from the disposal of wastes containing heavy metals and organic micropollutants like PCBs, and from the use of pesticides in agriculture. Water-quality changes are in fact part of a disturbance of total element cycles (Chapter 8). This results in an accumulation of substances in some of the spatial compartments and depletion in others.

Actual versus potential water supply
The increase of water supply in this century has largely been made possible by an

increase of the potential supply through artificial reservoirs (*Figure 6.1*). Nevertheless, a declining difference between actual and potential water supply is a trend found nearly everywhere on Earth (Kulshreshtha, 1993). The problem for the next century is that water demand will keep on growing while the possibilities for extending the potential supply through new reservoirs will run short. To safeguard future water supply, one can attempt to further increase the potential supply, but cutting down the growing demand will be a necessity (Young *et al.*, 1994). water-demand policy may comprise proper water pricing and technological development to bring about a more efficient water use. Beside the construction of new supply infrastructure, policies for water quality, land and soil management and climate policy could form a part of a water supply policy. The trade-off between water-demand and water supply policy does not only depend on the region, but also on the development stage. In developing regions with high population growth rates and with a total demand still far below the potential supply, the main factor causing water shortage is often the lack of infrastructure. In regions where demand moves closer to potential supply, the major cause of water scarcity becomes the rapidly growing demand itself. Major driving forces behind the increase of water demand are urbanisation, industrialisation, the spread of irrigated agriculture, population growth and the rise in living standards. In solving the problem of increasing demands at the root, it would be appropriate to address these underlying mechanisms. However, phenomena like population growth, food production and economic development are

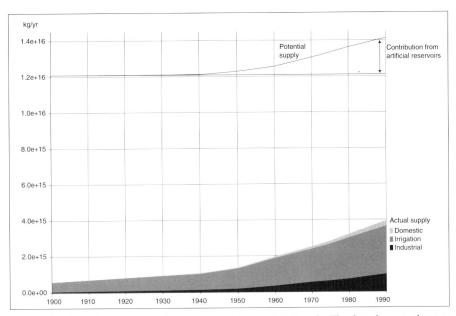

Figure 6.1 Actual versus potential water supply on a global scale. The data for actual water supply are taken from Shiklomanov (1993). The data for potential water supply are from this study (see Figure 14.3).

generally not considered to be a subject of water policy. This illustrates that water-related problems do not necessarily have water-related solutions and underlines the necessity of an integrated approach.

Water and food supply
The fivefold increase in global irrigated areas since the beginning of this century has gone along with the so-called Green Revolution in agriculture. At present, about 36% of the global harvest comes from the 16% of the world's cropland that is irrigated (Postel, 1992). Agricultural water use contributes to about 65% of the global human water use (Shiklomanov, 1993). These numbers indicate irrigation as a major cause of present water shortages. Given the projected population growth and increase in food demand, the demand for new irrigated land will keep growing in the future. The challenge is to assess how long the benefits of irrigation as a solution to growing food demand will outweigh the costs.

Sea-level rise
A global temperature increase, observed this century and projected for the next century, will result in thermal expansion of the ocean and enhanced melting of glaciers and ice sheets and thus in sea-level rise. Another mechanism that contributes to sea-level rise is via a long-term loss of water on land because of large-scale groundwater withdrawals, surface water diversions and land-cover changes (deforestation and wetland loss). By contrast, the construction of fresh-water reservoirs on land contributes to a sea-level decline. Sahagian *et al.* (1994) argue that the present rate of sea-level rise as a net result of groundwater withdrawals, deforestation, wetland reduction and reservoir construction is at least 0.54 mm/yr. Estimates of future sea-level rise vary considerably. According to the IPCC, we may expect a total rise of 13 to 94 cm over the period 1990-2100 (Warrick *et al.*, 1995). This uncertainty range does not yet include the very large uncertainties on the balance between cumulation and melting of the ice sheets of Antarctica and Greenland. According to Delft Hydraulics and RIKZ (1993), a future sea-level rise of 1 m would increase the number of people subject to annual flooding from the present 40-50 million to nearly 60-75 million, not accounting for population growth. It would also threaten half of the world's coastal wetlands. In anticipating the possible impacts of a sea-level rise, one has to consider what strategy would be most efficient: preventive (climate policy or land and water policy) or curative (coastal defence policy).

6.3 An integrated approach to water policy issues

Our hypothesis is that understanding today's water-related policy issues requires an integrated analysis of hydrology, water quality, water demand and water supply in relation to land-use changes, soil degradation, element cycles and climate change.

Such an analysis would also consider food and energy supply, human development and economics. Integration here includes different components (see Chapter 2). *Vertical integration* means an analysis of the complete cause-effect chain. A vertically integrated water analysis not only considers the dynamics of the water system (distinguishing a hydrological and a water-quality component), but also the nature of the various pressures on the water system (water demands, emission of pollutants, land-use changes, erosion, climate change) and the impacts in terms of a changed performance in the various socio-economic and ecological functions of the water system. *Horizontal integration* means that the interrelationships and feedback mechanisms are considered between the water-related processes and other environmental and socio-economic processes. A water analysis is horizontally integrated if it is embedded in a comprehensive analysis in which ecosystems, land use, soil, climate, energy, economics and human health are also included.

An integrated water analysis can provide the boundary conditions and priorities for more detailed analyses *(top-down approach)*. The other way round, detailed studies are necessary to guide the conceptualisation in an integrated analysis *(bottom-up approach)*. If a regional or short-term water policy analysis is not embedded in a more integrated analysis, there is a risk of suboptimisation.

Since interests have become more interwoven and problems more complex, there is a noticeable trend from problem-specific water analyses towards more integrated and more interdisciplinary analyses (Wisserhof, 1994).* This has resulted in comprehensive and elaborate computational frameworks, linking meta-models of originally separate (expert) models. Although these computational frameworks have proved to be very useful, they provide a poor basis for studying the coherence of water policies with other environmental and socio-economic policies. The reason is that these tools are *vertically* integrated to some extent but lack *horizontal* integration. The need for horizontal integration was the starting point for the development of AQUA.

AQUA has been designed as a *generic framework* so that it can be applied at different spatial scale levels, from the river basin to the global level. Genericity means that the different applications have the same modular structure using the same set of equations. An evident reason for applying AQUA at river basin level is that the physical aspects of water problems are typically river-basin specific (UN, 1958; 1970). Another reason is that considerable modelling experience is already available at this level, thus providing the possibility to link up with earlier experience; it will also be easier to validate the approach chosen. The rationale for applying AQUA at a global level has to do with our aim of including as many of the interrelationships and feedback mechanisms between water, environment and human beings as possible. Starting at a lower scale excludes the possibility of reckoning endogenously with global mechanisms like climate change and sea-level rise. Even an apparently regional mechanism like the water demand-supply mechanism has a global component due to long-distance exports of products from irrigated areas.

At present, the generic framework has been operationalised for the world as a whole and for two specific river basins: the Zambezi and the Ganges-Brahmaputra basins (Hoekstra, 1997). We will briefly discuss the river basin versions in section 6.5, but this chapter will focus on the global version of AQUA. The simulation of some processes, like runoff, have appeared to be possible for relatively easily validation at river-basin level, while this is very difficult at global level. The river basin versions of AQUA have served to provide some credibility to the simulation experiments with the global version.

6.4 Model description

6.4.1 Main model structure

The AQUA submodel follows the integrated systems methodology as described in Chapter 2. Following the Pressure-State-Impact-Response (PSIR) ordering mechanism, the following modules are distinguished, and presented in *Figure 6.2*:

- a *pressure* module calculating water demand from determinants such as population size, gross world (or national) product, industrial production, demand for irrigated cropland and water-supply efficiencies. Besides, the fractions of the water returning to groundwater or surface water and of water lost through evaporation are simulated;
- a *state* module describing hydrological processes and fresh-water quality. The hydrological cycle is modelled by distinguishing different water reservoirs and by simulating the flows between these reservoirs. This yields estimates of, for example, net precipitation, river runoff, groundwater level decline, fossil groundwater depletion and sea-level rise. Water quality is described in terms of four water-quality classes;
- an *impact* module calculating actual water supply to households, agriculture and industry; water supply to terrestrial ecosystems; suitability of surface waters for aquatic life; and socio-economic risks of sea-level rise;
- a *response* module representing the societal response to negative impacts and providing the possibility to introduce water policy measures. Response options are partly autonomous and partly in the form of interactive changes in policy variables; these include investments in infrastructure, water pricing, and legislative and managerial measures.

The temporal resolution used varies from one month to one year (see also section 6.4.4). The simulation period is from 1900 to 2100. The present global version of AQUA does not distinguish socio-economic or hydrological regions in a geographically explicit way. The global land area is regarded as one large 'river basin'. Spatial varieties are taken into account by distinguishing different land-cover types and climate zones.

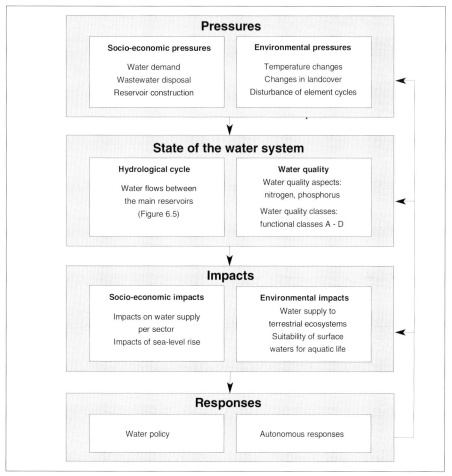

Figure 6.2 The AQUA Pressure-State-Impact-Response (PSIR) diagram.

6.4.2 Position within TARGETS

The AQUA submodel forms an integral part of the TARGETS model (*Figure 6.3*). The Population and Health submodel calculates the population size, one of the variables used in AQUA to calculate water demands. Demographic and health dynamics simulated in the Population and Health submodel also depend, amongst other factors, on the access to safe drinking water and proper sanitation, data provided by AQUA. Expansion of hydroelectric generation capacity from the Energy submodel determines the demand for new water reservoirs in AQUA.

The CYCLES submodel interlinks with AQUA in many ways. First, the hydrological cycle influences the element cycles: the outflows of elements from groundwater and surface water are driven by the water outflows from these water

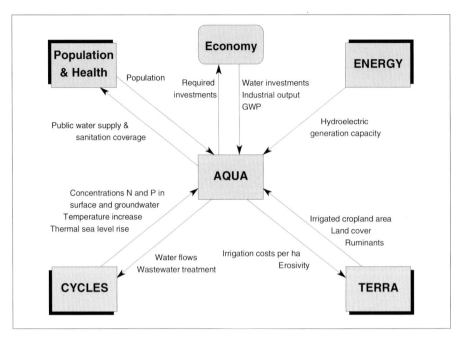

Figure 6.3 Interactions of AQUA with the other TARGETS submodels.

reservoirs. Second, AQUA simulates domestic and industrial waste-water treatment which determines the fate of nutrients. Third, the concentrations of various substances in fresh groundwater and surface water, calculated in CYCLES, are transported to AQUA to calculate the division of the fresh-water bodies into different water-quality classes. A fourth link concerns the effects of a temperature change on the hydrological processes.

Land-cover changes simulated in the TERRA submodel affect the processes of evaporation, infiltration, percolation and river runoff simulated in AQUA. The irrigation water demand within AQUA depends on the demand for irrigated cropland in TERRA. The other way round, actual irrigation and natural water availability influence the productivity of cropland. Finally, one of the factors determining erosion, the rain erosivity factor, is calculated in AQUA on the basis of the rainfall distribution throughout the year.

Gross World Product (GWP) and industrial output from the economic scenario generator are used within AQUA to calculate water demand. Conversely, increasing supply costs if water gets scarcer requires more investments and may hinder economic development. Traditional water models do not account for such feedbacks because they start from exogenous population and economic scenarios. In the TARGETS1.0 model we have begun to explore such feedback mechanisms.

6.4.3 Pressure module

Environmental pressures on the water system are: land-cover changes, soil degradation, water pollution and temperature changes. These environmental pressures, inputs to AQUA, are transported directly to the water system module. *Socio-economic pressures* on the water system as discussed below are driven by demographic, economic and technological developments.

Water demand

AQUA separately calculates water demand for the domestic, agricultural and industrial sectors. Demands are considered to be the driving force behind supplies, but actual supplies (simulated in the impact module) may be lower than demands due to allocation constraints (simulated in the response module) and are thus conceived as latent demands which are not necessarily met.

Domestic water demand includes demands for households, municipalities, commercial establishments and public services. Their determinants are: population size, type of water supply, gross world product per capita (GWP_{pc}), water price (WP) and water supply efficiency (Eff_{act}) (*Figure 6.4*). As for the type of water supply, the model discriminates between public and private supply. One reason is that the number of people having with water supply gives an indication of the number of people with *proper* water supply, which is a determinant of human health. Another reason is that public supply generally means a larger demand per capita. Public water supply systems include public hand pumps, standpipes and house taps. People relying on self-supply obtain their water via dug-wells, tube-wells or yard taps, or directly from rivers, canals, lakes or ponds. The water demand per capita (WD_{pc}) for public supply is calculated as:

$$\frac{dWD_{pc}(t)}{dt} = WD_{pc}(t) \times \left[E_G(t) \times \frac{dGNP_{pc}(t)/dt}{GNP_{pc}(t)} + E_P(t) \times \frac{dWP(t)/dt}{WP(t)} - \frac{dEff_{act}(t)/dt}{Eff_{act}(t)} \right] \text{ [l/yr/cap/yr]} \quad (6.1)$$

in which E_G represents the growth elasticity and E_P the price elasticity. The growth elasticity is assumed to be a function of GWP_{pc}. The actual water supply efficiency, Eff_{act}, depends on the water-saving technology available, i.e. the maximum possible efficiency Eff_{max}, and the extent to which this technology is actually being used. The value of Eff_{max} has been set exogenously. The actual efficiency is supposed to move towards the maximum possible one, as an autonomous process but possibly accelerated through policy measures. Although improving water supply efficiencies is often mentioned as an important policy instrument (Postel, 1992), knowledge on the effectiveness of this instrument is poor. A simple logistic curve with a diffusion rate d has been assumed to simulate the diffusion of water-saving technology:

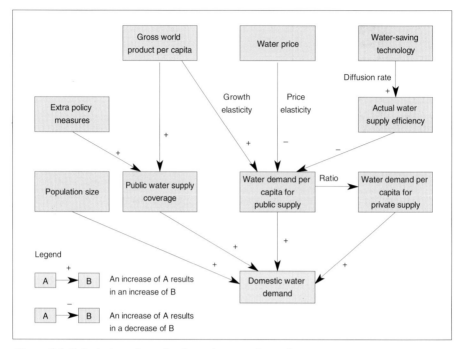

Figure 6.4 Calculation scheme for domestic water demand.

$$\frac{dEff_{act}(t)}{dt} = d \times \left(Eff_{act}(t) - Eff_{min}\right) \times \left(Eff_{max}(t) - Eff_{act}(t)\right) \qquad [1/\text{yr}] \qquad (6.2)$$

Irrigation water demand is calculated by multiplying the irrigated cropland area as simulated in the TERRA submodel by the water demand per hectare. The latter is calculated similarly to the domestic water demand per capita. Irrigation efficiency is defined as the fraction of the total water withdrawal that actually benefits the crop, i.e. the part taken up and transpired by the plant. The remainder consists of water losses through evaporation and infiltration and is often larger than the water actually used by the plant. The maximum possible efficiency for irrigation has a natural upper limit of 100%. Actual irrigation efficiency is simulated similarly to the actual domestic water supply efficiency.

Livestock water demand follows from the number of heads or flocks and the water demand per head or flock. The global version of AQUA reckons with only one livestock category, namely ruminants (including cattle, sheep and goats); the number of ruminants in 'ox equivalents' follows from TERRA. Average water demand per head is assumed to be constant.

Industrial water demand consists of demands from various industrial sectors, including cooling water demands from thermoelectric power plants and groundwater withdrawal requirements from mining industries. The aggregate demand is calculated as the product of the total industrial output, expressed in US dollars per year, and the average water demand per dollar. The latter is calculated analogously to the domestic water demand per capita.

The effect of water use on the hydrological cycle
The actual pressure of water use on the hydrological cycle follows from the overall balance of the water used. At present, the major portion of the water for human purposes in most parts of the world is derived from fresh surface water and shallow groundwater. However, other sources are fossil groundwater (e.g. in Libya) and saline water (e.g. in Saudi Arabia). The sinks of used water are the fresh groundwater and surface water reservoirs, and the atmosphere. The *water re-use potential* is defined as the part of the total fresh-water withdrawal that can possibly be reused. It is the complement of the *consumptive water use*, i.e. the part of the water withdrawal that does not return to one of the fresh-water reservoirs but is lost through evaporation. The water re-use potential for irrigation is calculated as a function of the irrigation efficiency; the water re-use potentials for domestic, livestock and industrial water use are assumed to be constant. The part of the domestic and industrial waste-water flow to fresh surface water, which is treated before discharge, is calculated on the basis of investments in wastewater treatment (response module) and average treatment costs per litre, which are assumed to be constant.

Reservoir construction
The effects of reservoir construction on evaporation and fresh-water storage capacity are simulated separately from the disturbance of the hydrological cycle by offstream water use. Artificial fresh-water reservoirs are planned for a variety of reasons: hydroelectric generation, water supply, flood control, navigation, recreation and aquaculture. The present version of AQUA considers hydroelectric generation as the main driving force, assuming the volume of artificial reservoirs to be proportional to hydroelectric generation capacity.

6.4.4 Water system module

Schematisation of the water system
Four spatial compartments are distinguished: land, atmosphere, oceans and ice sheets. The present global version of AQUA considers the land compartment as one large 'river basin'. One or more water reservoirs are distinguished in each compartment (*Figure 6.5*). The land compartment encloses: snow and glaciers, soil moisture, biological water, fresh surface water (including rivers and lakes), renewable and fossil groundwater. In order to account for different land characteristics, the model distinguishes two climate zones – tropics and temperate – and seven land-cover types – forest, grassland, desert, wetland, rainfed cropland,

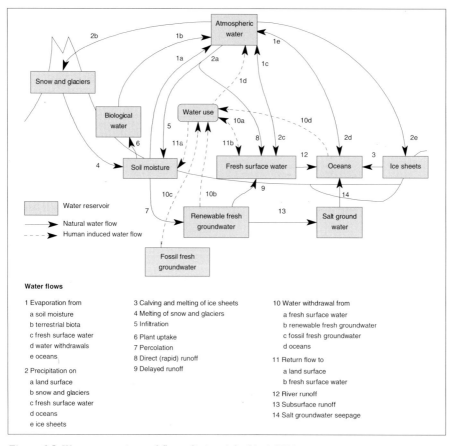

Figure 6.5 Water reservoirs and flows distinguished in AQUA.

irrigated cropland and urban area (see Chapter 7). The hydrological cycle is modelled by simulating the water flows between the water reservoirs and by calculating the water balance per reservoir. All fresh-water flows are simulated on a monthly basis. Such a temporal resolution is the minimal requirement to simulate seasonal variations in runoff and to assess stable fresh-water availability. The water flows between atmosphere, oceans and ice sheets are simulated on a yearly basis.

For each water reservoir, the model contains a mass balance of the form:

$$\frac{dS(t)}{dt} = \sum F_{in}(t) - \sum F_{out}(t) \qquad [kg/yr] \qquad (6.3)$$

in which S is the storage of the reservoir, $\sum F_{in}$ the sum of the inflows and $\sum F_{out}$ the sum of the outflows. Since inflows and outflows are not necessarily equal, reservoir storages change. However, three of the water reservoirs distinguished are assumed to be in dynamic equilibrium. The saline groundwater storage remains constant because the outflow into the oceans is supposed to equal the inflow from fresh groundwater. The atmospheric water content does not change because global precipitation is supposed to equal global evaporation. We are allowed to do so because we use a time -step of one month, while atmospheric water vapour has a renewal time of about 10 days (Speidel and Agnew, 1988). Hence, the model does not account for possible long-term changes in atmospheric water storage, which, however does not significantly influence the model results because atmospheric water amounts to only 0.001% of the total water storage on Earth (Shiklomanov, 1993). The biological water storage remains constant in the model by assuming that water uptake and transpiration by plants are equal. Although seasonal and long-term biomass changes are realistic (Houghton *et al.*, 1983) and important for the global distribution of *carbon* (Chapter 8), we may neglect them in our calculation of the global distribution of *water* because biological water forms only 0.0001% of the global water storage, equivalent to 3 mm sea-level rise.

A change in the oceanic water storage corresponds with a change in the average sea-level. The separate contributions to sea-level change from ice sheets, glaciers, fresh groundwater, fresh surface water and soil moisture are calculated by dividing the changes of these water storages by the global ocean surface (361×10^6 km^2). Total sea-level rise follows from the net increase of the oceanic water storage plus the increase in volume as a result of thermal expansion. An overview of different mechanisms affecting sea-level is given in *Figure 6.6*. The average decline of groundwater level is calculated by dividing the decrease of the renewable fresh groundwater storage by the total land area (133×10^6 km^2, excluding Antarctica and Greenland).

Evaporation and precipitation
Evaporation from *land* is calculated per climate zone and land-cover type on a monthly basis using the empirical relationships introduced by Thornthwaite (1948) and Thornthwaite and Mather (1957). These relationships imply that if precipitation exceeds potential evaporation, a possible soil moisture deficit is replenished up to the soil water holding capacity and the potential evaporation becomes actual. The rest of the precipitation is then available for runoff. However, if the precipitation is lower than the potential evaporation, there is nothing available for runoff. Evaporation from *fresh surface water* is calculated by assuming that the actual evaporation is equal to the potential evaporation. Evaporation from *oceans* is simulated on a yearly basis as a function of the initial oceanic evaporation and the global average temperature increase using a temperature response factor.

Global precipitation is assumed to equal global evaporation. The present spatial distribution of global precipitation is used as the *initial* distribution. The main distribution is: 20.1% above land, 79.4% above oceans and 0.5% above ice sheets

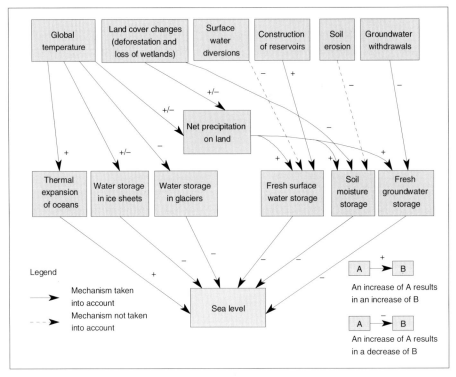

Figure 6.6 Mechanisms affecting sea level.

(Gleick, 1993). The distribution of land precipitation over the climate zones and land-cover types distinguished is derived from the IIASA climate database (Leemans and Cramer, 1991). The *change* of the distribution of land precipitation is simulated with a simple equation that relates a certain distribution to the global temperature increase and has been derived from results of the IMAGE2.0 model (Alcamo, 1994). AQUA scales the precipitation distribution to fit the total land precipitation change simulated. This method differs from the method outlined by Santer *et al.* (1990), which links a certain GCM-based precipitation distribution directly to the simulated global average temperature. The advantage of the AQUA method is that not only are the effects of temperature changes on evaporation and precipitation taken into account, but also the effects of land-cover changes and increased human water consumption. For a quantitative notion of the comparative importance of the driving forces behind evaporation and precipitation changes, one could consider two scenarios: (i) for doubling atmospheric CO_2, GCMs estimate an equilibrium increase of global precipitation of 2.5% to 10% (Houghton *et al.*, 1992); and (ii) for doubling human-induced evaporation, one may expect an increase in global precipitation of 0.4% (calculated from Shiklomanov, 1993). Although the estimated change under the latter scenario is small compared to the effect of global warming, its occurrence is more plausible.

Calving and melting ice sheets

Ice sheets grow as a result of precipitation and shrink through calving and melting. Evaporation from ice sheets is negligible and therefore ignored. In Antarctica, melting is also negligible because of the extremely cold climate. The complex dynamics of the Greenland and Antarctica ice sheets are represented by simple relationships in which the ice sheets respond to certain climate sensitivity parameters (Warrick *et al.*, 1995). The simulated change is superimposed on the initial imbalance.

Melting of snow and glaciers

The storage of snow and ice in mountainous regions is simulated by distinguishing short-term and long-term dynamics. The short-term dynamics are driven by the temperature variation within a year: in winter, addition by snowfall dominates and in summer, depletion by melting. Melting is calculated by reckoning with an average time lag between snowfall and snow-melt. The long-term change of the yearly average storage of snow and glaciers is a function of climate change, calculated according to Oerlemans (1989):

$$\frac{dS_{glac}(t)}{dt} = \alpha \times T_{incr}(t) \times \left[S_{glac,i} \times e^{-T_{incr}(t)/\beta} - S_{glac}(t) \right] \qquad \text{[kg/yr]} \qquad (6.4)$$

where S_{glac} represents the actual storage, $S_{glac,i}$ the initial storage and T_{incr} the global average surface temperature increase since the initial year in °C. Parameter α involves the characteristic response time in per °C per yr; parameter β is the global temperature increase in °C for which the storage of ice becomes 1/e of the initial value.

Runoff

The precipitation available for runoff is calculated per climate zone and land-cover type as the total precipitation (including melt and irrigation water) minus the evaporation and the soil moisture change. The precipitation for runoff goes partly as *direct runoff* to the fresh surface-water reservoir and partly *percolates* into the fresh groundwater reservoir, determined by a land-cover-specific partition factor. *Infiltration* of water from the Earth's surface into the unsaturated zone is calculated as the total precipitation minus the direct runoff. The infiltration water replenishes the soil moisture, which is depleted again by evaporation and percolation. The outflow of the fresh groundwater reservoir is assumed to relate linearly to the storage in the reservoir, with a constant lag time of the groundwater reservoir. Using a constant partition factor, the total groundwater outflow is divided into two components: *delayed surface runoff* and *subsurface runoff*. Finally, the *river runoff*, R_{riv}, is calculated as follows:

$$R_{riv}(t) = R_{dir}(t) + R_{del}(t) + P_{sw}(t) - E_{sw}(t) - W_{sw}(t) + D_{sw}(t) \qquad \text{[kg/month]} \qquad (6.5)$$

in which R_{dir} and R_{del} are the direct and delayed runoff, P_{sw} the precipitation on surface water, E_{sw} the evaporation from surface water, W_{sw} the surface water withdrawal and D_{sw} the waste-water disposal. Because rivers have an average renewal time of about 16 days (Shiklomanov, 1993), we assume that the runoff generated in one month will flow into the ocean during the same month. The *total runoff* from land to oceans is the river runoff plus the subsurface runoff. The *stable* runoff, i.e. the runoff on which one can rely throughout the year, is equal to the lowest monthly runoff in a year.

Water use

In the model, water for human use is obtained from fresh surface water, renewable fresh groundwater, fossil fresh groundwater and ocean water. In the last case, the water is desalinated before use. The distribution of the total water withdrawal over the different sources is considered as an exogenous scenario, possibly changed by the user of the model (see response module). Water used in the domestic, livestock and industrial sector is partly lost by evaporation and partly discharged into the fresh surface water reservoir. Irrigation water arrives at irrigated cropland as precipitation.

fresh-water quality

The quality of a water body is determined by its physical properties and the content of various chemical substances and micro biological agents. Which aspects of water quality are considered, strongly depends on one's interest (Chapter 14). In AQUA, the main aim is to evaluate water quality in functional terms: is water suitable for aquatic life, drinking, irrigation, etc. Four water-quality classes are distinguished here:

- class A is suitable for all functions (high quality);
- class B does not meet ecological requirements but is suitable for human purposes;
- class C is unsuitable for both aquatic life and drinking; and
- class D is also unsuitable for agricultural and industrial purposes (low quality).

The set of water quality variables reckoned with corresponds with the substances distinguished in CYCLES (Chapter 8). Average surface-water concentrations of the following substances are transported from CYCLES to AQUA: dissolved inorganic nitrogen (DIN), dissolved organic nitrogen (DON), particulate organic nitrogen (PON), dissolved inorganic phosphorus (DIP), dissolved organic phosphorus (DOP), and particulate organic phosphorus (POP). Only DIN and DON are available for groundwater. AQUA processes the data on average concentrations, both for surface water and groundwater, in three steps to arrive at a distribution of the water stock over the different quality classes. First, log-normal distributions are projected on the average concentrations. Standard deviations have been derived from statistical analyses of observed data (Meybeck and Helmer, 1989; WHO/UNEP, 1991). Second, how the fresh-water stock is distributed over the four water quality classes is calculated per water quality variable. For that purpose, each class is characterised by

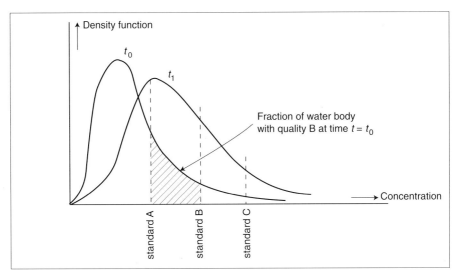

Figure 6.7 Characterising the quality of a water body by calculation using a log-normal distribution and water-quality standards.

maximum concentrations for all water-quality variables. Finally, the variable-specific distributions over the four water-quality classes are translated into one general distribution (*Figure 6.7*). The water-quality variable indicating the worst quality distribution is decisive for the general water-quality distribution.

6.4.5 Impact module

The impact module simulates the performance of the following socio-economic and ecological 'functions' of the water system: domestic, irrigation, livestock and industrial water supply; natural water supply to terrestrial ecosystems; and aquatic ecosystem maintenance. It further simulates population and capital at risk as a result of sea-level rise. Hydroelectric generation is simulated in the Energy submodel (Chapter 5).

Water supply
The yearly 'fresh-water availability' is defined as the natural net precipitation in a year, i.e. total precipitation minus natural evaporation. It includes water consumed by humans. The 'stable fresh-water availability' is the part of the total fresh-water availability that would form stable runoff if there were no human withdrawals. It is calculated on the basis of the partition factors that divide the precipitation available for runoff into a direct runoff component and a groundwater recharge component (see Section 6.4.4). The contribution of artificial reservoirs to stable runoff is

accounted for by taking a fixed percentage of the reservoir volume (compare Postel *et al.*, 1996). The 'potential water supply' is defined as the maximum possible amount of water that can be withdrawn from the total fresh-water reservoir from a long-term point of view. It is calculated by taking the fraction of the stable fresh-water availability that runs off in inhabited areas. As inaccessible remote areas we only take percentages of river basins that are thinly populated throughout, such as the Amazon basin (Postel *et al.*, 1996) because inaccessibility of a mountainous upstream part of a river basin does not imply that the water from this area cannot be used in the more densely populated area downstream.

Total actual water supply is the sum of domestic, livestock and industrial water supply and irrigation. Public domestic water supply is calculated by multiplying the annual expenditure in this sector by the average supply costs per litre. The expenditure is a function of the demand, the available budget and allocation priorities (Section 6.4.6). Private domestic water supply is supposed to equal the demand (Section 6.4.3). Irrigation, and livestock and industrial water supply are calculated similarly to public domestic water supply. Water supply costs follow from supply-cost curves which differ per type of water source, and per sector and water-quality class. For withdrawals from surface water and groundwater, supply costs are assumed to be a function of the ratio between consumptive water use and potential water supply (*Figure 6.8*). In Chapter 14 we will show that this is just one way of defining the supply-cost curve; also alternative definitions have been implemented (*Figure 14.1*). As shown in *Figure 6.8* the supply costs also depend on the quality of the intake water: the worse the quality of the intake water, the higher the costs. This is due to additional treatment requirements. In the model we assume that the quality of water withdrawals from surface water is distributed in accordance with the quality distribution of fresh surface water as simulated in the water system module. However, it is assumed that water of quality D is not used for water supply. The same holds for withdrawals from renewable groundwater. The costs of water supply from fossil groundwater and saline ocean water are assumed to remain constant. In all cases, calculated costs include investments to extend or renew supply infrastructure (fixed costs), and operation and maintenance costs (variable costs).

Public water supply and sanitation coverage
The number of inhabitants provided with public water supply is calculated by dividing the total public water supply by the average water demand per capita. The fraction of the population provided with public water supply is considered to have access to safe drinking water. The remainder, relying on private supply, run a higher risk of being exposed to bacteriologically contaminated water. Another health determinant is the lack of proper sanitation. The fraction of the population with access to sanitation facilities is calculated from the sanitation expenditures ($ per year), the size of the population and the average costs of sanitation ($ per capita per year). The latter are supposed to remain constant.

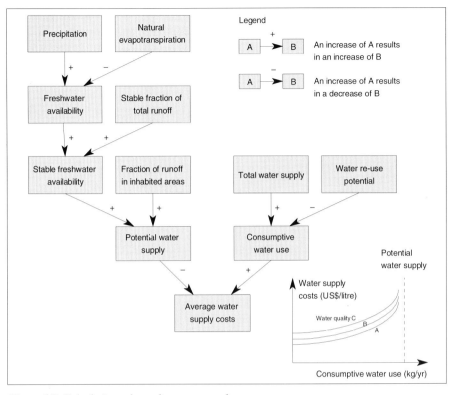

Figure 6.8 Calculation scheme for water supply costs.

Water for ecosystem maintenance

Several methods can be used to obtain a measure of the water availability for plant growth. For example, FAO uses the length of the growing period as a measure (Oldeman and Van Velthuyzen, 1991). This period is defined as the time within one year that precipitation exceeds half the potential evaporation and the temperature exceeds 5 °C. A measure of the natural water availability between zero and 1 can be obtained by taking the ratio between the actual and the potential evaporation (Leemans and Van den Born, 1994). Finally, the soil moisture content can be used as a measure of the natural water availability. This method, however, is very sensitive to the way soil moisture is simulated and is less robust than the methods based on precipitation and evaporation data. In AQUA, both the Length of Growing Period (LPG) and the ratio between actual and potential evaporation are calculated per climate zone and land-cover type. At present, these outcomes are not yet linked to the TERRA or CYCLES submodel.

One of the determinants of surface erosion (simulated in the TERRA submodel) is rain erosivity. Many different indices for rain erosivity have been developed (Bergsma, 1981). Because of its simplicity, the rain erosivity factor proposed by

Arnoldus (1977, 1978) has been chosen, which only depends on monthly precipitation data.

The functioning of aquatic ecosystems depends on both hydrological and water-quality variables. As a measure of the hydrological situation, the minimum river runoff is taken. As for the water quality, the model calculates which fraction of the fresh surface water storage meets the ecological water-quality criteria (water-quality class A).

6.4.6 Response module

The response module simulates societal responses to the impacts calculated in the impact module distinguishing *autonomous* and *policy-driven* responses. For the latter, the model contains a set of policy options that can be adjusted by the user; these measures vary from infrastructural, technological and financial to legislative and managerial.

Expenditures
The annual expenditure required to meet a certain water demand is calculated as the product of the water demand and the average water-supply costs. Required expenditures are calculated separately for the domestic, irrigation, livestock and industrial sectors. In a similar way, the model calculates required expenditures for sanitation, and domestic and industrial waste-water treatment. The calculated expenditures include both investments, and operation and maintenance costs. The required expenditures are added and compared with the maximum possible expenditure in the water sector, determined at a higher level of allocation within TARGETS (see Chapter 3). If demands cannot be met, the user of the model can allocate the shortage to the sectors.

Increase of the public water supply and sanitation coverage
Analysis of a large number of national data shows a weak statistical correlation between public water supply coverage and gross national product per capita (Hoekstra, 1997). However, the correlation is not strong enough to postulate a causal relationship between the two, with the latter as determinant of the former. The same holds for sanitation coverage. Both water supply and sanitation coverage can be strongly influenced by specific policy measures, as witnessed in the results of the International Drinking Water Supply and Sanitation Decade, an intensive investment programme carried out in the eighties (Christmas and De Rooy, 1991). The model offers the possibility of using the statistical relationships or exogenous scenarios, which can be adjusted.

Waste-water treatment

The demand for domestic waste-water treatment is calculated as a function of the total amount of domestic waste water discharged onto surface water, the actual treatment fraction and a policy variable, indicating the desired growth rate in treatment. The demand for industrial wastewater treatment is similarly simulated. In the absence of any specific policy, it is assumed that the existent wastewater treatment plants will be maintained and extended at a rate proportional to the increase in total wastewater discharge.

Water resources management and technology

In the absence of any specific policy, the distribution of water withdrawals from the different sources distinguished will remain constant. This means that most of the total water supply is derived from fresh surface water and groundwater. A policy option is to change the ratio between surface-water and groundwater withdrawals or to derive a larger fraction from fossil groundwater or saline water. The model also provides the policy option of setting a minimum monthly river runoff below which no further surface-water withdrawals are allowed. In this case, other (possibly more expensive) sources will be taken. In this way, a minimum dry season river runoff is

Impacts of sea-level rise

The impacts of sea-level rise are calculated according to the method developed by Delft Hydraulics and RIKZ (1993) and Resource Analysis (1994). The world's coasts have been classified into 60 clusters, each one characterised by a certain combination of hydraulic, demographic and economic properties. There are five hydraulic, three demographic and four economic classes. The impacts are calculated per coastal cluster and expressed in terms of people and capital at risk. Each cluster is characterised by a critical water level, H_{crit}, at which flooding of the coastal lowlands will occur. The probability that the sea-water level will exceed the critical water level is calculated by:

$$P_{crit}(t) = 10^{[c_1 - H_{crit}(t) + slr(t) + ss(t)]/c_2} \quad [1/yr] \quad (6.6)$$

in which slr represents the sea-level rise since the initial year and ss a possible aditional storm surge. The parameters c_1 and c_2 are cluster-specific constants in the water level-probability relationship. The total area is, per coastal cluster, subdivided into n subareas falling within different elevation classes. For each subarea and probability, the number of people and capital to be flooded are calculated. The number of people at risk is defined as the number of people in a certain area multiplied by the flooding probability for that area. Similarly, the capital at risk is the product of the capital in an area and the flooding probability for that area.

Coastal defence

Actual or expected impacts from sea-level rise will result in a demand for coastal defence measures such as raising dikes and dune reinforcements. It is assumed that the basic demand for additional coastal protection is to retain, per coastal cluster, the probability that the critical water level is exceeded under a certain maximum level, for which the model assumes the flooding probability of today. As a policy option, the basic demand for coastal protection can be adjusted The investment costs needed for an increase of the critical water level are calculated in three steps. First, the length of the coast that has to be strengthened is calculated, followed by the average rise and, finally, the protection costs, which follow from the length to be protected, the average rise, and the cost per unit length and unit rise. The model calculates three types of coastal protection (stone-protected sea dikes, clay-covered sea dikes and sand dunes), for which the costs of increasing the critical water level vary (Hoekstra, 1997).

guaranteed for maintaining ecosystems. The model further provides the possibility of deep-well injections to artificially recharge the groundwater reservoir.

Maximum possible water-use efficiencies in the domestic, livestock, irrigation and industrial sectors are implemented as exogenous scenarios. Since development of water-saving technology is considered one possibility to reduce water demand, the exogenous scenarios can be adjusted to analyse the effects of technological development on future water supply.

Water pricing
Worldwide, the price of water – the tariff charged to the consumer – has traditionally been lower than the actual water supply costs (Gleick, 1993; Serageldin, 1995). According to economic theory this may lead to inefficient water use. Therefore, water pricing – increasing the water price – can be an effective technique in demand management (Munasinghe, 1990). It is assumed that, without any specific policy, the fraction of the actual costs charged to the consumer remains constant. As a policy option, the model provides the possibility to change this fraction.

6.5 Calibration and validation

Data have been collected for three purposes: obtaining the required model input, calibration and validation. The model uses as input: initial values for a number of variables, model parameters and exogenous scenarios for developments outside the water sector. The latter are provided by the other submodels within the TARGETS1.0 model and not discussed here. For calibration and validation, we have used observed data for some key variables for the period 1900-1990 and independent estimates of model parameters from the literature. The data and the primary sources used are discussed in Hoekstra (1997). Some of the parameters have been taken directly from the literature as input for the model; other parameters have been determined by calibration. As validation of the model, the parameter values obtained by calibration have been compared to independent values in the literature. Another type of validation has been the application of AQUA at the river basin level. The reliability of the global version of AQUA is considered partly a derivative of the reliability of the river basin versions because the simulation of processes like runoff can better be validated at the more detailed level.

A sensitivity analysis in which we varied the most important model parameters, showed that the land-cover factors for simulating land evaporation and the temperature-response factor for oceanic evaporation considerably influence key variables such as river runoff, actual and potential water supply and sea-level rise. The land-cover factors have been calibrated such that total land evaporation equals 72×10^{15} kg/yr at present; this is the value given by Shiklomanov (1993). The temperature-response factor for oceanic evaporation is based on experiments using the General Circulation Model of the Max Planck Institute for Meteorology in Hamburg (Cubasch et al., 1992).

Annual river runoff is the most sensitive to the land-cover factors, but after that to the partition factor that determines the ratio between delayed surface runoff and subsurface runoff. This partition factor has been calibrated such that the subsurface runoff presently amounts to 2.3×10^{15} kg/yr (Shiklomanov, 1993). However, uncertainties are large, illustrated by the fact that estimates for subsurface runoff range between zero and 12×10^{15} kg/yr (Speidel and Agnew, 1988). Monthly river runoff values are also sensitive to the partition factors that determine the ratio between percolation and direct runoff. Finding these partition factors specifically per land-cover type is difficult due to a lack of land-cover-specific runoff data. Therefore an assumption has first been made with respect to the relative values of the partition factors for the different land-cover types, and secondly, the values have been calibrated such that the global average percolation is 25% of the total runoff, as estimated by Shiklomanov (1993). The literature generally suggests values between 25% and 30% (Postel et al., 1996; Ambroggi, 1977, 1980; Speidel and Agnew, 1988).

Actual water supply is sensitive to both hydrological parameters and parameters that determine demand. The most important hydrological parameters are the land-cover factors and the partition factors that determine the ratio between percolation and direct runoff. These parameters have been calibrated as described above. water demand is most sensitive to – in order of magnitude – growth elasticities, the diffusion rate of water-saving technology and price elasticities. A calibration dilemma exists in the estimated historical data on water demand being reproduced via more than one set of parameter values, that is, the historical water demand can be explained in more than one way. However, here we present the results of only one calibration. We calibrated separately for domestic, irrigation and industrial water demand and used the estimates of actual supplies given by Shiklomanov (1993). The calibration strategy has been to assume values for the parameters for which demand is least sensitive on the basis of the literature and to calibrate the most influential parameters, i.e. the growth elasticities.

Empirical data on price elasticities of domestic water demand per capita show a range between zero and -0.6 (Keller and Van Driel, 1985; Kooreman, 1993; Nieswiadomy, 1992; Renzetti, 1992). We have assumed a global average value of -0.2. For industrial water demand per dollar industrial production, we assumed a price elasticity of -0.4, assuming that water use in industry can be more easily reduced than water use for domestic purposes. The price elasticity for irrigation water demand per hectare has been estimated at nihil, assuming that water prices determine the irrigated area rather than the irrigation water demand per hectare. The diffusion rate of water-saving technology has been set at 0.5 % per yr. Calibration of the growth elasticities of water demand in the domestic, irrigation and industrial sectors has resulted in the values 1.2, 0.1 and 0.5, respectively. The low-growth elasticity of irrigation water demand per hectare can be understood by realising that growth and not so much the water demand per hectare affects the area of irrigated cropland. *Figure 6.9* shows the simulated historical water supply per sector in comparison to the estimated values from the literature.

Sea-level rise
The sea-level rise simulated appears to be sensitive to a gamut of parameters due to the fact that sea-level rise consists of different components, each with its own dynamics (*Figure 6.6*). The oceanic water storage is as it were the sink of the global water balance; changes in the amount of water stored in glaciers, soils, fresh surface water, groundwater or ice sheets all result in a change in the sea-level. *Figure 6.10* shows the different components of historical sea-level rise to be simulated. The sea-level rise as a consequence of glacier melting has been fitted with the values of Meier (1984) by calibrating the glacier melting parameters (Equation 6.4). The sea-level rise from the Greenland ice sheet follows from the initial imbalance (zero in this case) and the climate sensitivity assumed (Warrick *et al.*, 1995). The same holds for the Antarctica ice sheet for which a negative initial imbalance has been assumed. The sea-level rise from groundwater loss on land has not been calibrated, but follows from the groundwater depletion simulated. The model gives a rise of about 66 mm in the period 1900-1990. As a reference, Sahagian *et al.* (1994) give a conservative estimate of 9 mm. For the same period, Warrick *et al.* (1995) give a range for direct anthropogenic contributions to sea-level rise (including the groundwater contribution) of -50 to 70 mm. Our estimate is based on the assumption that the groundwater outflow linearly relates to the groundwater storage (Section 6.4.4). If we assume a quadratic relationship, i.e. outflow proportional to square storage, we simulate a weaker response of the groundwater storage to human withdrawals and thus a smaller contribution to sea-level rise of 61 mm in the period 1900-1990. The simulated sea-level rise from deforestation and loss of wetlands is small, corresponding to estimations in the literature (Warrick *et al.*, 1995). The simulated sea-level *decline* as a result of artificial reservoirs is 10 mm.

For the period 1900-1990 we simulate a total sea-level rise of 132 mm (assuming a linear relation between groundwater outflow and storage). This value fits with the observed sea-level rise during the last 100 years, which is estimated at 100-250 mm (Warrick *et al.*, 1995).

Validation from applications at river basin level
As mentioned in section 6.3, the generic framework of AQUA has been applied not only to the world as a whole, but also to specific river basins: the Zambezi basin in southern Africa and the Ganges-Brahmaputra basin on the Indian subcontinent. The river basin models can be better validated with empirical data than the global version of AQUA. At river basin level, we can also compare the results of AQUA with the outcome of other types of models. A comparison of the AQUA Ganges-Brahmaputra

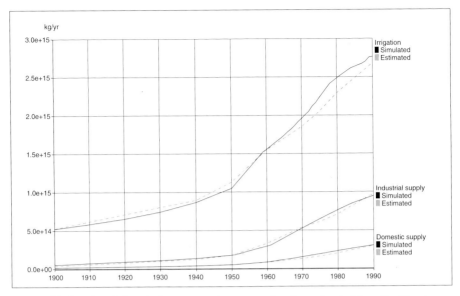

Figure 6.9 Simulated historical water supply compared to that estimated from the literature (Shiklomanov, 1993).

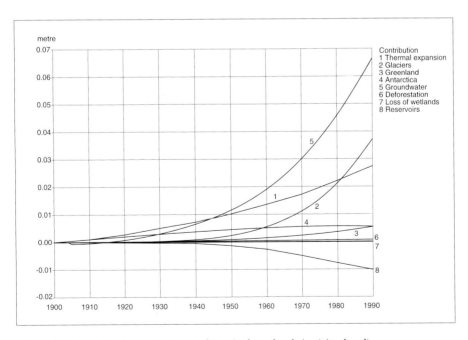

Figure 6.10 Individual contributions to historical sea-level rise (simulated).

model with the grid-based runoff model of Van Deursen and Kwadijk (1994) has shown that despite its relative simplicity, AQUA can reproduce observed river runoff as accurately as a more complex model (Hoekstra, 1995). Calibration of the AQUA Zambezi model, with respect to monthly river discharges for several subcatchments of the basin, shows that here too the model can reproduce observed runoff adequately (Vis, 1996; Hoekstra, 1997). The adequacy of the river basin models in simulating runoff is considered as support for the validity of the global AQUA version.

Applications of AQUA at river basin level

The AQUA Zambezi and the AQUA Ganges-Brahmaputra models include only the equations for the land compartment, leaving out those for atmospheric, oceanic and ice sheet processes. Further, the river basin models only differ from the global AQUA version in terms of input data, i.e., the initial values and parameters. The basins have been schematised into major sub-catchments and into different country areas. In each river basin model, the water system submodel describes the major subcatchments separately, including the interactions with neighbouring catchments. The pressure, impact and response modules describe each country area separately. For the translation from national pressures to pressures in the separate subcatchments, distribution matrices are used. The same holds for translating subcatchment changes back to impacts for the different country areas.

6.6 Conclusions

The AQUA submodel describes the global hydrological cycle as well as mechansisms of water demand and supply. The model has been calibrated such that observed trends in the past century can be reproduced. The historical increase in water supply has been explained in terms of different driving forces: population and economic growth, increase of the irrigated cropland area, increasing water costs and actual prices, and increasing water use efficiencies. Historical sea-level rise has been explained in terms of different contributing components. The results of experiments with the AQUA model demand a reconsideration of one of the basic IPCC assumptions. According to IPCC's best estimate, climate change is the only significant force behind sea-level rise in the past century (Warrick *et al.*, 1995). Mechanisms of long-term water loss on land have been recognised, but it is generally assumed that possible groundwater losses at some places are gains at other places, most importantly through artificial fresh-water reservoirs on land. In our explanation of twentieth century sea-level rise, however, the contribution of water loss on land is 42%. Although further research on this issue is required, we can conclude on the basis of our experiments that the least we can do is not to pre-empty the land component from studies of sea-level rise beforehand.

7 The land an food submodel: TERRA

"Der Mensch ist was er isst."
Ludwig Feuerbach

7 THE LAND AND FOOD SUBMODEL: TERRA

Bart J. Strengers, Michel G.J. den Elzen and Heko W. Köster

The aim of the land and food submodel is to simulate the key features of the global changes in land use and land cover that result from demand for food and the requirements of forestry. The submodel can reproduce the major historical trends in land use and land cover, food demand and supply, fertiliser use, etc. This is done, to a large extent, by employing of exogenous policy scenarios. The interaction with the other submodels, in particular CYCLES, allows the exploration of linkages between population growth, water availability and climate change on the one hand and food production on the other.

7.1 Introduction

The Earth's vegetation patterns have always changed in response to natural changes in, for example, geology, biology and climate. However, over the last few centuries human activities have made a considerable contribution to such changes. Natural ecosystems, forests, savannahs and wetlands have all been severely affected. The combination of growing populations and higher per capita food consumption has led to the gradual expansion of the land area used for food production and grazing. Increasing population density has led to forms of permanent agriculture which make more intensive use of land and this trend towards intensification is likely to continue in the decades to come. The growing demand for food may cause an imbalance between what can be produced and what is needed. This has happened many times in the past on a local scale (Braudel, 1981; Ponting, 1993; Turner II *et al.*, 1990a), but increasing global food trade is causing such imbalances to develop into a problem of global proportions.

People have generally responded in two ways to increases in population density: either by extensification of land or intensification of agriculture or both. Usually, this is accompanied by changes in tenure and demographic behaviour such as declining fertility and out-migration (see, for example, Bilsborrow and Okoth Ogendo (1992) for a discussion). Land extensification is achieved by clearing more of one's own land, appropriating neighbouring land or migrating to other areas to develop new land for agriculture. In all these cases it often leads to deforestation. *Figure 7.1* shows estimates of forest, grassland and agricultural areas for both the developed and the developing regions since 1900. Despite large uncertainties in these estimates (Leemans and van den Born, 1994), the declining area in forests is clear. Intensification has involved increasing cropping intensity by shortening fallow periods and multi-cropping, supported by additional inputs such

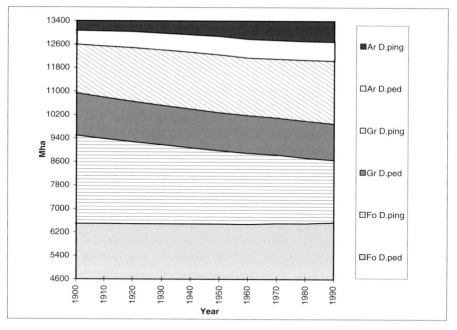

Figure 7.1 land-cover changes, 1900-1990. Sources: HYDE (Klein Goldewijk and Battjes, 1995) and AGROSTAT (FAO, 1992). land-cover type 'other' not shown (=4600 Mha). (D.ped = Developed region, D.ping = Developing region, Ar= Arable land, Gr= Grassland, Fo= Forest).

as human and animal labour, and organic manure, as well as irrigation devices and new skills. Labour was increasingly substituted in many regions for power from water, wind and fossil fuels. Organic manure has been supplemented with or replaced by inorganic fertilisers. Views on the causes of these developments and the underlying dynamics differ.

The diversity of agricultural systems does not lend itself to generalisation – either in terms of the past or of the future. Pastoral nomads and shifting cultivators have gradually been forced into a few of the world's regions, with an estimated 60-70 million people still living as shepherds and shifting cultivators (Grigg, 1974; 1982). Of the more advanced agricultural systems, rice culture is of great importance, requiring intensive land use with high labour inputs, but producing high yields with limited soil degradation. Livestock plays only a minor role, with much of the harvested rice being consumed on the farm. This is often referred to as subsistence farming. Wet-rice systems have been able to keep pace with rising population densities and although the Green Revolution has produced much higher yields with the use of hybrid varieties, these often require abundant irrigation water and a liberal application of fertiliser.

In Europe and North America, as well as other temperate areas where Europeans have settled, the main agricultural system is mixed farming. The three major trends shaping mixed farming in Europe between 1300 and 1800 were: reduction of fallow land by intercropping, the disappearance of common rights ('enclosure') and the increasing importance of livestock (Grigg, 1974, 1982). The rise in real per capita incomes caused a shift from cereals to vegetables and livestock. Scientific methods, cheap power, and transport and refrigeration technologies led to further advances. At present this mixed-farming, 'western' system is highly commercialised, with the integration of crops and livestock, and high yields as its main characteristics.

Numerous models have been developed in recent decades to assess the world's land and food situation. Some focus on the agro-technological aspects, while others take food markets as their starting point and still others take a biogeochemical perspective. One of the first global models was the World3 model (Meadows *et al.*, 1972, 1974). This captured in a crude and aggregate way the essential dynamics of land clearing and agricultural intensification, as well as feedbacks from both pollution to land yield and intensification to land degradation. An economically oriented modelling framework is the Basic Link system, that has been developed at IIASA over the last 15 years (Fischer *et al.*, 1988). It consists of a set of 26 national macro-economic simulation models which are linked through food demand and supply, and food price relationships.

Several analyses of land use, land cover and regional food supply and demand have been made within the RIVM (Bouwman *et al.*, 1992; Rotmans and Swart, 1991; Zuidema *et al.*, 1994), as parts of models which have a broader environmental focus such as IMAGE1.0 (Rotmans, 1990), ESCAPE (Hulme *et al.*, 1994) and IMAGE2.0 (Alcamo, 1994). The Land and Food submodel TERRA, which is strongly based on these models, is not meant to duplicate or improve upon them. Instead, it has two fairly straightforward objectives: (i) to simulate the key features of global changes in land use and land cover as a consequence of food demand and forestry, and (ii) to function as an integrated part of the TARGETS1.0 framework with a limited set of relations to other submodels.

7.2 Model description

7.2.1 PSIR approach

The TERRA submodel is structured along the Pressure-State-Impact-Response (PSIR) scheme as described in Chapter 2 (*Figure 7.2*). The *pressure* module (Section 7.2.4) calculates the demand for vegetable and animal products and for roundwood, both as a function of population size and income. Other (environmental) pressures considered are: water availability for irrigation, climate change and land degradation. The *State* module (Section 7.2.5) simulates the changes in the physical state of the land in terms of land areas and their characteristics, and the food production. Three main modules can be distinguished in the state system: (i)

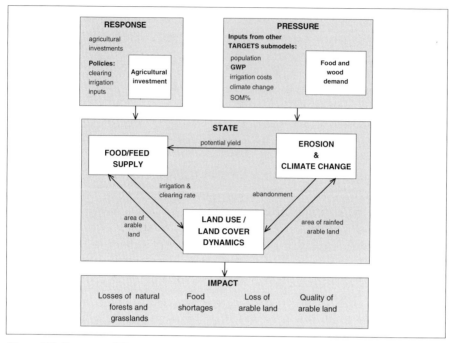

Figure 7.2 Structure of TERRA according to the PSIR approach.

land-use and land-cover dynamics, (ii) food and feed supply, and (iii) erosion and climate change. The *impact* module (Section 7.2.6) describes the impacts in terms of food shortages, the loss of (tropical) forests, and land degradation. This affects the quantity and quality or productivity of arable land and, hence, the food production potential. The *response* module (Section 7.2.7) represents different (agricultural) policy options: land clearing, extension of the irrigated arable land area, intensification on irrigated land (i.e. use of fertilisers and other inputs), land and/or soil conservation and reforestation. Total agricultural investments, and expenditures for intensifying the use of rainfed arable land are derived from these policies.

7.2.2 Position within TARGETS

The TERRA submodel is linked to the other submodels in TARGETS1.0 through the interactions shown in *Figure 7.3*. Changes in land-cover patterns, erosion and agricultural practices affect the element fluxes in CYCLES, resulting, among other things, in changes in Soil Organic Matter (SOM). This in turn has an impact on soil fertility and food production. Climate change resulting from rising concentrations of the greenhouse gases can affect the potential yield of arable land in TERRA both positively (CO_2 fertilisation) and negatively (heat stress). Changes in land cover affect the hydrological processes of evapotranspiration and runoff, which form part of the AQUA submodel. The irrigation water demand in AQUA depends on the area of irrigated arable land in TERRA. On the other hand, costs of irrigation are influenced by water availability and quality as calculated in the AQUA submodel. The rain erosivity factor as calculated in AQUA affects erosion; the size of livestock in TERRA influences water demand. The economic scenario generator, indicated by 'Economy' in *Figure 7.3* drives growth in Gross World Product (GWP) and, together with the Population and Health submodel, per capita income; this determines food demand. Conversely, available food per capita affects people's nutritional status and thereby human mortality. The Energy submodel indicates land requirements for biofuel plantations. Finally, desired agricultural investments are determined by the TERRA submodel. The economic scenario generator allocates the actual investments, which might be lower than the desired investments (Chapter 18).

7.2.3 Aggregation into land reservoirs and initialisation

The global land area in TERRA comprises 82 land reservoirs or stocks, differing with respect to economic, land-cover, climate and soil characteristics. Following the aggregation as used in FAO statistics, two economic world regions are distinguished: the developing and the developed (*Figure 7.4*). The economic regions differ with

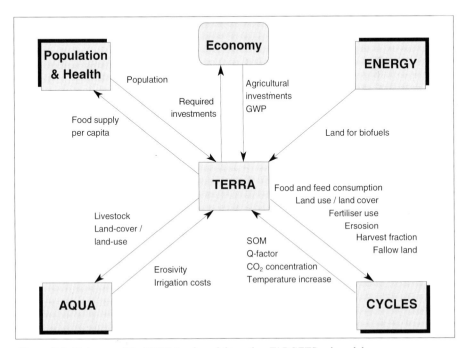

Figure 7.3 Linkages of the TERRA submodel to other TARGETS submodels.

respect to the stage in the 'agricultural transition' from low yields and fast cropland expansion to high yields and slow cropland expansion or even a shrinkage. This transition often coincides with a transition from low incomes and low, mainly vegetable, per capita food consumption, and high population growth to a situation of high incomes, high consumption of animal products and a stabilising growth in population. The clustering of countries into these two regions is fixed for the simulated period (1900 to 2100). In the CYCLES and AQUA submodels (Chapter 6, 8, 14 and 16), we also use the terminology 'tropical' and 'temperate' instead of 'developing' and 'developed'. Although these subdivisions are not fully interchangeable, the chosen economic subdivision shows similarity, with a division into tropical and temperate regions. This applies especially to the natural vegetation types 'forests' and 'grassland', which are most important to the dynamics in the CYCLES submodel. More than 95% of the natural vegetation in the developed region is located in the temperate region. In the developing region, more than 80% consists of tropical vegetation types.

Seven land-cover types are distinguished in TERRA. These are forests, grasslands (incl. pastures), rainfed arable land, irrigated arable land, degraded land, reforested area and other land (incl. wetlands and deserts), all based on the categories distinguished by the FAO (*Table 7.1*). The historically (1900-1990) estimated land-cover types for the two economic regions represented in TERRA are shown in *Figure 7.1*. As previously mentioned, these estimates are rather largly uncertain.

Information on plant growth potential and land-use possibilities is aggregated into the Length of Growing Period (LGP) (e.g. FAO, 1978-1981; Kassam *et al.*, 1991; Oldeman *et al.*, 1991). The LGP of an area denotes the estimated number of consecutive

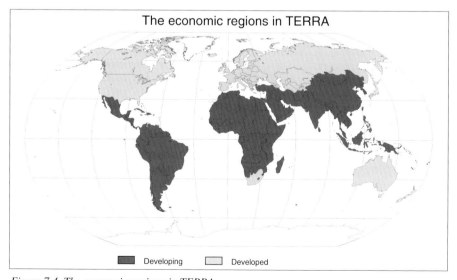

Figure 7.4 The economic regions in TERRA.

FAO	TERRA
Forest and woodland	Forests Reforested land
Permanent meadows and pastures	Grassland Degraded land
Arable land minus irrigated land	Rainfed arable land
Irrigated land	Irrigated arable land
Other land Land under permanent crops	Other land

Table 7.1 land-cover types: FAO vs. TERRA.

days that the standard soil will support rainfed plant growth without interference by extended dry or extremely cold periods. Three LGP classes are distinguished in TERRA:

- short (<90 days), it is generally not possible to grow rainfed crops;
- medium (90-240 days), it is generally not possible to grow more than one main crop per year; and
- long (>240 days), it is often possible to grow two main crops per year.

Soil differences can be aggregated using the soil productivity-reduction factor Q (ranging from 0 to 1), as introduced by Leemans and van den Born (1994). Q depicts the reduction of the potential yield due to less than optimal inherent soil conditions and calculated from the following reduction factors: fertility, salinity, acidity, drainage and rooting depth. In TERRA, three soil productivity classes, low ($Q < 0.5$), medium ($0.5 \leq Q < 0.75$), and high ($Q \geq 0.75$), are distinguished.

The land-use data from AGROSTAT (FAO, 1992), HYDE (Klein Goldewijk and Battjes, 1995) and IMAGE2.0 (Alcamo, 1994; Leemans and Cramer, 1991) are combined to obtain the historical land reservoir areas. The economic and land-cover classification is derived from AGROSTAT and HYDE, since they include historical data (1900-1990) for most land-cover types in both economic regions. The agro-ecological and soil classification is derived from IMAGE2.0 data. The IMAGE data base gives land-cover data for a reference year (1970) only, and is used to estimate the relative distribution of the land-cover classes in the developing and developed regions over the LGP and Q classes in 1900. The initial areas of the TERRA land reservoirs are estimated by combining the latest with the initial (1900) land-use areas, as derived from AGROSTAT and HYDE. Significant uncertainties are introduced in this procedure, but it seems the best solution for the moment. The combination of two economic regions, seven land-cover types, three LGP classes and three Q classes would result in a total of 2×7×3×3=126 reservoirs. However, not all combinations are possible (for example, rainfed arable land with a short growing period does not exist) so that 82 land reservoirs remain.

7.2.4 Pressure module

Food and wood demands are calculated in the pressure module. Food demand is divided into a vegetable and an animal component expressed in kcal per capita per day. This division is useful since it permits investigation of the effect of different diets on the vegetable production. If food crops are used for feed, then a diet containing animal products requires a higher vegetable production than a vegetarian diet (1 kcal of animal food requires up to 8 kcal feed). Demand for both vegetable and animal products are assumed to depend linearly on population size and log-linearly on per capita income. As in Zuidema (1994), we used a semi-log Engel curve: an increasing income per capita results in a slowly declining growth rate of demand for food, with a relatively (and absolutely) higher demand for animal products:

$$CNS_{i,r}(t) = CNS_{i,r}(t_0) \times \left(1 + x_{i,r} \ln \frac{GWP_{pc,r}(t)}{GWP_{pc,r}(t_0)}\right) \qquad [\text{kcal/cap/day}] \qquad (7.1a)$$

where CNS is per capita food demand and GWP_{pc} is Gross World Product per capita. Strictly speaking, the term 'Gross World Product' is inaccurate since in TERRA it has been split into two parts corresponding with the two economic regions as described previously. The variable x is the income elasticity parameter, t the index for time (1900 to 2100), i the index for commodity (vegetable or animal products) and r the index for region (developed or developing). The income elasticity parameter x is not the same as income elasticity. In a semi-log Engel curve income elasticity ε varies inversely with the quantities consumed. In equation form:

$$\varepsilon_{i,r}(t) = \frac{CNS_{i,r}(t)/\Delta CNS_{i,r}(t)}{GWP_{pc,r}(t)/\Delta GWP_{pc,r}(t)} = x_{i,r} \frac{CNS_{i,r}(t_0)}{CNS_{i,r}(t)} \qquad (7.1b)$$

from which the elasticity parmeter is seen to be equal only to the initial elasticity. Finally, in the reference scenario (see Chapter 15) we assume a saturation level of 3470 kcal/cap per day (FAO, 1993a) for the total food demand per capita and a maximum animal food demand per capita of 1260 kcal/cap per day, which is comparable with the diet of a West German in 1990. As to wood we restrict ourselves to the roundwood demand from the developing countries. This is because large-scale deforestation associated with roundwood production occurs mainly in this region. Based on FAO statistics, the roundwood demands are assumed to depend linearly on population size.

7.2.5 State module

land-use and land-cover dynamics
The land-use and land-cover dynamics are governed by eight processes (*Figure 7.5*):

- *Clearing and conversion of forest and grassland into rainfed arable land (1)*. Driven by an exogenous clearing policy scenario, this leads to an increase in the food and feed supply to meet the animal and vegetable food demands.
- *Clearing of forest to grassland (2)*. This part of deforestation is caused by the need of the ruminant livestock for pasture land (developing region only).
- *Deforestation associated with roundwood production (3)*. It is assumed that some fixed fraction of the total roundwood production causes permanent or long-lasting deforestation resulting in grassland. The remaining fraction is assumed to be produced 'sustainably' in the sense that the harvested forest regenerates to or remains natural forest.
- *Abandonment of rainfed arable land due to erosion (4)*. Rainfed arable land will be abandoned and left behind as 'degraded land' if the SOM in the topsoil is below 0.5%. In this case, the soil productivity is too low for any rewarding agricultural activity (Pieri, 1989; Rozanow *et al.*, 1990; van der Pol, 1993).
- *Irrigation (5)* refers to the conversion of rainfed arable land into irrigated arable land. It is introduced by an exogenous irrigation policy scenario. In TERRA irrigation will increase the average yield by 50% compared to rainfed arable land, assuming the same fertiliser input (Euroconsult, 1989; FAO, 1993a).
- *Natural reforestation (6)*. This is caused by natural recovery of secondary vegetation (e.g. grassland) reverting to forest. An average of 0.02% per year of the total area of grassland reverts to natural forest; this is an approximation based on FAO (1993a, 1993b).
- *Anthropogenic reforestation (7)* is the result of an exogenous reforestation policy scenario as determined in the response module (developed region only).
- *Recovery of degraded land (8)*. It is assumed that most of the degraded land will naturally recover to become grassland, which can then be used as cropland again.

Some possibly important flows, including urbanisation[1] and drainage of wetlands[2], are not included in TERRA. This is because, to date, we do not have enough insight into the dynamics and data to describe these processes satisfactorily.

---

[1] In several Western Europe countries it is estimated that 2% of agricultural land is being lost per decade to urban growth; the USA is losing 2.5 Mha of prime farmland per decade. In developing countries cropland is steadily being consumed by urban expansion (Tolba and El-Kholy, 1992).

[2] Since 1900 the world may have lost half of its wetlands to agriculture, clearance for forestry and urbanisation (Harrison, 1992). Today, total remaining wetlands in the world number between 530 and 860 Mha (Aselmann and Crutzen, 1989).

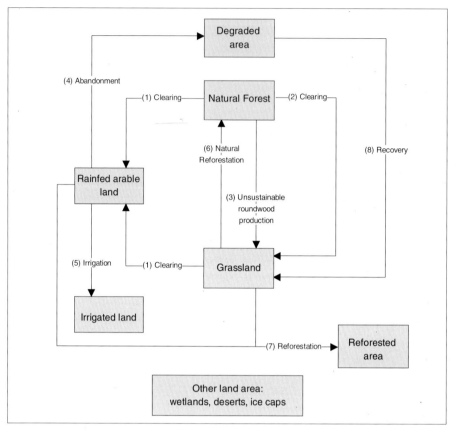

Figure 7.5 Land-use /land-cover dynamics in TERRA.

Food supply

The main structure and dynamics of food supply can be explained by the causal loop diagram shown in *Figure 7.6*. It contains three negative causal loops. In the first, the change in irrigated area is the result of an irrigation policy. The irrigation investment flow is calculated as the product of irrigated area expansion and irrigation costs per hectare. Average costs to irrigate land are assumed to increase due to, for example, declining water availability. When the limits of water availability (in AQUA) are reached, these costs will lead to increasing irrigation investments and, ultimately, expansion of irrigation comes to a halt.

The second loop includes expansion of the agricultural land area by land clearing. Clearing investments are the result of a clearing policy, and the marginal clearing and development costs per hectare. This leads to a decrease in the remaining potential arable land, implying higher costs assuming that potential arable land with easy access is cleared first *(Figure 7.7)*. Another criterion for clearing new land is land fertility or soil productivity (this is not indicated in *Figure 7.6*). As a result, at the

7 THE LAND AND FOOD SUBMODEL: TERRA

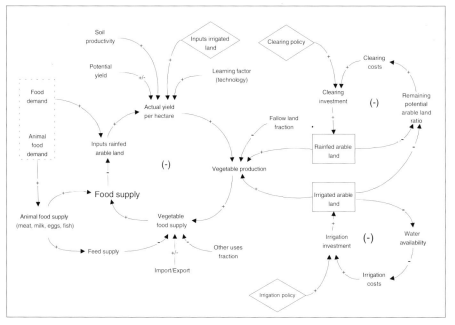

Figure 7.6 Food supply. Exogenous policy scenarios are enclosed in diamond-shaped boxes.

beginning mainly highly productive land is cleared; however increasingly newly developed lands will be less productive.

The third loop concerns the *Actual Yield* per hectare, *AY*, which is defined as the average yield actually harvested by the farmer. It is calculated for all classes of irrigated and rainfed arable land. For the latter, two LGP classes (medium and long) and three inherent soil productivity classes are taken into account, resulting in six corresponding actual yields. Since no irrigation is needed on arable land with a long growing period, only the medium LGP class is considered for irrigated arable land, resulting in three actual yields corresponding to the three soil classes. The general expression for the *Actual Yield* is (see also Zuidema *et al.*, 1994):

$$AY_t = PY_t \times Q_t \times \frac{MEI_t}{MEI_{max}} \times LF_t \qquad \text{[tonne CCE/ha/yr]} \qquad (7.2)$$

where *t* is the index for year. The *AY* is the average yield for the range of edible crops (cereals, roots and tubers, pulses, oilcrops, vegetables and sugar), expressed in Cereal Consumption Equivalents (CCEs)[3]. *PY* is the average *Potential Yield* with

---

[3] Based on AGROSTAT, we define one kg CCE to contain 3017 kcal in the developing and 2725 kcal in the developed region, based on the average caloric value of 1 kg cereals. The difference arises from a varied contribution of the separate cereal crops. The rationale for using different CCEs is to get a close fit to the actual dry weight harvested from the different crops.

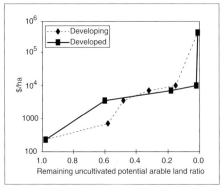
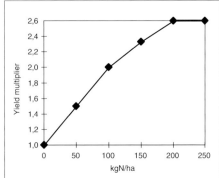

Figure 7.7 Clearing or development costs as a function of the remaining uncultivated potential arable land ratio (Buringh et al., 1975; Meadows et al., 1974).

Figure 7.8 N-fertiliser response curve with $MEI_{max} = 2.6$. Based on FAO (1987), Pimentel et al. (1973), and Wild (1988).

the same dimension as *AY* and based on the weighed average of the potential crop yield for seven main crops as computed by the crop model (see Box). The potential crop yield is the theoretical maximum yield obtainable from the crop considered. It is equal to the product of the potential net dry-matter biomass production throughout the growing period and a crop-specific harvest index. The potential biomass production is defined as the total dry-matter production by a green, healthy crop surface, as determined entirely by the amount of available light, prevailing temperatures and crop characteristics, assuming optimal conditions with respect to water, nutrient availability and crop coverage and absence of weeds, pests, diseases,

The crop model

We determine the potential production using a simple photosynthetic crop model based on the crop models of de Wit (1965) and the FAO (1978-1981), which were constructed to simulate the annual path of crop growth. We have adapted them to compute a mean growth rate during the growing period (Leemans and van den Born, 1994) in each of the two major agro-climates: temperate and (sub)tropical. The potential crop yields in the temperate and (sub)tropical regions are used as an estimate for the developed and developing countries, respectively (*Figure 7.4*). Cold climates are excluded because in boreal regions, few crops are grown successfully. Temperate regions are characterised by relatively mild winters and higher growing-season temperatures with adequate moisture. The (sub)tropical zones are characterised by temperatures high enough to support growth during the entire year. Here, crop growth is limited by the presence and amount of available water. Within each agro-climate, the potential crop yield is calculated for two LGP classes (medium and long), and for three temperature classes (temperate: 5-10 °C, 10-15 °C and 15-18 °C; and tropics: 18-20 °C, 20-25 °C and >25 °C), assuming constant harvest indices. The results are expressed in ton of CCE per hectare per year and aggregated for the seven main crops (wheat, rice, millet, sorghum, maize, cassava and potatoes, which cover more than 70% of the vegetable production in the world). This takes into consideration the relative contribution of each crop, which is assumed to be constant over time (as confirmed by AGROSTAT data for the period 1960-1990). The resulting Potential Yield (*PY*) is considered as an average yield for the whole range of edible crops. Finally, the *PY* can be affected by climate change, changes in the aggregated harvest index and irrigation. We assume that the aggregated harvest index changes proportionally to the technological development. As we calculated in *annual yields*, irrigation doubles the *PY* if temperatures allow double cropping.

erosion and strong winds etc. The harvest index accounts for the difference between the biomass production and the usable or edible part of the crop. Harvest indices can increase over time as (bio)technology develops.

The next factor in equation 7.2 is the *actual soil productivity* factor or Q factor ($0 \leq Q \leq 1$, see Section 7.2.3). Changes in Q are directly related to the fraction of SOM in the topsoil and the formation of gullies. Since SOM is an integral part of the element cycles, it is calculated in the CYCLES submodel, taking into account all relevant inputs from TERRA. In general, erosion will decrease the soil productivity. Other processes, for example, the accumulation of non-harvested plant parts and the conversion of forest and grassland into arable land, have a positive effect on the average SOM fraction of all arable land.

MEI in equation 7.2 is the *Multiplier Effect of Inputs* ($1 \leq MEI \leq MEI_{max}$), indicating yield increases from mechanisation, fertilisers and other intensification investments on irrigated and rainfed arable land. The total inputs on irrigated land have been implemented as an exogenous policy scenario. The level of inputs on rainfed arable land is determined by the ratio between the total food supply and total food demand. The multiplier effect of the inputs on the actual yield is determined through the N fertiliser response curve (see *Figure 7.8*) assuming that N fertilisers cover a fixed fraction of the total inputs and that all the remaining inputs in terms of P and K fertilisers, mechanisation, etc. are available.

Finally, *LF* is the Learning Factor ($0 < LF \leq 1$), which is dependent on technology development and its application, i.e. management, plant treatment, storage improvement, mechanisation, biotechnology, etc. For calculations in the past, the value of *LF* accounts for the difference between the computed feasible yields without the learning factor and the actual yields observed. In other words, it accounts for all factors not modelled explicitly in TERRA. Given the land-cover changes and the food demands in some future scenario, the learning factor is determined by calibration (see Section 7.3).

The total vegetable production for each economic region is equal to the product of the actual yields as described above and the harvested area of rainfed and irrigated arable land in the different LGP and Q classes (*Figure 7.6*). The fraction of fallow land, defined as all arable land not used for edible crops, is implemented as an exogenous time-series. Fallow land fractions tend to decrease as a result of decreasing fallow periods (Section 7.3.3). The total vegetable supply per region is derived from the vegetable production by taking into account a net trade fraction from the developed to the developing region. The *utilisation* of the vegetable supply is divided in three components: food, feed and other uses such as seeds, waste and processed items. In a normal situation, the vegetable *food* supply covers the vegetable food demand. The vegetable *feed* supply is calculated from the animal food supply, which is equal to animal food demand. This is based on a set of assumptions on animal diets, caloric content of animal products, conversion factors of feed to animal products and the extent to which animals graze.

Region	LGP	Erosion class		
		High	Medium	Low
Developing	90-240	0.25	0.25	0.50
	>240	0.30	0.25	0.45
Developed	90-240	0.10	0.10	0.80
	>240	0.15	0.15	0.70

Average loss of topsoil: High = 50.4 ton/ha per yr = 3.6 mm per yr,
Medium = 16.8 ton/ha per yr = 1.2 mm per yr
Low = 5.6 ton/ha per yr or 0.4 mm per yr.

Table 7.2 Relative distribution of potential erosion classes on rainfed arable land in 1970 [4].

Erosion

In the present version of TERRA, erosion by rain is estimated for rainfed arable land only. It starts with the potential erosion class distribution for 1970 (*Table 7.2*). Potential erosion is defined as the erosion that can occur assuming 'standard' cropping practices. The distribution in other years has been obtained by assuming that all erodible arable land in 1900 was in the 'low' class, that an exponential increase of both medium and highly erodible arable land occurred between 1900 and 1970 and that this increase levelled off after 1970 towards a more or less stable distribution in 2100. The erosion class distribution is combined with the erosivity change factor (from AQUA) and an additional conservation policy to arrive at the average actual erosion level for each LGP class. These conservation practices are defined through an exogenous conservation policy scenario, expressed in Mha. Actual erosion causes the loss of relatively fertile topsoil, resulting in a decreasing fraction of SOM in the topsoil. This has two effects:

- less organic matter results in a decreasing soil productivity Q and thus decreasing actual yields per hectare; the annual return of non-harvested plant parts will also decrease, which tends to lower the SOM;
- decreasing SOM in the topsoil increases the amount of rainfed arable land where the productivity is too low for farming practices, resulting in abandonment. It is assumed that cropland is abandoned once the SOM in the topsoil is less than 0.5%.

The effect of land-use and land-cover changes on SOM can be either positive or negative thus affecting soil productivity Q. For example, if potentially highly

---

[4] A rough estimate of cropland distribution over the TERRA erosion classes in 1970 was obtained by using the Universal Soil Loss Equation (USLE) (Lal, 1994; Troeh *et al.*, 1991; Wischmeier and Smith, 1978). Here it was assumed that 70% of the global cropland is on land with a slope of 0-8%, 15% with a slope of 8-15%, 10% with a slope of 15-30% and 5% with a slope of >30%. Global rainfall distribution characteristics and selected agronomic default values were used for other USLE parameters under 'standard' cropping practices.

productive forests and grasslands are taken into production, this will result in a higher average productivity of arable land. Conversely, if marginal or potentially lower productive areas are taken into production, the result is a lower average productivity.

Climate change
The crop module is used to calculate the effects of climate change on potential crop yields. The first effect accounted for in our model is heat stress. To arrive at a multiplication factor for temperature effects on potential yield, the crop-specific changes in yield due to temperature changes are averaged and weighed with respect to their relative contribution to total yield. The results are aggregated into four multipliers (one for each economic region and LGP class), accounting for the change in the potential yield (*Figure 7.9*). Besides the direct multiplier effect of a temperature increase *within* a LGP class, there is an indirect effect caused by changes in the LGP *distribution*. Based on experiments with IMAGE2.0, the LGP changes have been estimated as a function of global temperature increase: an increase of 1 °C will result in a shift of the area with low LGP to medium LGP by about 2% in the developing and 12% in the developed regions. The change from medium to high LGP is modelled as a fixed fraction of the change from low to medium (0.6 and 0.1 for the developing and the developed regions, respectively).

The second effect related to climate change considered here is CO_2 fertilisation. Many studies suggest that high atmospheric CO_2 concentrations enhance plant growth (for a review see, for example, IPCC (1995a)). CO_2 increases photosynthesis, dry-matter production and yield; it decreases stomatal conductance and transpiration, and improves water efficiency. C_3 species are more sensitive to atmospheric CO_2 concentrations than C_4 plants (Bazzaz and Fajer, 1992). However, many interactions between CO_2 fertilisation and temperature, soil moisture, nutrient availability and increases in UV-B radiation (due to depletion of the ozone layer) may be present (Klein Goldewijk *et al.*, 1994; Teramura *et al.*, 1990). Adaptation to elevated CO_2 concentrations may take place, resulting either in an increase or decrease of the photosynthetic rate (Sage *et al.*, 1989). Given the complexity of the interactions mentioned above, and the often insufficient scientific knowledge of the basic processes underlying these mechanisms, we took the effects of the CO_2 fertilisation into account by multiplying the net biomass production with a single fertilisation factor (F_{CO_2}):

$$F_{co_2} = 1 + \beta_r \times \ln \frac{[CO_2]_t}{[CO_2]_o} \tag{7.3}$$

where β_r is the actual β-factor for arable land[5]. In Alcamo (1994, p. 209), a mean

---

5   β-factors for the other land-cover types are used in CYCLES to simulate the CO_2 fertilisation effect on the C cycle.

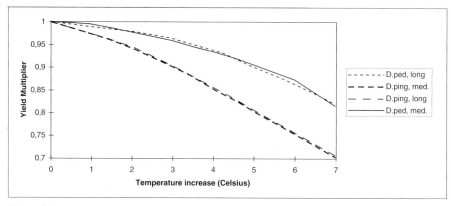

Figure 7.9 Aggregated multiplier effect of temperature increase on potential yield. (D.ped = Developed region, D.ping = Developing region, long= LGP>240 days, med = LGP<240 days)

gobal value of 0.28 is presented. The β-factor for arable land will be higher since agriculture is assumed to be practised on more or less suitable soils and using growth-promoting inputs (fertilisers, biocides, etc.). Therefore in the reference scenario, which is presented in Chapter 15, we have used the 'best guess' or 'central estimate' which is 0.40 in the developed and 0.37 in the developing regions.

7.2.6 Impact module

Impacts in the context of the TERRA submodel can be considered as changes in the food situation, the size and quality of the land reservoirs, and changes in the hydrological cycle. With respect to the food situation, we cannot make statements about changes in food availability and distribution because these factors heavily depend on non-modelled aspects like world food markets, poverty, income distribution, political situation, etc. Instead, we use global food shortages, that is, a mismatch between supply and demand, and combine it with a food distribution function within the Population and Health submodel. An unequal food distribution can result in a situation where no global food shortages exist, and yet a significant part of the world population suffers from malnutrition.

The size of the land reservoirs, forest and grassland (as far as it refers to natural grassland), can be considered as the first proxy for the loss of ecosystems and biodiversity. Eutrophication of groundwater and surface water as modelled in AQUA, deposition of NO_x and NH_3, and climate change as modelled in CYCLES can also have a considerable effect on the functioning of ecosystems. However, as ecosystem dynamics have not been modelled in TARGETS1.0, we can only give an indication of the various pressures on the land and water reservoirs.

An important impact indicator of the quality of arable land is soil productivity, as modelled by the *Q*-factor (see Equation 7.2). Changes in soil productivity are related to changes in the fraction of SOM in the topsoil (see Section 7.2.5). Below a certain fraction, arable land is abandoned, leaving behind infertile degraded land. Impacts on the hydrological cycle due to irrigation, such as increased evapotransporation and decreased river runoff, which leads to water scarcity, are considerable. At the moment, water scarcity results only in higher irrigation costs in TERRA; no other impacts are considered.

7.2.7 Response module

The response module allows for the introduction of various exogenous time paths for some crucial variables reflecting a variety of micro- and macro-behaviour. Within the TERRA submodel, these time paths are interpreted as policy scenarios. Three types of agricultural policies can be distinguished. First of all, we distinguish policy measures for each of the economic regions aimed at increasing the amount of arable land in terms of the annual clearing rate. Secondly, there is a set of policy measures focusing on the increase of crop yields per hectare. This can be achieved in two ways: (i) conversion of rainfed arable land into irrigated arable land, and (ii) additional inputs, including fertilisers, on irrigated arable land. A third option, additional inputs on rainfed arable land, is not modelled as a policy but is used to balance food supply and demand endogenously within the constraints of the available investment funds. Thirdly, land conservation policy measures can be initiated to protect current arable land from erosion and soil degradation. Conservation measures in the form of subsidies, education, etc. are expressed as the total amount of rainfed arable land to be conserved, where 'conserved' is defined as establishing an erosion rate lower than 1 tonne of soil per hectare per year. Besides these agricultural policies, a reforestation policy is also included. This option allows for reforesting rainfed arable land and/or grassland.

7.3 Calibration

Calibration has been carried out in three consecutive steps for both economic regions for the period 1900-1990. We started with land-use and land-cover changes as shown in *Figure 7.1*, then food demand and supply and, finally, vegetable production and yields.

land-use and land-cover changes
Calibrating the irrigated area is done by equating the policy scenarios to the historical data. Between 1961 and 1990, the irrigated area is assumed to be equal to the irrigated area in the AGROSTAT database (FAO, 1992). Irrigation areas from Gleick (1993) are used for the period 1900-1960 (*Table 7.3*). We have compared the area of *rainfed arable land* between 1900 and 1960 with historical data on cropland from the

Calibration variable	Source(s)	Period
Irrigated arable land	(Gleick, 1993)	1900-1960
	(FAO, 1992)	1961-1990
Rainfed arable land	(Klein Goldewijk and Battjes, 1995)	1900-1960
	(FAO, 1992)	1961-1990
Forest	(Klein Goldewijk and Battjes, 1995)	1900-1960
	(FAO, 1992)	1961-1990
Grassland & degraded land	(FAO, 1992)	1961-1990
Abandonment	(Oldeman et al., 1991)	1946-1990

Table 7.3 Calibration variables and sources with respect to land-cover changes.

HYDE data base (Klein Goldewijk and Battjes, 1995), which equal the sum of the AGROSTAT categories of (rainfed) *Arable land* and *Permanent crops*. To estimate the area of rainfed arable land in HYDE, we assume the fraction of rainfed arable land as derived from the AGROSTAT database in 1961 to be applicable to the cropland category in HYDE for the period 1900-1960. Between 1961 and 1990 this fraction hardly changed (±6% world-wide). The data from AGROSTAT are used for the period 1960 to 1990. Reproducing the historical time-series was achieved mainly by determining an appropriate clearing policy scenario, taking into consideration the effect of abandonment.

The simulated area of *forest*, i.e. natural forest plus reforested area, between 1961 and 1990 should be equal to the area of *Forest and Woodland* in AGROSTAT. The simulation results for between 1900 and 1960 are compared with the trends of Forest in the HYDE data base because its definition of *Forest* applies to a larger area than *Forest and Woodland* in AGROSTAT. Keeping the clearing policy more or less fixed, the calibration of forest in the developing region is done by varying the parameters for deforestation from roundwood production.

The historical data on forest area for the developed region are reproduced by introducing a reforestation rate increasing linearly from 0 in 1900 to 0.88 Mha/yr in 1925 to 1.1 Mha/yr in 1990. In the real world recent reforestation rates were much higher since in Europe alone (excluding the former USSR) the reported area of new or renewed forest and other woodland between 1980 and 1990 was 3.8 Mha/yr (UN-ECE/FAO, 1992, p. 7). The difference is caused by the fact that deforestation in the developed countries is simulated only as far as clearing for agricultural purposes is concerned. Commercial logging and any other causes of deforestation in the developed countries are not simulated in the current version of TERRA. In the real world, reforestation is not limited to the developed countries as is assumed in our simulations. The reported area of forest plantations in 90 developing countries (excluding China) at the end of 1990 was 30.7 Mha after 30% deduction of mortality and failures (FAO, 1993b); this amounts to 5% of the deforested area.

The sum of *Grassland* and *Degraded land* in TERRA for between 1960 and 1991, is compared with *Permanent meadows and pastures* in AGROSTAT. Assuming the area of *other land* to be fixed (see Section 7.2.1), the historical time-series of *Grassland* and *Degraded land* between 1900 and 1960 could be estimated. *Degraded land* in TERRA, which is equal to abandoned rainfed arable land, was calibrated by using the GLASOD data (Oldeman et al., 1991)[6]. In TERRA, it is assumed that a large part of the strongly and extremely degraded cropland in the GLASOD inventory will be or will have been abandoned – probably 60% in the developing countries and possibly around 90% in the developed countries. This difference is based on the assumption that farmers in the developing countries will usually decide to abandon their land at a later stage than in the developed countries. Based on these assumptions and the GLASOD information, historical land abandonment rates are estimated for the period 1900-1990 and can be reproduced by TERRA on the basis of SOM values from the CYCLES submodel.

Food demand and supply

The elasticity parameter values describing the relation between GWP per capita and animal and vegetable food demand have been chosen so that the simulated food demands between 1960 and 1990 equal the animal and vegetable food intake data of AGROSTAT. Historical food demands in the period 1900-1960 are derived by assuming fixed elasticity parameters. The simulated vegetable food supply reproduces the historical data fairly well (*Figure 7.10*). Animal food supply and historical intake (1960-1990) are equal to animal food demand and therefore not shown.

Vegetable production, supply and yields

In TERRA the total vegetable supply is derived from the total vegetable production by taking into account an exogenous net trade fraction for trade from the developed to the developing region (*Table 7.4*). As described in Section 7.2.5, the utilisation of the vegetable supply is divided in three components: food, feed and other uses. Based on AGROSTAT, the 'other uses' are assumed to claim 12% and 15% of the vegetable supply in the developing and the developed region, respectively. The fallow-land percentage of rainfed arable land in the developed countries is rather constant at ±45%[7] for the period 1960-1990. We assume the same value for the

---

6 The most consistent effort to estimate the state of global soil degradation has been completed in UNEP's project 'GLobal Assesment of SOil Degradation' (GLASOD). In this project, soil degradation induced by humans from 1946 up to 1990 was assessed. The inventory covered all types of soil degradation and confirmed erosion by rain to be the most widespread form of soil degradation.

7 This high value is related to the fact that the harvested area in AGROSTAT does not include arable land used for fodder crops or leys in crop rotation schemes. In the developing countries cultivation of fodder crops on arable land is negligible. Futhermore, crop failures in the developed countries make harvesting more unrewarding than in the developing countries where subsistence farming is more common.

Parameter	Region	Value
Net trade percentage[a] for trade from the developed to the developing region	–	0 in 1900 0.8% in 1961 5.2% in 1990
Other uses fraction[b]	Developing Developed	12% 15%
Fallow-land fraction of rainfed arable land[c]	Developing Developed	28% in 1900 10% in 1990 45%
Learning factor	Developing Developed	0.15 in 1900 0.17 in 1990 0.27 in 1900 0.29 in 1990

a 1900-1960: own estimate. 1961-1990: AGROSTAT.
b Constant between 1961-1990 in AGROSTAT, therefore also applied to the period from 1900 to 1960.
c In many farming systems arable land is not harvested or cropped in certain years; in this period the land is considered as fallow in TERRA.

Table 7.4 Parameter settings for vegetable production.

periods before 1960 and after 1990. In the developing countries the fraction has declined between 1960 and 1990; a trend which we have extrapolated.

The historical vegetable production (in tonne CCE/yr), supply and utilisation between 1961 and 1990, as derived from AGROSTAT, are reproduced by TERRA if the relevant parameters are set as shown in *Table 7.4*. No other determinants of vegetable production can be used for the calibration because there are a number of constraints:

- areas of rainfed and irrigated arable land must equal historical estimates;
- vegetable and animal food demands are related to GWP/cap;
- simulated N fertiliser use, which is strongly dependent on the learning factor, must be equal to the historical data in AGROSTAT[8] ; and
- the share of different animal product types (milk, eggs, ruminant and non-ruminant meat, and fish) in the total animal food demand is kept constant.

---

8 Since no data are available, the initial use of N fertiliser is based on the assumption of it being equal to 1/20 of the 1970 level. In the developing countries the 1970 value is 15 kg NPK per ha. Assuming an N:NPK ratio of 1:2, an initial amount of 0.25 kg N per ha results. For calibration, no distinction was made between irrigated and rainfed arable land because no data are available at this level of aggregation. Between 1900 and 1961 a linear increase is assumed between the initial values as mentioned above and the 1961 AGROSTAT values.

7 THE LAND AND FOOD SUBMODEL: TERRA

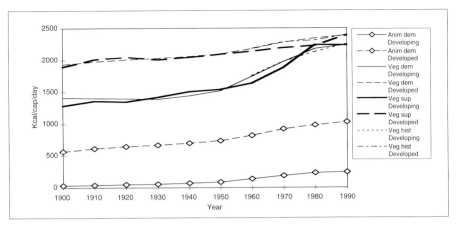

Figure 7.10 Simulated vegetable food demand (Veg dem) and supply (Veg sup), historical vegetable food intake (Veg hist, 1961-1990) and simulated animal food demand (Anim dem).

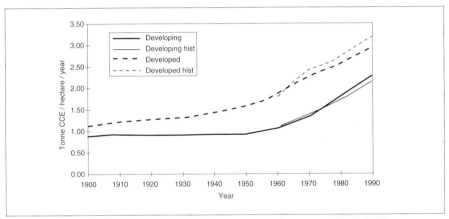

Figure 7.11 Comparison of simulated yields with historical estimates (1960-1990).

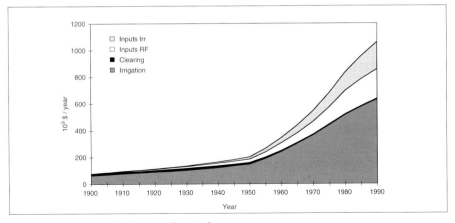

Figure 7.12 The simulated costs of agriculture.

Given the calibration steps, it is possible to simulate some variables which can be used for model validation. Yields are calculated by dividing the total vegetable production minus the production of permanent crops (around 3% of the total vegetable production) by the surface area of the harvested land. Clearly, if the total surface area of arable land and the total vegetable production are calibrated correctly, the historical yields are also reproduced (*Figure 7.11*).

Another variable which can be derived is the total annual costs of agriculture. These are subdivided into operational costs to clear land, irrigation costs and inputs on rainfed and irrigated arable land (labour excluded) (*Figure 7.12*). Conservation costs are not shown as they are set to zero in the period 1900-1990. The calculated costs of agriculture are about 5% to 8% of GWP, which is in the range of World Bank estimates of the world's agricultural share in GWP for the years 1980 and 1990 (World Bank, 1996). However, as historical data are not available, we have not attempted to calibrate the costs at global level. For this reason, we consider cost trajectories to give only an indication of trends and relative importance.

7.4 Conclusions

The TERRA submodel is able to reproduce the major agricultural trends in land use and land cover, food demand and supply, fertiliser use and irrigation in both developed and developing regions. This is done to a large extent by using exogenous parameters and policy scenarios which can be chosen in such a way that the simulation results are in agreement with historical data or future scenarios. The TERRA submodel does not yet contain the dynamics needed to simulate food demand-supply imbalances in a meaningful way. A more dynamic and realistic model formulation would require the inclusion of economics (world food markets and prices) and the social dynamics of agricultural processes in more detail (such as the development of agricultural systems from shifting cultivation to permanent agriculture as population pressure increases).

The value of the TERRA submodel in its present form lies in its interactions with the other submodels, which make it possible to explore linkages between population growth, water availability and climate change on the one hand and food production on the other.

The biogeochemical submodel: CYCLES

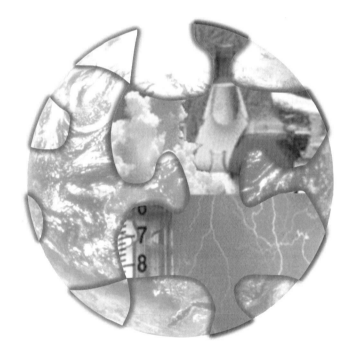

"The balance of evidence suggests a discernible human influence on global climate."
IPCC, Climate Change 1995

8 THE BIOGEOCHEMICAL SUBMODEL: CYCLES

Michel G.J. den Elzen, Arthur H.W. Beusen, Jan Rotmans and Heko W. Köster

The CYCLES submodel describes the long-term dynamics of the global biogeochemical cycles of carbon (C), nitrogen (N), phosphorus (P) and sulphur (S), their interactions and their impacts on climate change. The model analysis balances past carbon and nitrogen budgets – emphasising the importance of the N fertilisation feedback – and supports the future projections of the fate of anthropogenic emissions of both carbon and nitrogen compounds in the global environment presented in Chapter 16. This chapter focuses on the link between the global cycles of C and N and their feedbacks, providing calculations of global flows of these basic elements and their related compounds within and between the major reservoirs.

8.1 Introduction

Carbon, hydrogen and oxygen, together with the basic nutrient elements nitrogen, phosphorus and sulphur, are essential for life on Earth. The term 'global biogeochemical cycles' is used to describe the transport and transformation of these substances in the global environment. In recent decades detailed studies have been carried out on the global biogeochemical cycles of the basic elements, in particular carbon (C), nitrogen (N), phosphorus (P) and sulphur (S) (Bolin *et al.*, 1979; Bolin and Cook, 1983; Schlesinger, 1991; Butcher *et al.*, 1992; Wollast *et al.*, 1993). *Figure 8.1* depicts how anthropogenic disturbances of the global cycles of the basic elements of C, N, P and S lead to a variety of global environmental consequences.

Research has hitherto mainly been focused on separate global cycles rather than on interactions between the various global cycles. In spite of earlier attempts to model interactions between the global cycles of C, N, S and P (Jörgensen and Meijer, 1976), it is only in the last decade that a start has been made on quantitative studies of the interactions between the global cycles using sophisticated compartment models (Raich *et al.*, 1991; McGuire *et al.*, 1992; Keller and Goldstein, 1994; Hudson *et al.*, 1994; MacKenzie *et al.*, 1992). What is lacking so far, however, is an integrated modelling framework which describe the global cycle of carbon, nitrogen, phosphorus and sulphur: where each originates, where it remains and how the various global budgets can be balanced.

The CYCLES submodel captures in a comprehensive, albeit simplified, way the entire cause-effect chains of the basic elements C, N, P and S and their interactions (den Elzen *et al.*, 1995; 1997). These cause-effect chains have been aggregated into

8 THE BIOGEOCHEMICAL SUBMODEL: CYCLES

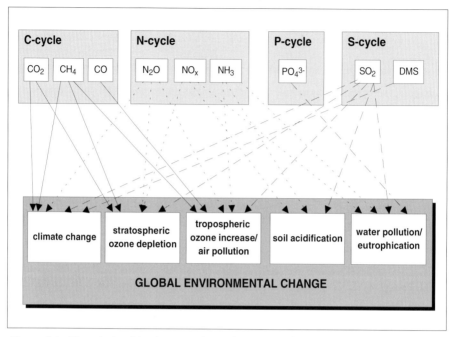

Figure 8.1 The relationships between the main compounds of the global biogeochemical cycles of C, N, P and S and the environmental themes. For example: soil acidification is a consequence of perturbations in the N cycle (NO_x, NH_3) and the S cycle (SO_2) (The main chemical compounds are: CO_2: carbon dioxide, CH_4: methane; CO: carbon monoxide, NO_x: NO and NO_2, NH_3: ammonia, PO_4^{3-}: phosphate; SO_2: sulphur dioxide; and DMS: dimethyl sulphide).

pressure, state, impact and response modules, using simple, one-dimensional box models to represent the physical, chemical and biological fluxes between the atmosphere, hydrosphere and terrestrial biosphere. In addition to the basic elements, the CYCLES submodel comprises such chemical compounds as halocarbons and tropospheric ozone and its precursors.

Within the context of this study, the CYCLES submodel is used to project the future fate of anthropogenic emissions of C and N compounds given the condition that the carbon and nitrogen budgets are balanced. This chapter therefore does not offer an exhaustive description of the whole CYCLES submodel but will merely describe those modules of CYCLES relevant to the analysis presented here: the coupled carbon and nitrogen cycle – and climate module. A full description of the CYCLES submodel which includes the model details and the sulphur and phosphorus cycles can be found in den Elzen *et al.* (1995).

8.2 The global carbon and nitrogen cycles and related feedbacks

8.2.1 The global carbon cycle

All life forms on Earth are primarily composed of carbon, so that studying the global carbon cycle in the past and present gives an indication of the comparative state of the biosphere. The cycles of other elements are closely tied to those of carbon through oxidation and reduction reactions. The natural global carbon cycle encompasses exchanges of CO_2, carbonates, organic carbon, etc. between three reservoirs (the atmosphere, the hydrosphere and the terrestrial biosphere) of several billions of tonnes of carbon per year. The anthropogenic increment due to the burning of fossil fuels and changing land use is relatively small compared with most of the natural exchanges of carbon between the reservoirs. However, there is strong evidence that the anthropogenic increment of carbon is disturbing the balance of the global carbon cycle, leading to an increase in the atmospheric CO_2 concentration. An increase which, as the subject of numerous studies during the past decades indicate, will change the Earth's climate (Intergovernmental Panel on Climate Change (IPCC), 1990, 1992, 1994, 1995b).

However, there are considerable uncertainties in our knowledge of the present sources of and sinks for, the anthropogenically produced CO_2. In fact, the only well-understood source is fossil fuel combustion, while in contrast, the source associated with land-use changes is less understood. The amount of carbon remaining in the atmosphere is the only well-known component of the budget. With respect to the oceanic and terrestrial sinks, the uncertainties are likely to be in the order of $\pm 25\%$ and $\pm 100\%$, respectively, mainly resulting from the lack of adequate data and from the deficient knowledge of the key physiological processes within the global carbon cycle (IPCC, 1994).

The carbon balance concept
The uncertainties concerned can be expressed explicitly by the formulation of a basic mass conservation equation which reflects the global carbon amount, as defined in:

$$\frac{dC_{CO_2}}{dt} = E_{fos} + E_{land} - S_{oc} - E_{for} + I \qquad [GtC/yr] \qquad (8.1)$$

where dC_{CO_2}/dt is the change in atmospheric CO_2, E_{fos} the CO_2 emission from fossil fuel burning an cement production, E_{land} the CO_2 emission associated with changing land use, S_{oc} the CO_2 uptake by the oceans and E_{for} the CO_2 uptake through forest regrowth. The best available current knowledge on the sources and sinks of CO_2, which comprises a mixture of observations and model-based estimates, does not permit us to obtain a balanced carbon budget. To balance the carbon budget, another

8 THE BIOGEOCHEMICAL SUBMODEL: CYCLES

Component [GtC/yr]	1980-1989
Emissions from fossil fuel burning and cement production (E_{fos})	5.5 ± 0.5
Net emissions from landuse change (E_{land})	1.6 ± 1.0
Change in atmospheric mass of CO_2 (dC_{CO_2}/dt)	3.2 ± 0.2
Uptake by the oceans (S_{oc})	2.0 ± 0.8
Uptake by northern hemisphere forest regrowth (E_{for})	0.5 ± 0.5
Net imbalance ($I = (E_{fos} + E_{land}) - (dC_{CO_2}/dt + S_{oc} + E_{for})$)	1.4 ± 1.6

Table 8.1 *Components of the carbon dioxide mass balance (Schimel et al., 1995).*

term, I, is introduced, which represents the missing sources and sinks ('missing carbon sink'). I might therefore be considered as an apparent net imbalance between the sources and sinks. The analysis of the net imbalance in the global carbon cycle has become a major issue in the last decade (Tans *et al.*, 1990), particularly the question of how to account for very large carbon sinks in the terrestrial biosphere in the northern temperate latitudes. *Table 8.1* presents the global carbon balance over the 1980s in terms of anthropogenically induced perturbations to the natural carbon cycle, as given by the IPCC in 1994 (Schimel *et al.*, 1995). Schimel *et al.* stated that the imbalance of 1.4±1.6 GtC/yr may be inappropriate, since sink mechanisms (CO_2 fertilisation [0.5-2.0 GtC/yr for the 1980s], N fertilisation [0.2-1.0 GtC/yr] and climatic effects [0-1.0 GtC/yr]) would account for it.

8.2.2 The global nitrogen cycle

Although the atmosphere consists of 78% nitrogen (N), most biological systems are N limited (Schlesinger, 1991; Vitousek and Howarth, 1991) because these systems are unable to use unreactive N_2. Therefore N must first be converted to reactive forms of N by the process of N fixation by bacteria so that it can be used by biological systems until it is converted back to N_2 by denitrification. Through anthropogenic perturbation of the cycle, whether associated with agricultural, industrial and household activities or with the burning of fossil fuels and biomass, the N compounds play an important role in a wide range of environmental issues such as climate change, soil acidification and nitrate pollution of groundwater and surface water. Nitrous oxide (N_2O) is a long-lived greenhouse gas. NH_3 influences the alkalinity and acidity of the atmosphere and soils. NO_x (NO and NO_2) plays an important role in tropospheric chemistry. At low NO_x concentration levels, tropospheric O_3 is chemically destroyed, although at high NO_x levels this leads to O_3 formation and thereby control of the concentration of OH (the most important tropospheric oxidising agent) by NO_x (Crutzen, 1988).

Component [TgN /yr]	Present (1990)	
Anthropogenic nitrogen inputs ($E_{fos}+I_{fert}+I_{crop}$)	145	
fossil fuel combustion (E_{fos})		25
fertiliser inputs (I_{fert})		80
legumes (I_{crop})		40
Ocean ($S_{N,oc} = D_{N,oc} + R_{N,oc}$)	59	
deposition ($D_{N,oc}$)		18
river inputs ($R_{N,oc}$)		41
Atmosphere (dC_{N_2O}/dt)	~4	
N_2O-accumulation (dC_{N_2O}/dt)		~4
Net imbalance ($I = (E_{fos}+I_{fert}+I_{crop}) - (dC_{N_2O}/dt + S_{N,oc})$)	~82	

Table 8.2 Components of the nitrogen mass balance (Galloway et al., 1995).

In the absence of human activities, N fixation is the primary source of reactive N, providing about 90-130 TgN/yr on the continents, which is in balance with the natural denitrification. Human activities have resulted in the fixation of an additional ~145 TgN/yr (Galloway *et al.*, 1995). The whereabouts of only a part of this anthropogenic N it is known. N_2O accumulates in the atmosphere at a rate of about 3.9 TgN/yr (Prather *et al.*, 1995). Coastal oceans receive another 59 TgN/yr via rivers, or by atmospheric deposition (*Table 8.2*). The remaining 82 TgN/yr is either stored on continents in groundwater, soils or vegetation, or denitrified to N_2 (atmosphere) ('missing nitrogen sink').

The nitrogen balance concept
The uncertainties in the fate of the anthropogenic inputs can be expressed in a mass conservation equation for the terrestrial biosphere, oceans and the atmosphere. As an example, the yearly accumulation on land consists of: (i) net inputs from deposition from anthropogenic emissions minus the terrestrial anthropogenic emissions, (ii) N inputs from fertiliser use and legume production (legumes: soybeans, groundnuts, pulses and forage) and (iii) anthropogenic N losses to surface water:

8.2.3 Interactions between the global biogeochemical cycles and climate change

On a global scale the biogeochemical element cycles not only have common inputs, but are also interlinked through mutual physical, chemical and biological interactions and feedbacks and through common environmental impacts. To illustrate this, in this section we will seek to demonstrate how the problem of global climate change can be described in terms of interlinkages and disturbances between the different disturbed global element cycles.

Geophysical and biogeochemical climate feedbacks

The various global element cycles are interlinked via a whole range of climate-related feedback mechanisms. Such mechanisms can amplify (positive feedback) or dampen (negative feedback) the response of the climate system due to enhanced concentrations of the elements of C, N, P and S resulting from anthropogenic perturbations. Our understanding of these feedbacks is surrounded by considerable uncertainties, chiefly because of the lack of knowledge of their characteristics under both present and future circumstances.

In general, two classes of climate feedbacks can be discerned: geophysical and biogeochemical feedbacks. Geophysical climate feedbacks are caused by physical processes directly affecting the response to radiative forcing. The most important ones are the water vapour and cloud feedbacks, and snow-ice albedo. We will further concentrate on the main biogeochemical climate feedbacks which are those related to the response of the marine and terrestrial biosphere and geosphere, thereby indirectly altering the radiative forcing. These feedbacks, which are all incorporated in the CYCLES submodel are shown in *Figure 8.2*, will be discussed below, except for the feedbacks related to the P and S cycle (P fertilisation, temperature feedback on marine production and the DMS feedback) (see IPCC, 1995c).

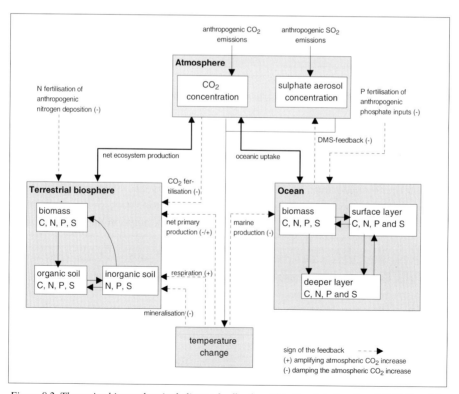

Figure 8.2 The major biogeochemical climate feedbacks within the global cycles of C, N, P and S.

Carbon cycle

The major biogeochemical feedbacks within the global carbon cycle are the terrestrial feedbacks (Schimel et al., 1995): CO_2 and N fertilisation and temperature feedbacks on the net primary production (net photosynthesis minus respiration of non-photosynthetic plant parts) and on soil respiration.

Short-term experiments under optimal conditions of water supply, temperature and nutrients (non-stressed conditions) show that most plants achieve an increase in photosynthetic rate and plant growth at enhanced CO_2 levels. This process is called the 'CO_2 fertilisation effect'. However, the final overall impact of this feedback on the net primary production depends very much on the availability of water and nutrients in these natural ecosystems (Schimel et al., 1995). So far, model studies show that nutrients limit the response of terrestrial ecosystems to CO_2 fertilisation to 20-40% of the potential with unlimited nutrients (Comins and McMurtrie, 1993).

Temperature feedbacks involve the combined effects of the stimulation of photosynthesis and respiration at increased temperatures (Harvey, 1989; Kohlmaier et al., 1990; Schimel et al., 1995). The effect of rising temperature on photosynthesis will, up to an optimum level, be to stimulate plant growth and thus C storage by vegetation (negative feedback). Another feedback involves the temperature effect on soil respiration, resulting in an increase in the release of CO_2 to the atmosphere (positive feedback). The overall impact of such temperature-related feedbacks may become a significant positive climate feedback in the future (Schimel et al., 1995).

Nitrogen cycle

The major biogeochemical feedbacks within the nitrogen cycle are nitrogen fertilisation and temperature feedbacks on nitrogen mineralisation. N fertilisation is related to anthropogenic releases of NO_x and NH_3. The consequent depositions of their oxidation products increase the level of nutrients in the soils of natural ecosystems, which increases the productivity of the ecosystems. This increases the CO_2 uptake by vegetation by about 0.2-1.0 GtC/yr over the 1980s, with a proposed upper limit of about 2 GtC/yr (Peterson and Melillo, 1985; Schlesinger, 1993; Schindler and Baley, 1993). However, high levels of N addition are also associated with acidification, which may damage ecosystems.

Global warming could increase decomposition and nitrogen mineralisation and thereby increase the net primary production, especially in the northern latitudes (Rastetter et al., 1991; Hudson et al., 1994). Te temperature feedback could possibly offset the increased CO_2 releases from the northern soils (temperature feedback on soil respiration).

8.3 Model description

8.3.1 PSIR approach

Following the PSIR approach introduced in Chapter 2, the CYCLES submodel is subdivided into pressure, state, impact and response modules, which are briefly described below and represented in *Figure 8.3*. The response module is not considered in this submodel, since the anthropogenic disturbances can be counteracted by measures within in the energy, land and water response modules:

- The *pressure* module describes the driving forces underlying the human perturbations of the element cycles, i.e. emissions and flows of compounds of C, N, P and S from (i) the energy and industrial sectors (fossil fuel combustion, cement production, etc.), (ii) the agricultural sector (biomass burning, fertiliser use, etc.) and (iii) other human-related activities (land-use changes, erosion, harvesting and grazing). In addition to these, changes in the hydrological cycle represent an indirect perturbation. The above pressures are all calculated in the other TARGETS submodels.
- The *state* module describes the physical, chemical and biological cycling of the basic elements C, N, P and S and chemicals (chlorofluorocarbons, carbon tetrachloride, methyl chloroform and their alternatives, tropospheric ozone and its precursors). The state module consists of one-dimensional box models for (i) the atmosphere (one single, uniformly mixed box), (ii) the terrestrial biosphere (vegetation biomass, two organic and two inorganic soil layers, groundwater and fresh surface water); and (iii) the oceans (a warm-water and cold-water column, with a surface layer including marine biota and four deep layers).

  The terrestrial biosphere is further subdivided into highly aggregated soil, economic[1] and land-use classes (see Chapter 7). The module describes the fluxes between the reservoirs (or compartments) and the most relevant internal processes: biological, chemical and physical (including the main terrestrial feedbacks). The initial 'steady state' of the subcompartments is described in mass-balance equations. Finally, after imposing the anthropogenic fluxes on this natural system, the 'disturbed' state is calculated.
- The *impact* module describes the impact of the disturbed cycles on the global environment. This module comprises a climate module and an ozone module. The climate module simulates the radiative forcing and global-mean surface

---

1 Here we use the terminology 'tropical' and 'temperate' instead of 'developing' and 'developed'. Although these subdivisions are not fully interchangeable, the chosen economic subdivision shows similarity with a division in a tropical and a temperate region, especially for the natural vegetation types 'forests' and 'grassland', which are most important to the dynamics in the CYCLES submodel. More than 95% of the natural vegetation in the developed region is located in the temperate region. In the developing region, more than 80% consists of tropical vegetation types.

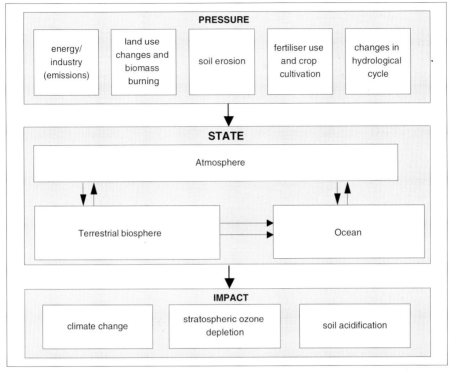

Figure 8.3 The Pressure-State-Impact-Response diagram of the CYCLES submodel.

temperature changes due to the changes in concentrations of greenhouse gases and sulphate aerosols (den Elzen *et al.*, 1995; 1997). The ozone module, which uses as input the atmospheric chlorine and bromine concentrations, calculates the stratospheric ozone depletion and increased UV-B radiation (den Elzen, 1993).

8.3.2 Position within TARGETS

The CYCLES submodel (den Elzen *et al.*, 1995) is closely interrelated with the other submodels of TARGETS. The role of CYCLES within the overall framework is to provide calculations of global flows of basic elements and related compounds within and between the major reservoirs. The main linkages between the CYCLES submodel and the other submodels of the TARGETS1.0 are depicted in *Figure 8.4*. As this figure shows, the CYCLES submodel is particularly interwoven with the TERRA submodel (see Chapter 7). The interlinkages include changes in land use; changes in soil fertility and quality; emissions as a result of agricultural activities (fertiliser use, burning and domestic animals); food and feed consumption, physical changes in concentrations and temperature; and the flow of organic matter and

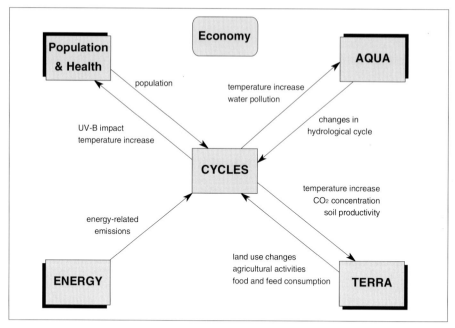

Figure 8.4 Input-Output diagram of the CYCLES submodel.

inorganic compounds. There are also multiple linkages with the AQUA submodel exist (see Chapter 6): water pollution compounds and temperature change patterns are exchanged, while, conversely the CYCLES submodel uses changes in the global hydrological cycle for the calculation of transport and movement of substances by water.

The following sections describe those modules of the CYCLES submodel which are of vital importance to an understanding of the global budget calculations presented next: the coupled carbon and nitrogen cycle and climate modules.

8.3.3 The global carbon cycle module

The carbon cycle module is an extended version of the model described in Goudriaan and Ketner (1984) and den Elzen *et al.* (1997). The main input is formed by the CO_2 emissions associated with fossil fuel combustion, and the production of cement, but also the direct effects of human activities themselves, such as soil erosion, land-use changes, etc. The systems diagram underlying the carbon cycle module (*Figure 8.5*), consists of three subsystems: the terrestrial biosphere, the ocean and the atmosphere.

The atmospheric system
The atmosphere is represented as a single well-mixed reservoir, with a mixing time

8 THE BIOGEOCHEMICAL SUBMODEL: CYCLES

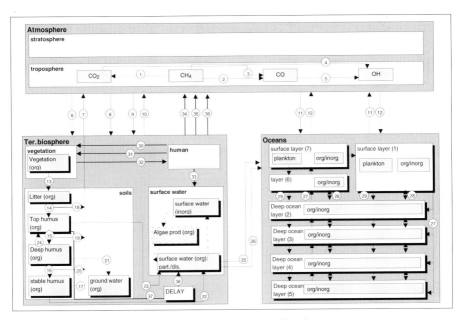

Natural processes (fluxes in GtC.yr⁻¹)

Atmospheric processes
1. oxidation of CO to CO_2 ($oxid_{CO}(t)$)[1]
2. oxidation of CH_4 ($oxid_{CH4}(t)$)
3. oxidation of CO by OH ($uptake_{CO,OH}(t)$)
4. OH uptake by oxidation of CO to CO_2
5. OH uptake by oxidation of CH_4

Processes between terrestrial biosphere and atmosphere
6. net primary production ($NPP_{cl}(t)$)
7. total soil respiration ($RC_{S\,cl}(t)$)
8. CH_4 uptake by soils ($uptake_{CH4}(t)$)
9. CO uptake by soils ($uptake_{CO}(t)$)
10. oxidation of NMHC ($emNMHC(t)$)

Processes between atmosphere and ocean
11. carbon uptake by the ocean (warm column) ($F_{oa,7}(t)$)
12. carbon uptake by the ocean (cold column) ($F_{oa,1}(t)$)

Terrestrial biospheric processes
13. decay of terrestrial vegetation ($DC_{B\,cl}(t)$)
14. decay of litter ($DC_{L\,cl}(t)$)
15. decay of top humus ($DC_{TH\,cl}(t)$)
16. decay of deep humus ($DC_{DH\,cl}(t)$)
17. decay of stable humus ($DC_{SH\,cl}(t)$)
18. respiration of litter ($RC_{L\,cl}(t)$)
19. respiration of top humus ($RC_{TH\,cl}(t)$)
20. respiration of deep humus ($RC_{DH\,cl}(t)$)
21. DOC leaching to groundwater ($leach_{DOC\,cl}(t)$)
22. DOC flux from groundwater to surface water ($outfl_{DOC\,gw}(t)$)
23. natural erosion POC flux into surf. water ($eros_{nat\,cl}(t)$)
24. erosion carbon flux from deep to top humus ($erosC_{DH\,cl}(t)$)

Terrestrial biosphere to ocean processes
25. river discharge of DOC to oceans ($riv_{DOC}(t)$)
26. river discharge of POC to oceans ($riv_{POC}(t)$)

Oceanic processes
27. massflow of inorganic/organic compounds ($mfl_i(t)$)
28. diffusion flow of inorganic/organic compounds ($tbm_i(t)$)
29. precipitation of dead organic matter ($prco_i(t)$)

Anthropogenic processes (fluxes in GtC.yr⁻¹)
30. landuse changes ($flandC_{B\,cl}(t)$, $flandC_{L\,cl}(t)$, $flandC_{TH\,cl}(t)$, $flandC_{DH\,cl}(t)$ and $flandC_{SH\,cl}(t)$)
31. yield improvement (by technology) ($yld_{cl}(t)$)
32. harvesting ($harv_{C\,cl}(t)$)
33. human consumption of harvested products to sewage (excretia consists of carbon ($sewage_{DOC}(t)$)
34. emissions of CO_2 (fossil fuel combustion and cement production ($E_{fos}(t)$); and biomass burning ($E_{land}(t)$)
35. emissions of CH_4 (fossil fuel combustion); biomass burning, agricultural activities, rice production, landfills, domestic animals
36. emissions of CO (fossil fuel combustion, transport, and biomass burning)
37. anthropogenic erosion flux of organic carbon from top humus ($erosC_{TH\,cl}(t)$)
38. flux of POC by anthropogenic erosion reaching the fresh-water system (consisting of the direct and indirect flows) ($erosPOC_{sw}(t)$)

[1] Expressions in parentheses represent the model variables which may be referred to in the text.

Figure 8.5 The carbon cycle model (the natural flows are represented by light solid lines and the anthropogenic flows by heavy solid lines).

delay of one year. The CO_2 concentration is calculated on the basis of the mass balance of CO_2, as represented by equation 8.1, where $E_{land}(t)$ are the net terrestrial biospheric emissions (i.e. direct emissions from land-use changes, minus the global net ecosystem production).

The terrestrial biosphere system

The terrestrial biosphere system includes a fresh-water system and is horizontally subdivided into the various land-use classes. Vertically, the biosphere is subdivided into five compartments: biomass, litter, topsoil (top humus, organic rich, 0-0.2 m), deep soil (deep humus, organic poor, 0.2-1.0 m) and stable humus. The following biogeochemical feedbacks are incorporated: the CO_2 fertilisation feedback, the temperature effects on net primary production and soil respiration.

Atmospheric CO_2 is initially consumed in the process of net primary production (process 6 in *Figure 8.5*). The atmospheric CO_2 is stored in the biomass until it decays, whereupon the carbon passes through the litter and subsequently through the top, deep, and stable humus reservoirs (13-17), thereby releasing carbon dioxide to the atmosphere (soil respiration) as part of the decay process (7, 17-20). The decay in biota, litter and humus soil layers is modelled as an exponential process which is essentially linear (natural decomposition) and much slower in the deeper soil layers. The temperature effect on the soil respiration is modelled as a function of the global surface temperature increase (ΔT_s) and parameterised using the Q_{10} value (Q_{10}^{res}) for the respiration of soils and litter (Rotmans and den Elzen, 1993b). The net ecosystem production, or the net growth of the biomass of the terrestrial biosphere, is calculated as the net primary production (see Equation 8.3) minus the soil respiration.

As an illustration of the mass balance equations, the equation for the top humus $C_{TH\ cl}(t)$ [GtC] for a specific land-use class (see Chapter 7), cl, as a function of the input and output fluxes is given below:

$$\frac{dC_{TH\ cl}(t)}{dt} = hf_{cl} \times DC_{L\ cl}(t) - \left[DC_{TH\ cl}(t) + erosC_{cl}(t) + flandC_{TH\ cl}(t) \right] \text{ [GtC/yr]} \quad (8.2)$$

where hf_{cl} is the humification fraction, $DC_{L\ cl}(t)$ and $DC_{TH\ cl}(t)$ the respective decays related to carbon in the litter and top humus layer, $erosC_{cl}(t)$ the erosion flux, including natural and anthropogenic erosion and $flandC_{TH\ cl}$, the changes in carbon content of the top humus due to land-use changes.

The total net primary production ($NPP_{cl}(t)$) of each land-use class, cl, is calculated as a multiplicative function which considers the effect of changes in land use and soil productivity (Q). A number of feedbacks are also considered, namely: CO_2 fertilisation (f_{CO_2}), temperature (f_T) and nutrient fertilisation (f_{nut}), namely:

$$NPP_{cl}(t) = npp_{cl\ i} \times A_{cl}(t) \times \frac{Q_{cl}(t)}{Q_{cl\ i}} \times f_{CO_2\,cl}(CO_2(t))$$
$$\times f_{T\,cl}(T_{cl}(t)) \times f_{nut\,cl}(NPP_{cl}(t), NUP_{cl}(t))$$
[GtC/yr] (8.3)

where $npp_{cl\ i}$ is the net primary production per hectare in the year 1700 for land-use class cl [tC/ha per yr]. $A_{cl}(t)$ is the area of land-use class cl at time t. $Q_{cl}(t)$ and $Q_{c\ i}$ express the actual and initial soil productivity of land-use class cl, respectively. $CO_2(t)$ is the amospheric CO_2 concentration at time t [ppmv], $T_{cl}(t)$ the temperature of land-use class cl at time t [°C] and $NUP_{cl}(t)$ the nutrient uptake of land-use class cl at time t [TgN/yr].

The soil productivity, $Q_{cl,}$ depends primarily on the soil fertility, which is directly linked to the proportion of soil organic matter. For the topsoil and deep soil layer, the amount of soil organic matter is calculated as a function of the carbon content through a fixed conversion factor between soil organic matter and carbon, soil bulk density and soil-layer depth (the top humus layer remains constant at 0.2 m, while the deep humus-layer depth decreases under severe erosion conditions). The CO_2 fertilisation feedback, f_{CO_2}, is modelled using a logarithmic relationship of the atmospheric CO_2 concentration (Gifford, 1980; Goudriaan and Ketner, 1984), using $\beta_{cl}(t)$ as the biotic growth factors (Klein-Goldewijk et al., 1994). The temperature feedback, f_T, is described by a parabolic function of temperature, with a maximum of 1 at the optimum temperature, $T_{opt\ cl}$, from which values decrease monotonically to 0 at $T_{min\ cl}$ and $T_{max\ cl}$, the minimum and maximum temperature (Raich et al., 1991). f_{Nut} represents the feedback of nutrient stress or N fertilisation on net primary production. If there is a shortage of nutrients (here for all land-use types of nitrogen), the function will be defined as $stress_{IN}(t)/stress_{IN\ i}$, with $stress_{IN}(t)$ as a parabolic stress curve scaled to decrease from 1 (no stress) to 0 (maximum stress). The latter is a function of the simulated C:N ratio; $1000 \cdot NPP_{cl}(t)/NUP_{cl}(t)$.

Since we do not distinguish different compartments for the vegetation biomass, no changes in the allocation of the net primary production under different degrees of nitrogen stress are assumed. Regarding N fertilisation, for high levels of nitrogen addition due to depositions of anthropogenic NO_x and NH_3 emissions (no stress), net primary production follows the increases in the nitrogen uptake by vegetation using the simulated C:N ratio between net primary production and N uptake. The net primary production of agricultural land is calculated with the TERRA submodel (Chapter 7).

Anthropogenic processes
land-use changes, as simulated with the TERRA submodel, affect the various biomass and soil pools of organic and inorganic matter, both directly and by transferring soil organic matter and inorganic forms from one landaus type to another. If land which contains a considerable amount of organic matter changes to land with little organic matter, as is the case with deforestation, this results in a net atmospheric release of gases such as CO_2, NO_x and SO_2. The direct net release of the

gases is assumed to depend on the area undergoing transition, the type of biomass burning, the density of organic matter and the amount of organic matter in biomass and litter that is converted to gases and vapours. Also included is the first-year carbon sequestering effect associated with the regeneration of new biomass according to Houghton *et al.* (1983; 1987). Besides these direct releases, we also take into account the delayed emissions due to the microbial decomposition of the remaining biomass and litter organic material after burning, as well as delayed carbon sequestering mechanisms through the regrowth of the new land-use type. Forms of biomass burning (other than forest clearance), which also lead to atmospheric releases, are not included in the model, since most of these activities have a zero net effect if regrowth is taken into account.

Topsoil losses by erosion lead to a decrease in the organic matter in the topsoil/humus, which in turn decreases soil productivity. The net loss of topsoil due to soil erosion consists of the net effect of organic matter loss from the topsoil (because of the assumed constant depth of the topsoil layer) discounted by the small compensating effect due to additional organic matter in the deep humus. The loss of organic matter from the topsoil is modelled as a multiplicative function of (i) an enrichment factor (expressing the fact that soil particles lost through erosion contain more organic matter than average topsoil), along with the total soil erosion loss per hectare, (ii) the fraction of organic matter for the soil in the top humus layer, and (iii) the area devoted to a specific land-use type. The carbon content of the flux is calculated using a conversion factor; for the nitrogen and phosphorus content, C:N and C:P ratios of the topsoil are used. A part of this erosion flux reaches the fresh-water system, which is modelled via a delay-box mechanism.

The fresh-water system

The fresh-water system consists of two compartments: groundwater and fresh surface water (lakes and rivers), each of which contains dissolved organic carbon (DOC) and particulate organic carbon (POC). The carbon fluxes are calculated in the CYCLES submodel, whereas transport mechanisms through water are calculated in the AQUA submodel (Chapter 6). The major water fluxes are: soil leaching, the groundwater outflow to the surface water and the river discharge. The relevant anthropogenic disturbances are the increased POC flux from soil erosion (runoff) and the flow of sewage containing faeces carrying POC and DOC.

The ocean system

The ocean system is a slightly modified version of the ocean model of Goudriaan and Ketner (1984). The ocean is considered as one stratified unit, subdivided into seven well-mixed layers: two surface layers (a 75-metre warm mixed layer and a 400-metre cold mixed layer), one 325-metre intermediate layer and four 900-metre layers. The distribution of CO_2 among these different ocean layers is described as a result of: the transport of dissolved carbon by upwelling and downwelling, and through turbulent mixing or vertical diffusion (physical processes); the CO_2 exchange between the atmosphere and the two surface ocean layers (chemical processes) and through precipitation of organic material (biological processes) and inflow from rivers.

This precipitation flux for the surface layers is modelled as a multiplicative function of a fixed fraction of the total primary production (the new production) and average ratios of C and N to P (Baes *et al.*, 1985). The precipitation flux of organic matter for the deeper layers is inversely proportional to the height of the water column and the new production (Suess, 1980).

Although the biological production, i.e. phytoplankton growth, is limited to short time-scales in most of the sea by nitrogen, phosphate is usually, and also here, used as the limiting nutrient in box models (Joos *et al.*, 1991). The production is modelled as a multiplicative function which considers not only the concentration of phosphate in the surface layers (Goudriaan, 1989), but also the temperature and UV-B radiation (den Elzen *et al.*, 1997)

8.3.4 The global nitrogen cycle module

The global nitrogen cycle module simulates the concentrations of the different nitrogen compounds in the atmosphere, the terrestrial biosphere and the ocean as is shown in *Figure 8.6* (den Elzen *et al.*, 1997). For the sake of simplicity no distinction is drawn between different forms of soil inorganic N (nitrate, nitrite and ammonium). The main input are the anthropogenic emissions of NH_3, NO_x and N_2O as previously discussed.

The atmospheric system
The atmosphere module simulates the atmospheric sources and sinks of NO_x and NH_3. Since these compounds have short atmospheric lifetimes, their deposition is virtually equal to the emissions, except in the case of oxidation of NH_3 to NO_x and N_2 (processes 1-2 in *Figure 8.6*). The mass balance of atmospheric N_2 is also calculated, although this hardly changes. The N_2O concentration is calculated using a mass balance equation, where the removal process in the stratosphere is directly proportional to the atmospheric concentration and inversely proportional to the atmospheric lifetime of N_2O (Prather *et al.*, 1995).

The terrestrial biosphere system
As *Figure 8.6* shows, the disaggregation scheme for representing the different stages of global cycling of nitrogen through each ecosystem is similar to that for global carbon cycling. However, in the case of nitrogen the soil compartment is further partitioned into an organic and inorganic reservoir. The following biogeochemical feedbacks are incorporated: N fertilisation feedback, and temperature effects on the nitrogen uptake of the vegetation and on soil mineralisation processes.

8 THE BIOGEOCHEMICAL SUBMODEL: CYCLES

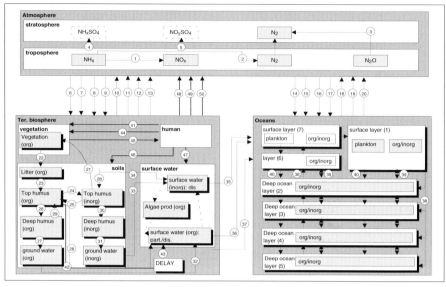

Natural processes (fluxes in TgN/yr)

Atmospheric processes
1. oxidation of ammonium to NO_x
2. oxidation of ammonium to N_2
3. stratospheric loss of N_2O to N_2
4. oxidation of ammonia emissions to aerosols
5. oxidation of NO_x emissions to nitrate aerosols

Processes between terrestrial biosphere and atmosphere
6. deposition of NO_x on land
7. deposition of NH_3 on land
8. lightning to land (NO_x deposition)
9. N_2 fixation of land ($Nfix_{cl}(t)$)[1]
10. volatisation of NH_3 from land ($volNHx_{cl}(t)$)
11. NO_x releases from soil ($soilNOx_{cl}(t)$)
12. denitrification of N_2 from land ($denitrN_{cl}(t)$)
13. denitrification of N_2O from land ($denitrN2O_{cl}(t)$)

Processes between atmosphere and ocean
14. deposition of NO_x on ocean
15. deposition of NH_3 on ocean
16. lightning to ocean (NO_x deposition)
17. N_2 fixation of ocean
18. volatisation of NH_3 from ocean
19. denitrification of N_2 of ocean
20. denitrification of N_2O of ocean

Terrestrial biospheric processes
21. terrestrial biospheric nitrogen uptake ($NUP_{cl}(t)$)
22. decay of terrestrial vegetation to litter ($DN_{B\,cl}(t)$)
23. decay of litter to top humus ($DN_{L\,cl}(t)$)
24. mineralisation of top humus ($INminer_{cl}(t)$)
25. mineralisation of inorganic nitrogen ($INminer2H_{cl}(t)$)
26. decay of top humus to deep humus ($DN_{TH\,cl}(t)$)
27. leaching of DON to groundwater ($leach_{DON}(t)$)
28. natural erosion flux of top humus ($erosN_{nat\,cl}(t)$)
29. infiltration of DIN in soil ($infN_{cl}(t)$)
30. nitrogen flux from deep to top humus ($DN_{TH\,cl}(t)$)
31. leaching of DIN from top to deep soil ($leachIN_{cl}(t)$)
32. leaching of DIN from deep soil to groundwater ($leachIN_{TH2gw}(t)$)
33. DON groundwater flow to surface water ($outfl_{DON\,gw}(t)$)
34. DIN flow in groundwater to surface water ($outfl_{DIN\,gw}(t)$)
35. rapid runoff of DIN to surface water ($runoffIN(t)$)

Terrestrial biosphere to ocean processes
36. river discharge of DIN to oceans ($riv_{DIN}(t)$)
37. river discharge of DON to oceans ($riv_{DON}(t)$)
38. river discharge of PON forms to oceans ($riv_{PON}(t)$)

Oceanic processes
39. massflow of inorganic/organic compounds
40. massflow of inorganic/organic compounds
41. precipitation of dead organic matter to deeper layers

Anthropogenic processes (fluxes in TgN/yr)
42. landuse changes ($flandN_{B\,cl}(t)$, $flandN_{L\,cl}(t)$, $flandN_{TH\,cl}(t)$, $flandN_{DH\,cl}(t)$ and $flandN_{SH\,cl}(t)$
43. erosion flux of PON from top humus
44. flux of PON by erosion reaching the fresh-water system (consisting of direct and indirect flows)
45. yield improvement
46. harvesting
47. fertiliser use and legume production
48. human consumption of vegetable and animal products to sewage (excretia), as well as animal excretia of organic and inorganic nitrogen)
49. emissions of NH_3 (fossil fuel combustion, biomass burning, domestic animals, fertilisers, consumption)
50. emissions of NO_x (fossil fuel combustion and transport, biomass burning)
51. emissions of N_2O (fossil fuel combustion and transport, biomass burning, fertilisers, acidic and nitric production

[1] Expressions in parentheses represent the model variables which may be referred to in the text.

Figure 8.6 The nitrogen cycle model (the natural flows are represented by light solid lines and the anthropogenic flows by heavy solid lines).

The biological nitrogen cycle starts with nitrogen uptake by vegetation (21), which primarily depends on the available pool of inorganic N in the soil. This uptake is stored in the biomass, where it is followed by exponential decay involving organic nitrogen through the litter, and top and deep humus layers (22-26). In the top humus, the decay process is followed by volatisation, in which inorganic N is produced (24: mineralisation); this is modelled as a linear function of the organic N in top humus. This production of inorganic N is partly immobilised by decomposer organisms to become part of its biomass (25), modelled as a fraction of the mineralisation rate. Other outputs of the organic soil compartments are the natural constant erosion flux (28) of particulate organic nitrogen (PON) (top humus) and the soil leaching flux of dissolved organic nitrogen (DON) (27) from deep humus to groundwater.

In addition to these organic soil processes, there is a variety of inorganic soil processes. The processes of N fixation, NO_x releases from the soils, NH_3 volatisation, denitrification (N_2 and N_2O, based on a constant N_2:N_2O ratio (Galloway et al., 1995; Weier et al., 1993)) and infiltration of the natural deposition of NH_4^+ and NO_3^- are modelled as a linear function of the surface area. A proportion of the deposition flux is wet deposition, which reaches the Earth's surface by precipitation and is then transported by runoff to the fresh surface water, or infiltrates into the topsoil. The other part of the deposition, dry deposition, accumulates directly in the topsoil. Leaching of dissolved inorganic nitrogen (DIN) (mainly nitrate) from top humus to deep humus is modelled as a fraction of the inorganic nitrogen in the top humus, which is dissolved by the leaching water flow divided by the lag time during which the water remains in the topsoil. The percolation flux of DIN from deep humus to the groundwater (*31*) is modelled in a similar way.

The nitrogen uptake ($NUP_{cl}(t)$) for land-use class, *cl*, is calculated in a manner analogous to net primary production, as a multiplicative function. This takes into account the effect of changes in land use and soil productivity (*Q*) but also feedbacks, namely, the nitrogen fertilisation feedback (f_{IN}) and the temperature feedback (f_T). Thus:

$$NUP_{cl}(t) = nup_{cl\,i} \times A_{cl}(t) \times \frac{Q_{cl}(t)}{Q_{cl\,i}} \times f_{IN\,cl}(IN_{cl}(t)) \times f_{T\,cl}(T_{cl}(t)) \qquad [\text{TgN/yr}] \qquad (8.4)$$

where $nup_{cl\,i}$ is the initial N uptake per m² for land-use class *cl* [gN/m² per yr]. $A_{cl}(t)$ is the area of land-use class *cl* [1000 Mha]. The variable, $IN_{cl}(t)$, represents the inorganic N pool in the soil [gN.m⁻²]. The N uptake of the agricultural systems is calculated using a constant C:N ratio for net primary production. The temperature feedback (f_T) is modelled by reference to the Q_{10} value (Q_{10}^{nup}) for the N uptake. f_{IN} is the N fertilisation feedback, which is modelled in a manner similar to that employed for the CO_2 fertilisation effect: a logarithmic relationship between the elevated content of inorganic nitrogen in the soil attributed to the deposition of ammonium and nitrate, on the one hand and the N uptake on the other, using $\beta_{IN\,cl}$ as the nitrogen biotic growth factors.

Anthropogenic disturbances

The fraction of the fertiliser (47) which is taken up by agricultural biomass can be calculated on the basis of the yield response curve for fertiliser of Addiscott *et al.* (1991). The remaining part of the fertiliser is lost in three ways (i) gaseous losses in the form of NH_3 and NO_x; (ii) the denitrification flux (N_2 and N_2O, based on a constant $N_2:N_2O$ ratio) and (iii) leaching and runoff (used to balance the nitrogen budget), all calculated as fractions of the fertilisers actually applied and based on IPCC (1995b).

The additional N fixation by legumes is modelled as an exogenous variable (based on the data of Galloway *et al.* (1995)). Part is lost by denitrification, modelled as a function of the N_2O emissions and a constant $N_2:N_2O$ ratio.

The impact of anthropogenic releases of NH_3/NO_x (49-50) and the subsequent depositions is modelled analogous to the modelling of the natural depositions, namely, by reference to the infiltration and the runoff fluxes. The infiltration flux is a linear function of the wet and dry depositions, which are modelled as a fraction of the total deposition flux of NH_4^+ and NO_3^-. The depositions are assumed to depend on the area of the land-use type but also on constant deposition ratios for the different land-use types, based on Hudson *et al.* (1994). The remainder (after runoff losses) is lost by denitrification (about 5%), vegetation uptake and leaching.

Harvested products (48) are consumed by (i) animals, resulting in excrement, recycled fixed organic nitrogen (via fertilising using the manure excretions of livestock), or NH_3, which can volatise, or (ii) humans (including meat consumption), normally resulting in sewage or NH_3 volatisation. The pathways of this nitrogen to the atmosphere, or to fresh surface water is modelled on the basis of specified fractions (see model analysis).

The fresh-water system

As discussed for the carbon cycle module, the water fluxes from AQUA are used to determine the flows between the three water reservoirs of dissolved and particulate organic nitrogen (DON and PON) and dissolved inorganic nitrogen (DIN). The four anthropogenic perturbations of the global nitrogen cycle impinge upon the fresh-water system in various ways: (i) anthropogenic releases of NH_3 and NO_x, which reach fresh surface water directly by runoff (34) and indirectly via the groundwater by infiltration and leaching; (ii) fertiliser use (DIN), via leaching (via groundwater) and sewage (28, 36, 47); (iii) the human-induced erosion flux of PON via runoff; (iv) the flow of domestic sewage containing DIN and DON via faeces. For nitrogen we use the concentration of nitrate in the fresh surface water as one of the main determinants for the water quality, as calculated in the AQUA submodel (Chapter 6).

The ocean system

In the ocean module the amount of nitrogen (mainly nitrate) for the different ocean layers is calculated on the basis of physical and biological processes, similarly to the carbon cycle module. Over and above these processes, there are the inflow processes from the river (carrying PON, DON and DIN) and from the atmosphere, namely, deposition of NO_x and NH_3 (14-15). The latter is modelled as a fraction of the total atmospheric NO_x and NH_3 releases (Galloway et al., 1995) and the deposition from NO_x production by lightning (16). Other fluxes are the denitrification flux by the oceans (19-20) and N_2 fixation by the oceans (17), both of which remain constant.

8.3.5 The climate module

The climate module is an upwelling-diffusion energy-balance-box model based on Wigley and Schlesinger (1985) and Wigley and Raper (1992), as described in den Elzen et al. (1997). The input to the module is the induced radiative forcing due to the changes in the concentration of the different greenhouse gases, stratospheric ozone, DMS (dimethylsulphide) and sulphate aerosols derived from emissions of sulphur dioxide from fossil fuel combustion and biomass burning.

8.4 Calibration

The main input data of the submodel are composed of the following human-induced perturbations of the global C and N cycles (i) land-use changes and soil degradation, food production/consumption and fertiliser use, as simulated in the TERRA submodel, and (ii) energy-related emissions of C and N compounds, simulated with the Energy submodel.

In the following calibration analysis, a number of simulation experiments have been performed to compare historical (atmospheric, biospheric and oceanic) concentration values of carbon and nitrogen with simulated values. A key principle in this is the use of the modelled feedbacks and fluxes to balance the past and present carbon and nitrogen budgets. In the steady-state case, the human-induced disturbances and feedbacks are switched off. Then the human-induced disturbances are switched on serially (per cycle) and imposed on the steady-state case. In addition, the various terrestrial feedbacks, i.e. CO_2 fertilisation effect, temperature feedback on net primary production and soil respiration (C), nitrogen vegetation uptake and mineralisation (N) are activated sequentially. Finally, the main interaction mechanism between both cycles, the N fertilisation feedback, is activated.

8.4.1 Past carbon budget

Human-induced disturbances
For the first experiment (run 1, no feedback case), all feedbacks are switched off and only the impact of the human factors, fossil fuel combustion and land-use changes are imposed on the system. The resulting simulated global carbon fluxes of the fossil carbon emissions (fossil fuel and cement production), carbon uptake by the oceans and emissions from the land-use changes are given in *Figure 8.7*. The CO_2 emissions are comparable with the trend as presented by Schimel *et al.* (1995), especially over the period 1940-1990. The cumulated land-use flux over the period 1850-1990, amounting to 130 GtC, is within the range of 122 ± 40 GtC reported by Houghton (1994). The rather high oceanic carbon uptake of about 2.6 GtC/yr is due to the simulated atmospheric CO_2 concentrations, which are much higher than the observed values (in 1990: 383 ppmv, as opposed to the 353 ppmv observed (IPCC, 1990)). The modelled imbalance of about 1.2 GtC/yr over the 1980s, is comparable to the remaining imbalance of Schimel *et al.* (1995) (see also *Table 8.3*).

The global terrestrial fluxes are hardly influenced by the other human disturbances (soil degradation, fertiliser use, other agricultural inputs and food production), as shown in *Table 8.3* (run 2-6). This is not only because agricultural production is a small component of the global net primary production (5-10%) but also because agricultural harvesting and other losses (grain and straw) lead to a near balance between the soil inputs and decay rates for the agricultural systems.

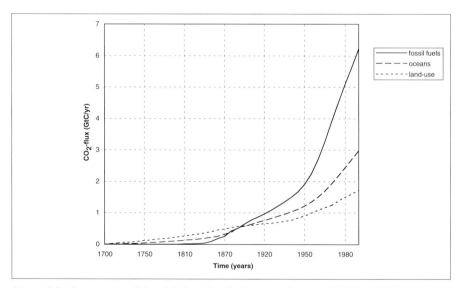

Figure 8.7 Components of the global carbon budget over the period 1900 to 1990 (run 1, no feedback case). The simulated values for runs 2-6 are not shown here, since they hardly differ from the results of run 1.

Component	Without feedbacks		With feedbacks		
	run 1	run 2-6	run 7	run 8	run 9
Emissions from fossil fuel burning and cement production (E_{fos})	5.5	5.5	5.5	5.5 (5.5)[2]	5.5
Net terrestrial biospheric emissions (E_{land})	1.6	1.4	0.4	0.4 (-0.1)	0.3
Observed change in atm. CO_2 (dC/dt^*)	3.2	3.2	3.2	3.2 (3.2)	3.2
Change in atmospheric CO_2 (dC/dt)[1]	4.4	4.3	3.7	3.7 (3.4)	3.2
Uptake by the oceans (S_{oc})	2.7	2.6	2.2	2.2 (2.0)	2.0
Imbalance ($I = [E_{fos}+E_{land}] - [dC/dt^*+S_{oc}]$)	1.2	1.1	0.5	0.5 (0.2)	0.0

1 In the model calculations this net imbalance is stored in the atmosphere, leading to higher CO_2 concentrations than the observed values over the historical period. This also explains the rather high oceanic uptake, which decreases in the subsequent runs through decreasing atmospheric carbon storage.
2 Values in parentheses represent results of model experiment, in which all CO_2 fertilisation feedback parameters are set at their maximum value.

Table 8.3 Components of the modelled carbon dioxide mass balance over the period 1980-1989 with and without feedbacks (in GtC/yr).

Terrestrial feedbacks

Terrestrial feedback mechanisms are introduced to explain the missing carbon sink and to balance the past and present carbon budget, namely, CO_2 fertilisation feedback and temperature feedback on the net primary production and on soil respiration. Thus *Figure 8.8* primarily indicates the relative contribution of the different feedbacks in explaining the apparent net imbalance. Firstly, the experiment with the CO_2 fertilisation effect (run 7) clearly shows the importance of the CO_2 fertilisation feedback on the net primary production. The net biospheric carbon emissions are now reduced by about 1.0 GtC/yr averaged over the 1980s (compare run 6 and run 7) and the apparent mismatch is now -0.5 GtC/yr over the 1980s (see *Table 8.3*). Temperature feedbacks (run 8) have a minor effect on the net biospheric emissions over the period 1900-1990, since the temperature increase has been rather small. A combination of both factors, CO_2 fertilisation effect and the temperature feedbacks, is not sufficient to obtain a good match between the observed and simulated atmospheric CO_2 concentration values. The imbalance over the 1980s in the submodel has now been reduced from 1.1 GtC/yr (no feedback case, run 6) to 0.5 GtC/yr.

Even when the CO_2 fertilisation parameters are all set at their maximum values, an imbalance of 0.2 GtC/yr is created. The above experiments illustrate that a model in which only terrestrial carbon feedbacks are incorporated cannot create a balanced past carbon budget. The last part of this imbalance must therefore be explained by the fertilisation of the anthropogenic nitrogen accumulated on land, i.e. temperate forests (industrialised world). This will be shown in the following analysis but, to do this, we first have to analyse how much and where anthropogenic nitrogen accumulates on land.

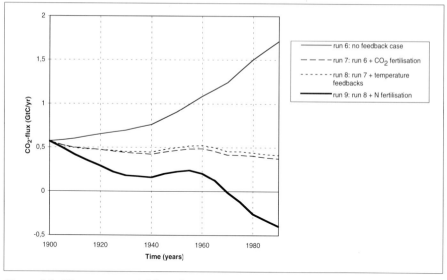

Figure 8.8 The net terrestrial biospheric emissions of CO_2 (direct emissions from landuse changes, minus the net ecosystem production) over the period 1900 to 1990 (the emission flux of run 6 corresponds with the land-use flux of Figure 8.7).

8.4.2 Past nitrogen budget

Human-induced disturbances
To analyse the fate of the anthropogenic nitrogen emissions and flows, all the processes discussed previously in relation to the carbon budget are switched on successively. Starting with the steady-state case, land-use changes (run 1) are imposed on this system; the resulting changes in the components of the global nitrogen budget are given in *Figure 8.9*. The increase of nitrogen on land results from the dominating effect of the global denitrification decrease due to the losses of forested ecosystems with high denitrification rates compared to the nitrogen losses from biomass burning. The effects of soil erosion (run 2) are a decrease in soil organic matter (including N) and an increase in the PON flow of 20 TgN/yr to the oceans, which is about the same as the pre-industrial flux of 21 TgN/yr (Meybeck, 1982).

The effect of fertiliser use (run 3) is the 1990 vegetation uptake (40 TgN/yr) of about 50-55% of the total fertiliser use (75 TgN/yr), somewhat lower than the 65% in 1900. The remainder of the applied N fertiliser is lost due to volatisation, denitrification, leaching and runoff. Agricultural inputs (run 4) increase global N fixation (43 TgN/yr in 1990) but also the denitrification flux. The total harvesting flux of vegetable products (run 5: food consumption) is about 1.25 GtC/yr, or 62.5 TgN/yr. Based on the 1990 anthropogenically dissolved nitrogen river transport of

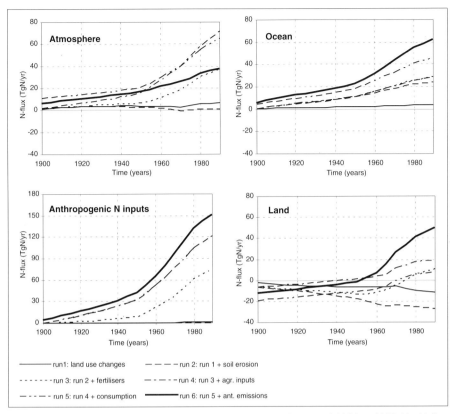

Figure 8.9 Components of the global nitrogen budget over the period 1900 to 1990 (the N-flux of anthropogenic N inputs for each run is distributed over the different components of the global nitrogen budget: atmosphere, oceans and land).

20 TgN/yr (Duce *et al.*, 1991), it is estimated that about 35% is lost via sewage and river transport to the oceans. A small part (about 10%) is lost to the atmosphere (denitrification, cattle). The rest mainly denitrifies, while a part is lost through NH_3 emissions (mainly from domestic animals) (Schlesinger and Hartley, 1992). The anthropogenic nitrogen inputs from energy production and other minor industrial sources (run 6) of about 25 TgN/yr, including NH_3 (2 TgN/yr in 1990), NO_x (22 TgN/yr) and N_2O (1.3 TgN/yr) are almost all redeposited on the land surface. A small proportion is lost to the oceans via runoff and deposition (NH_3 and NO_x) and to the atmosphere (N_2O emissions).

Terrestrial feedbacks

The temperature feedbacks on the terrestrial processes do not affect the exchange fluxes between the compartments and thereby hardly affect the distribution of the nitrogen over the components.

In summary, human activities in 1990 have resulted in the fixation of an additional 148 TgN/yr through fertiliser, legume and energy production and other industrial sources. The oceans receive in total 64 TgN/yr and in the atmosphere, N_2O accumulates at a rate of 4 TgN/yr. These results merely correspond with the observed values (Table 8.2). The key issue in the missing nitrogen sink is now where the remaining ~80 TgN/yr accumulates? According to our calculations, it accumulates in the atmosphere (~38 TgN/yr) and partly on land (~42 TgN/yr), which results in a balanced past nitrogen budget in the model. The atmospheric accumulation is mainly caused by denitrification from fertilisers and animal wastes and to a lesser extent biomass burning. In the soils nitrogen remains at about 42 TgN/yr, whereas for the biomass there is a loss of nitrogen through biomass burning of about 12 TgN/yr.

8.4.3 Linkage between past carbon and nitrogen budget: N fertilisation

The N accumulation in the soils of temperate forests in the industrialised world is of major importance for the magnitude of the N fertilisation feedback and thus also for the past carbon budget. The issues of the missing nitrogen and carbon sinks are thus coupled as shown below.

The present global anthropogenic depositions of about 60 TgN/yr (about 40 TgN/yr for the temperate forests) result in an additional nitrogen uptake of about 28 TgN/yr in 1990 (70% of deposition). This uptake increases net primary production by 1.9 GtC/yr in 1990. The overall net ecosystem production increases by 0.6 GtC/yr over the 1980s and balances the past carbon budget (run 9, see Table 8.3). As a result, the relative differences between the observed and simulated atmospheric CO_2 are less than 2-3 ppmv over the period 1958-1990. Thus, by incorporating the N fertilisation feedback, a balanced past carbon budget can be created. The N fertilisation feedback constitutes a significant part of the carbon imbalance in the submodel, about 20-50%. The final net terrestrial biospheric emissions are in good agreement with the results of those determined by deconvolution of historical atmospheric CO_2 records (Siegenthaler and Oeschger, 1987; Hudson *et al.*, 1994).

8.5 Conclusions

Although the biological, physical and chemical processes representing the biogeochemical cycling of basic elements in the CYCLES submodel are highly aggregated representations of complicated, geographically dependent processes that are only partly understood, the submodel seems to be a reasonable first step towards describing the global biogeochemical cycles in an adequate, quantitative way.

Simulation experiments with the CYCLES submodel show that incorporating terrestrial carbon feedbacks, i.e. the CO_2 fertilisation and temperature feedbacks, is not enough to create a balanced past carbon budget. The N fertilisation of the anthropogenic nitrogen depositions on temperate forests is necessary to account for the final part of this imbalance and much depends on the fate of anthropogenic nitrogen inputs. Only for a proportion of these inputs is it known where they remain. By varying the N fluxes due to river transport, depositions and denitrification a balanced past nitrogen budget can be obtained and the imbalance of nitrogen of about 80 TgN/yr is almost equally distributed over the atmosphere and land. A major cause of the N accumulation on land is the anthropogenic depositions of 40 TgN/yr for the temperate forests, which results in an additional N uptake of about 28 TgN/yr (~70% of deposition). By incorporating the N fertilisation feedback, the net primary production increases by about 1.9 GtC/yr, which leads to a balanced past carbon budget. We found that the N fertilisation feedback constitutes a significant part of the carbon imbalance in the model (about 20-50%). The above analysis shows that the issues of balancing the past carbon and nitrogen budgets are strongly linked and cannot be considered in isolation.

Indicators for sustainable development

9 INDICATORS FOR SUSTAINABLE DEVELOPMENT

Jan Rotmans

The most widely used social, economic and environmental indicators are scale, sector or subject-specific. Indicators for sustainable development, however, need to address the linkages between different aspects of global change and this requires a systemic approach. One way of systematically structuring the interlinkages between indicators is by using integrated assessment models. In this chapter we discuss indicators from a modeller's point of view, including their use for communicating model results. A hierarchical framework is introduced for models in general and TARGETS in particular.

9.1 Introduction

Indicators are pieces of information designed to communicate complex messages in a simplified, (quasi-)quantitative manner so that progress in the field of decision-making can be measured. Social and economic indicators have been used for decades at both the national and international level. More recently, environmental indicators have been developed, which are not yet as widely adopted as socio-economic indicators. The most widely used social, economic and environmental indicators are scale, sector or subject-specific. Indicators for sustainable development, however, need to address the interlinkages between the social, economic and environmental aspects of sustainable development. Because there are so many different linkages at different levels, this requires a systemic approach. One way of systematically structuring the interlinkages between indicators is by using models, in particular integrated assessment models.

Chapter 40 of Agenda 21 (UNCED, 1992) calls for the development of indicators for sustainable development, at multiple levels. Indicators for sustainable development are needed in order to provide decision-makers with information on sustainable development that is simpler and more readily understood than raw or even analysed data (Billharz and Molda, 1995). In particular there is a need for highly aggregated and composite indicators, here defined as indices, in which condensed information is assembled. Hitherto, economic and social indices such as GNP (Gross National Product) and HDI (Human Development Index) have been widely adopted and accepted as useful one-dimensional measures, and these have had considerable influence on both national and international policy debates. However, comparable environmental indicators and indices with broad international acceptance are lacking.

Because sustainable development, as a cross-linking concept, requires a coherent approach to social, economic and environmental aspects (Chapter 1), indicators for sustainable development have to address the interlinkages between them. One of the major problems in designing indicators for sustainable development is that there are a variety of linkages at different aggregation levels. Integrated assessment models can be helpful in analysing a number of dynamic interlinkages between social, economic and environmental aspects of sustainable development. In this chapter it is proposed to design a hierarchical framework of indicators which includes different levels of aggregation, varying from aggregated indices to measurable quantities. Such a hierarchical framework can be readily linked to models in general, and to the TARGETS1.0 model in particular.

9.2 Indicators and indices

We frequently use indicators to measure processes that affect our daily lives, although we may call them by other names such as signals or pointers. Socio-economic indicators have been used for decades at both the international and national level. The history of environmental indicators, however, is rather short, as they have only been developed since the 1970s. The path towards their implementation is strewn with obstacles, as they are not yet as widely accepted and used as socio-economic indicators. An even more recent field, which is still in an embryonic stage, is that of indicators for sustainable development (Bakkes *et al.*, 1994; Kuik and Verbruggen, 1991).

It is rather difficult to give an overall definition of an indicator. Ott (1978) puts it as follows: "Ideally an indicator is a means devised to reduce a large quantity of data to its simplest form, retaining essential meaning for the questions that are being asked of the data." In more general and practical terms, indicators describe complex phenomena in a (quasi-)quantitative way by simplifying them in such a way that communication is possible with specific user groups. So key words are: *(quasi-)quantification, simplification and communication*. The additive 'quasi' indicates that, although indicators are mostly quantitative in nature, in principle indicators could also be qualitative (e.g. changes in the water or soil quality conditions are often expressed in terms of improvement, no change or deterioration). Qualitative indicators may be preferable to quantitative indicators where the underlying quantitative information is not available, or the subject of interest is not inherently quantifiable (Gallopin, 1996).

The distinction between indicators and indices is based on a difference in aggregation level. An index is defined here as a multi-dimensional composite made up from a set of indicators. An index is primarily designed to simplify and combine information at a high abstraction level, whereby the inevitable loss of information should be kept to a minimum (Ott, 1978). In practice, indicators and indices are the result of a compromise between scientific accuracy, being concise and informative and usefulness for strategic decision making (Opschoor and Reijnders, 1991).

Indicators can be used for many different purposes at many different levels, for many different user groups. Ultimately, the user determines whether an indicator is relevant or not. Therefore, the process of selecting and designing indicators should be participatory, involving all stakeholders, including decision-makers, scientists, data producers and the general public. It is important to realise that, like models, indicators are context-bound and without specifying the context in which they should operate, indicators are meaningless. In view of the various possible perspectives on sustainable development we maintain that there is no simple set of indicators for sustainable development that can be used in all contexts.

Within the context of our research an indicator is defined as a characteristic of the status and the dynamic behaviour of the system concerned. From this systems-based definition of an indicator, it follows that an indicator is a one-dimensional systems description, which may consist of a single variable or a set of variables. An indicator may be absolute or relative, where the latter is an absolute indicator per unit or relative to a reference state. Generating and analysing time-series of indicator values can be an important aid in describing and explaining the complex dynamics of interacting processes in a system.

Sustainable development can be portrayed as a dynamic process of trade-offs between social, economic and ecological interests, which can be described in terms of capitals (reservoirs or stocks) and flows (fluxes). In the long-term capitals are more relevant than flows, while in the short-term the latter are important as a measure for the rate of change of capital forms. In operationalising sustainable development, it is therefore important to develop indicators representing capital forms as well as flows. Unfortunately, most indicators developed so far have focused on flows rather than on capital forms. This underlines the need for developing indicators representing capital forms. In *Table 9.1* some examples are given of social, economic and ecological indicators representing capital as well as flow items. Both capital and flow indicators are composites, with the difference that the composite flow indicators are made up of measurable quantities, whereas the capital indicators are abstract, characterised by dimensions (in the form of issues). This illustrates that the concept of capital indicators still has to be crystallised out (see e.g. Serageldin (1996)).

The institutional indicators have hardly been fleshed out in the international indicator arena. For this reason we don't take into account the institutional indicators here, although we realise that these indicators are of vital importance.

In the course of time all kinds of rather theoretical requirements and academic criteria have been formulated for indicators. This list varies from observability and predictability to unambiguity and reproducibility (Adriaanse, 1993; Bakkes *et al.*, 1994; Opschoor and Reijnders, 1991). Instead we use a core set of practical conditions for indicators which have to (i) have added value vis-à-vis data sets or observations, (ii) be policy relevant with a scientific basis, (iii) be based on quality of

	Capital indicators	**Flow indicators**
Social	SOCIAL CAPITAL • quality of life • education • human health • poverty • equity • production/consumption patterns • space	HDI * life expectancy * literacy * GNP
Economic	CAPITAL GOODS • means of production • houses • infrastructure	GNP * economic transactions * goods im/exported * rate of inflation
Ecological	NATURAL CAPITAL • air • water • soil • fishing grounds	RATE OF POLLUTION * air emissions * water discharge * soil uptake forests

• = dimension/issue
* = indicator

Table 9.1 *Capital and flow indicators for the three types of capital.*

data available, (iv) be sensitive to human-induced changes, and (v) be clear and recognisable.

Single indicators, or combinations of indicators as indices, can be useful tools in assessing an overall situation and evaluating strategic decisions. A prerequisite for indicators for sustainable development is that they represent the status and dynamic behaviour of the human-environment system as a whole, i.e. they portray the pressure on, the state of, and the impact on the human-environment system. In other words, they should capture as much as possible of the cause-effect chains they represent, and relate pressure and impacts to criteria for sustainable development. This causal chain approach is comparable with the Pressure-State-Impact-Response (PSIR) mechanism outlined in Chapter 2. Usually, indicators representing the pressures on the human-environment system are of a socio-economic character, whereas state and impact indicators can be both socio-economic and biological/chemical/physical in nature. This implies that indicators for sustainable development are inherently hybrid. Whether these indicators for sustainable development will be completely new, or combinations of existing indicators defined for a particular system, remains to be seen. If possible, these indicators need to be aggregated into a number of overall sustainability indices, which may prove to be useful tools for measuring sustainable development (Liverman *et al.*, 1988). Examples of composite, integrated indices are given in section 9.5.

9.3 Linkages between indicators and models

The relation between models and indicators has two sides. On the one hand, indicators can help in improving the communication between modellers and decision-makers. Decision-makers are used to formulating policies and related policy targets in terms of indicators. As regards models, however, decision-makers generally are on unfamiliar ground, considering model outcomes opaque and difficult to interpret. Formulating model results in terms of a selected set of key indicators makes it easier for decision-makers to interpret model projections. In this respect indicators may serve as vehicles for communication of the model results, on the basis of which promising policy routes can be mapped out for decision-makers. On the other hand, models can help in identifying and analysing the dynamic interlinkages between indicators. This is of vital importance, because, as stated above, indicators for sustainable development should address the *interlinkages* between social, economic and environmental aspects. Analysing the *dynamic behavioural patterns* of those indicators yields essential information for analysing sustainable development. In particular integrated assessment models are suitable tools for systematically structuring the interlinkages between indicators.

The main advantages of linking a set of indicators to an integrated modelling framework are: (i) it shows how the various indicators are interlinked (linkages within the cause-effect chain of an issue (vertical integration) and between different issues (horizontal integration)); (ii) it yields insights into the relevance and dynamic behaviour of indicators (behavioural patterns of social, economic and environmental systems); (iii) it enables projections for sustainable development (long-term trends for social, economic and environmental indicators); (iv) it identifies critical system variables and offers a guide for the selection and aggregation of indicators; (v) it may result in a more comprehensive set of indicators, where model variables which appear to be of pivotal importance for trend projections are not yet part of the existing set of indicators; and (vi) it may serve as a guide for the further development of the integrated modelling framework. Such coherent and integrative information can only be generated by an interconnected framework of indicators.

After having addressed the question *why* we should link indicators and models, the question now arises *how* to link them. In particular how could we describe a set of indicators in such a way that they have a logical connection to the model used? A logical way would be to use the same organising principle for indicators as for models. This leads us to use the PSIR framework as an organising principle for both our models and indicators (Chapters 2 and 3). By doing so, we use an identical ordering mechanism, which is based on the concept of causality, for both our models and indicators. In fact, the purpose of integrated assessment models is to quantify linkages between parts of the PSIR indicator framework, as depicted in *Figure 9.1* (see Swart *et al*, 1995).

It is important to realise that we are not yet able to link all indicators dynamically

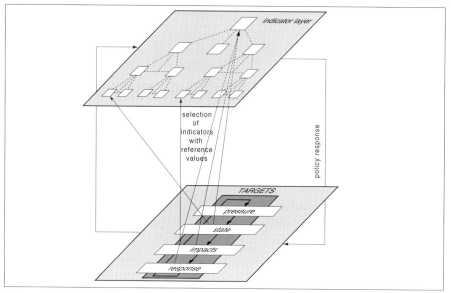

Figure 9.1 Indicator layer versus model layer.

through one type of model. First because knowledge of existing causal relationships between and within the human and environmental system is limited. And secondly, as models are only reflections of a part of reality, the set of indicators enclosed by models is limited. Nevertheless it is possible to transfer the current system of static indicators into a dynamic system of mutually dependent indicators. In *implementing* this transferal between indicators and models, there are various strategies that can be followed. It is possible to build a set of indicators within a model. Models can also be built around an existing set of indicators. The implementation strategy followed in our research is explained in the following sections.

9.4   A model-based indicator framework

Indicators and models can be linked at various levels of aggregation, varying from highly aggregated indices to measurable indicators directly based on observed data or statistics. Here it is proposed to design a hierarchical framework of indicators encompassing different levels of aggregation, varying from highly aggregated indices to measurable, statistics-oriented indicators. The starting-point is to link the hierarchical framework of indicators to the TARGETS1.0 model. The indicator framework can then be considered as a layer on top of the model layer (*Figure 9.1*). Within the hierarchical framework of indicators, four levels of aggregation are distinguished. The first level represents the model-aggregated indices; the second level denotes aggregated indices for the various submodels; the third level denotes relative indicators (absolute indicators per unit or relative to a reference state); while the fourth layer represents observable indicators. In this way a tree diagram of pressure, state, impact and response indicators and indices is built up (*Figure 9.2*), with at the top-level an overall sustainability index, capturing the pressure, state, impact and response dynamics for the TARGETS1.0 model as a whole. One level below the sustainability index, indices for the various submodels are constructed.

The ultimate goal of linking the hierarchical indicator framework with the TARGETS1.0 model is to show the dynamic behaviour of the model and its submodels in a transparent but simplified way. The framework of indicators makes it possible to capture the key social, economic and ecological characteristics of the human-environment system. Furthermore, it demonstrates the dynamic interrelations between abstract indices and the real-world, observable indicators. Depending on the interests of the user, the framework can be entered at any of the information levels. The more aggregated levels are appropriate for giving (horizontal) integrated information, which suffices when there is consensus about a specific issue. The observable indicators, however, are used in the case of single issues, where the underlying dynamics may be poorly understood.

Using the hierarchical framework of indicators, insights derived from model experiments can be communicated, and made accessible to decision-makers, thus enabling the formulation and evaluation of strategies for sustainable development. The interesting scientific question to be answered then is what the minimum set of indicators is that still represents the TARGETS1.0 model adequately. This question can only be answered by performing numerous experiments with the TARGETS1.0 model. In designing the hierarchical framework of indicators the following steps should be taken into consideration:

- *selection:* based on the knowledge gleaned from a range of systematic simulation experiments, representative indicators can be selected and formulated for each of the subsystems;
- *scaling*: transforming the various components of indicators down to a single scale and rendering the units of measurement dimensionless;
- *weighing*: valuation of the various components of the indicators;
- *aggregation*: aggregating results to bring indicators up to one (integrated) index;
- *visualisation*: multi-dimensional representation of indicators and indices.

The above steps should be treated with great care, because they are value-based and depend on subjective perceptions. In particular the processes of scaling, weighing and aggregating are value-loaded. One should avoid the pitfalls of comparing chalk with cheese. Value-based decisions will always be subject to criticism. Methods that can be used for underpinning the above processes are expert judgement, delphi-techniques, multi-criteria analysis, public opinion polls, and model experiments (Bakkes *et al.*, 1994; Hope and Parker, 1993; Hope *et al.*, 1992). In this book we mainly rely on the first and the last, i.e. expert judgement and model experiments.

9 INDICATORS FOR SUSTAINABLE DEVELOPMENT

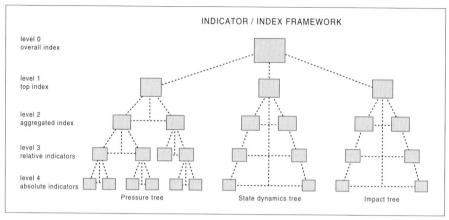

Figure 9.2 Hierarchical framework of indicators.

The abstract level of information generated by indices might create a barrier, hampering the communication of model results to potential user groups. An alternative to the scaling, weighing, and aggregation procedures – resulting in possibly indigestible aggregation levels – is to visualise indicators in a multidimensional starlike figure. This enables us to represent more than three dimensions in one picture. Such a visualisation has been used under the name of 'amoeba' for water pollution and ecosystems (RIVM, 1991). It is also used in orientor theory (Bossel, 1987) and known as eco-compass in industrial efforts to reduce resource intensity. An example is presented in *Figure 9.3*, which represents four indicators of the human-environment system from TARGETS1.0. The idea is to represent visually the state of the human-environment system in 2100 compared to 1990. Overall, the visualisation shows that, for this scenario, we are better off in 2100 in terms of more people who live longer, consuming more food, with no additional detrimental climate impact. The picture illustrates two important aspects of visualisation. On the one hand, it facilitates the provision of linked information in a concise way, without loss of any information. On the other hand, the resulting pictures are more complex and may need more explanation.

9.5 Implementation for TARGETS1.0

The process of designing and implementing an indicator framework is not only time-consuming and complicated, but also an iterative, highly participatory process. Each step in this process should be discussed thoroughly with the stakeholders involved. This requires frequent and intensive interaction between different types of stakeholders: modellers, disciplinary scientists, data experts, and decision-makers. So far we have not yet fully explored this participatory process. The process of

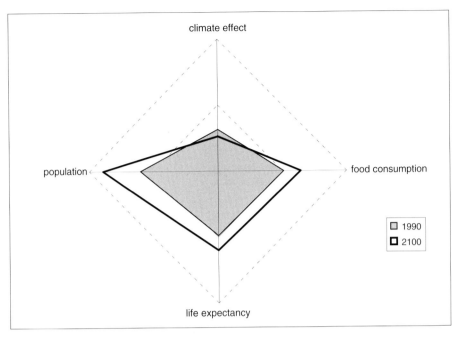

Figure 9.3 An illustrative example of visualisation: the state of the human-environment system.

designing the indicator framework and linking it to the TARGETS1.0 model, took place in our own experimental (model) garden. Owing to lack of time, we have not yet consulted widely about the basic assumptions underlying the indicator framework. Because we realise that we are walking on thin ice, we consider the time not yet ripe for presenting and discussing the whole indicator framework linked to TARGETS in this book. Instead, we will present two characteristic examples of indicator frameworks that represent submodels of TARGETS1.0. The indicator frameworks selected are those for demographic dynamics and biogeochemical cycles. They can be considered as indicative for the human and environmental system and are formulated on the basis of expert judgement and model experiments. The indicator frameworks can be viewed as a particular meta-representation of the submodel.

9.5.1 Indicator framework for the Population and Health submodel

A prerequisite for indicators and indices for future population and health development is that they represent socio-economic and environmental pressures, including the state of population and the impact of the various health determinants on health levels as well as on future population size. In *Figure 9.4* the indicators and indices that are derived from the integrated Population and Health submodel of TARGETS 1.0 (as described in Chapter 4) are represented in the form of a hierarchical indicator-tree.

At the highest aggregation level of the indicator framework, an aggregated population & health index expresses the health risk, steady-state and health status. This is comparable to the UNDP human development index (UNDP, 1994). Weighing of health risk against actual health loss will depend on the level of risk one is prepared to accept. Weighing fertility and population indices the pressure and impact indices will depend on the kind of policy chosen. In order to construct a meaningful overall index, the indices defined above for the different components of the system are scaled between 0 and 1. By simply adding up the scaled indices one overall index is created describing population and health in time. One level below, aggregated indices represent the pressure on, the state of, and the impact on the population and health system. This has been done in line with the human and social indices established by e.g. the UNDP (1994). Similarly, social policy can be evaluated, taking into account access to and distribution of water, food and economic resources. At the lowest aggregation level the regular model output figures are depicted. At this level, pressure and impact indicators – in relation to the stage of the demographic and epidemiological transition – are collected from data sets, generated by international monitoring systems as established by WHO, UNICEF, FAO, UNESCO and UNDP.

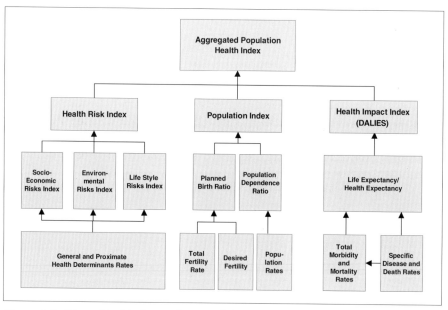

Figure 9.4 Hierarchical indicator framework for the Population and Health submodel of the TARGETS1.0 model.

Pressure

Risk levels for the population level are expressed in terms of people exposed to certain health risks. Danger threshold have usually been defined based on research results found in experimental settings. Taking the relative risk associated with a particular exposure, one can calculate the population-attributive risk which indicates the proportion of disease incidence that is explained by the level of exposure. At the pressure level, an overall 'health risk' index is proposed which adds up the risks related to all determinants categories while weighing for their contribution to the incidence of diseases. A discount rate can be used to correct for the moment in time when the disease incidence occurs. In this version there is no weighing attached to the age at which the disease incidence occurs. Because of the delays involved, the overall pressure index will provide an early warning signal for potential future health changes.

State

Two state indices are proposed: one related to fertility and one related to population structure. The fertility index expresses the percentage of children born that were planned: the 'planned birth ratio'. The ratio depends on the total fertility rate and desired family size, for which there are a number of empirical data sources such as censuses, demographic surveys and family planning programmes. This fertility index demonstrates the potential for fertility change and, hence, the momentum of population growth. The second index proposed is the dependence ratio describing the number of dependants within a population, i.e. those under 15 and those above 65 years of age as a proportion of the whole population. It indicates the proximity of the population to a possible steady state.

Impact

Health status assessment at the population level implies a quality-of-life measure. It is expressed in the health expectancy measure. Until the past decade, population health has been measured in terms of overall life expectancy, stressing the gains in mortality reduction and hence, gain in absolute life years. Recently World Bank's (1993) composite quality measures, including years spent in disease, have been made operational, making use of the health statistics that are available internationally. Weighing the total years lived for the time spent with and without disease produce disability-adjusted-life-years per 1000 persons (the DALY measure). The available data on DALIES is still insufficient, and if available, is based on modelling, i.e. life-table extrapolations and a Delphi-like method (World Bank, 1993). The severity of the disease is weighed according to a disability scale that consists of six categories. Using this measure, one can correct for the age groups in which life years are lost and also discount certain life years. In this way, time lived with a disability is made comparable with time lost due to premature mortality.

Response

This component is not shown in *Figure 9.4*. Customary indicators are the percentages of the available budgets that have been allocated for particular social or health policy areas. Other lower level indicators in relation to social and health regulations are also used (e.g. level of educational requirements).

9.5.2 Indicator framework for the CYCLES submodel

The hierarchical framework of indicators and indices which is linked to the CYCLES submodel is shown in *Figure 9.5*. At the highest aggregation level, an overall index provides information on the total pressure on, the actual state of, and the impact on the global environmental system. At the second level less aggregated indices which underpin the top indices represent the various modules of CYCLES. The two lowest information levels in the indicator framework consist of relative and absolute indicators, representing state variables taken into account in the simulation model. For the CYCLES submodel no response indicators have been formulated, because response options are not included in the model (see Chapter 8).

Environmental pressure index
Environmental pressure indices provide aggregate information on human activities that result in pressures on the functioning of the biogeochemical cycles: energy-related, industrial and agricultural activities, expressed in terms of emissions and flows. At the lowest level the indicators are emissions or flows of the main biogeochemical elements C, N, P and S, and a number of other chemicals. At the third level, the relative indicators are defined as absolute emissions or flows relative to 1990 levels.

At the second level the emissions of the different compounds are clustered in a theme-specific way and combined on the basis of their potential impacts. The environmental themes of climate change, stratospheric ozone depletion, acidification and eutrophication (water pollution) have thus been taken into account. The aggregation protocol used for designing an overall pressure index involves a normalisation and weighing method which is described in den Elzen *et al.* (1995). The weighing schemes are based on expert judgement and valuation as presented by the Dutch Ministry of Housing, Physical Planning and Environment (VROM, 1994).

Environmental state indices are designed to provide information on the changes in the various environmental compartments (reservoirs), expressed in changes in concentrations, in radiation and temperature. For the various compartments state indicators are formulated which represent changes in quality. For the atmosphere, concentrations of a number of chemical compounds are normalised and weighed to obtain aggregated concentration indices, as described in den Elzen *et al.* (1995). The soil compartment indicators include soil organic and inorganic matter in natural and

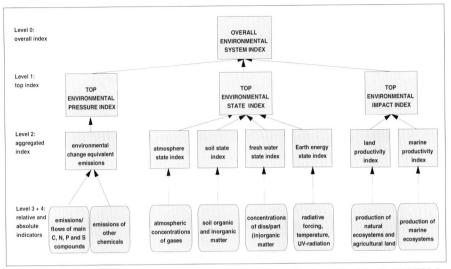

Figure 9.5 Hierarchical indicator framework for the CYCLES submodel of the TARGETS1.0 model.

cultivated ecosystems: the fresh-surface-water-quality indicators contain information on concentrations of dissolved and particulate organic matter (C, N, P and S) in surface waters, and on concentrations of dissolved inorganic matter (the acidifying compounds, SO_4^{2-}, NO_3^- and NH_4^+, and phosphate) in surface water and ground water. At the aggregated level the fresh-water-quality index as developed within the AQUA submodel (Hoekstra, 1995) is used. The various atmospheric, soil and fresh-water indices are weighed to create an overall environmental state index. For the environmental themes taken into account – global climate change, stratospheric ozone depletion, acidification, and deterioration of land and water resources – weighing schemes are used as presented by the Dutch Ministry of the Environment (VROM, 1994).

Environmental impact index
The environmental impact indices give information on the impact of global environmental changes on the terrestrial and marine biological productivity of ecosystems, and agricultural production. At the lowest aggregation level the direct model output figures are depicted, including the terrestrial productivity per unit area and marine productivity. At a higher aggregation level the normalisation procedure for absolute total land and marine production is based on the initial (1900) values of these indicators. At the highest aggregation level the top environmental impact index is created by weighing the land productivity and marine ecosystem productivity index, which have simply been averaged at this stage in the development of TARGETS.

9.6 Linkage with existing indicator programmes

Clearly, the spheres of measurable, policy-relevant and modelled indicators do not necessarily coincide. The limited availability of data will probably constrain the choice of indicators even more than the limitations imposed by models. In such cases the indicator framework should be reconciled with available data and statistics, which are external to the modelling framework used. With regard to policy relevance, the clear discrepancy between policy-relevant indicators and model-based indicators can be illustrated by adopting a list of indicators widely used in the policy arena. At the third session of the UN Commission on Sustainable Development, held in April 1995 in New York, a list of indicators was politically approved and endorsed (UN-DPCSD, 1995). As has been pointed out by others (Billharz and Molda, 1995), there is anything but a scientific consensus on the CSD list of selected indicators. But instead of waiting for a perfect, consistent and purely scientifically-based list, we prefer to use the existing CSD list of indicators. Meanwhile we can make an endeavour to improve on the current CSD list of indicators.

In *Table 9.2* the current list of CSD indicators for sustainable development, which is structured along the lines of the Driving Forces-State-Response framework, is presented. Those indicators of the CSD list that are included in the TARGETS1.0 version are printed in italics. A first screening shows that the economic and environmental indicators of the CSD list are largely covered, but that particularly the social and institutional indicators are poorly represented. In general, social and institutional indicators are hardly presented in the current generation of integrated assessment models. The CSD list of indicators at least gives an indication of current policy priorities in relation to the subject of sustainable development. This prioritisation could be used as a guiding principle for the further development of both the TARGETS1.0 model and the indicator framework linked to it.

	Driving force	**State**	**Response**
Social indicators			
Poverty	unemployment rate	female/male wage	
Demographic dynamics	*population growth* *fertility* net migration	population density	
Human health		life expectancy infant mortality *access to clean water* access to sanitation nutritional status	immunisation contraceptive prevalence *health expenditures* local health care expenditure
Education	*literacy* primary/secondary school enrolment	women/men in labour force male/female school enrolment	*GDP spent on education*
Human settlement	urban population growth *fossil fuel consumption by motor vehicle transport*	rural/urban population area of urban settlements	infrastructure expenditure
Economic indicators			
International cooperation	GDP growth net investment share in GDP exports and imports as % of GDP	environmentally adjusted GDP share of manufactured goods in merchandise exports	
Production/ Consumption patterns	*energy consumption* *energy production* *share of natural-resource intensive industry*	mineral reserves *fossil fuel reserves* resource intensity renewable energy consumption	
Financial mechanisms	resources transfer ODA as % of GDP	debt / GNP debt service / export	*environmental expenditures* additional funding for sustainable development
Technology transfer	capital goods imported foreign direct investments	environmentally sound capital goods	technical cooperation grants
Environmental indicators			
Fresh-water resources	*withdrawal of ground/ surface water* *domestic water consumption*	ground water reserves faecal coliform	*waste-water treatment coverage*
Oceans	population growth in coastal areas discharge of oil *release of nitrogen and phosphorus*	actual abundance versus max yield abundance stock of marine species versus maximum yield algae index	
Land			
Land management	*land-use change* *population*	land condition change	local natural resource management
Desertification	poor people in dryland areas		
Mountains	population dynamics in mountain areas		

Table 9.2 CSD list of indicators; those in italics are included in the TARGETS1.0 model.

	Driving force	**State**	**Response**
Environmental indicators (continued)			
Deforestation	*wood harvesting intensity*	*forest area change*	managed forest area ratio protected forest area ratio
Biodiversity		threatened species versus native species	protected area versus total area
Rural Development	use of pesticides use of fertilizers irrigated arable land energy use in agriculture	*arable land per cap salinisation*	agricultural area affected by research agricultural education
Biotechnology			R&D expenditure on biotechnology biosafety regulations
Atmosphere	*emissions of greenhouse gases emissions of sulphur oxides emissions of nitrogen oxides consumption of ozone depleting substances*	*concentrations of pollutants*	*expenditure on air pollution abatement*
Solid waste	generation of industrial and municipal solid waste		expenditure on waste management waste recycling rate
Hazardous waste	generation of hazardous waste import and export of hazardous waste	land contaminated by hazardous wastes	expenditure on hazardous waste treatment
Radioactive waste			
Toxic chemicals		acute poisoning	chemicals banned or restricted
Institutional indicators			
Decision-making			sustainable development strategies programmes for integrated accounting national councils for sustainable development
Institutional arrangements			
Legal instruments and mechanisms			ratification of global agreements implementation of ratified agreements
Capacity building			
Information for decision-making		telephone lines per 100 inhabitants access to information	programmes for national environmental statistics
Science		scientists/engineers per million population	expenditures on R&D
Major groups			representation of major groups contribution of NGOs

Table 9.2 continued

10 Uncertainties in perspective

"Es gibt keine Tatsachen, nur Interpretationen."
Friedrich Nietzsche

10 UNCERTAINTIES IN PERSPECTIVE

Marjolein B. A. van Asselt and Jan Rotmans

Any exploration of future developments inevitably involves a considerable degree of uncertainty and integrated assessment modelling is no exception. One of the major uncertainties has to do with the direction of policy-making. In order to accommodate a wide variety of world views and management styles, this chapter introduces the concept of multiple model routes. These are alternative ways of looking at model relationships, taking into account the bias and preferences of a number of stereotypical perspectives. These perspectives, which each represent a different attitude to nature and society, are typified as hierarchist, egalitarian, individualist and fatalist. Matching a consistent management style with the first three (active) perspectives permits an analysis of 'utopias', while the opposite case – when world view and management style are out of step – reveals the risk of a number of possible 'dystopias'.

10.1 Introduction

The future is inherently uncertain and thus unpredictable. Nevertheless, people in general, and decision-makers in particular, are interested in exploring future developments in order to make plans. One of the roles of science is to assist decision-makers by sketching images of the future of the planet and of humankind. Scientists are facing an increase in both the magnitude and the degree of complexity. The issues currently associated with global change differ from other scientific problems in several respects (Funtowicz and Ravetz, 1993):

- they are global in scale and long term in their impact;
- the available data is lamentably inadequate; and
- the phenomena, being novel, complex and variable, are themselves not well understood.

As a result, while science has hitherto been concerned with attempts to resolve uncertainties, they are now seen as intrinsic to many scientific investigations. Because of the complexity of global change, it can be approached from numerous valid perspectives. The idea that there are various perspectives, implying that each model is regarded as the reflection of merely one specific perspective, is an emerging theme within the modelling community, although outsiders have pointed out that subjectivity is inherent in models (Funtowicz and Ravetz, 1993; Keepin and Wynne, 1984).

Uncertainties play a key role in integrated assessment modelling. Because integrated assessment models are multi-disciplinary in nature, they include many different types of uncertainty. Various attempts have been made to classify the different types and sources of uncertainty. Morgan and Henrion (1990) distinguish uncertainty about empirical quantities and uncertainty about the functional form of models, which, *inter alia*, may arise from: inherent randomness; subjective judgement; systematic and random errors; approximation; and disagreement among experts. An alternative classification which is useful within the context of this study is the distinction between technical uncertainties (concerning observations versus measurements), methodological uncertainties (concerning the right choice of analytical tools) and epistemological uncertainties (concerning the conception of a phenomenon) (Funtowicz and Ravetz, 1989). A taxonomy of uncertainty is depicted in *Figure 10.1*. Because integrated assessment models are end-to-end approaches, they contain an accumulation of uncertainties. Furthermore, the highly abstracted, parameterised process formulations in those models often 'hide' uncertainties. In addition to the problem of gaps in knowledge about the past, present and future state of the world, there is inherent uncertainty with respect to the adaptive behaviour of both humans and natural systems, due to the occurrence of surprises.

Hitherto, there have been no adequate methods available to integrated assessment modellers for analysing uncertainty and various perspectives in a systematic way. Current uncertainty analysis techniques used in integrated assessment, e.g. Monte Carlo sampling and probability distribution functions, merely address technical uncertainties. These are not suitable for analysing methodological and epistemological uncertainties, which primarily arise from subjective judgements. At

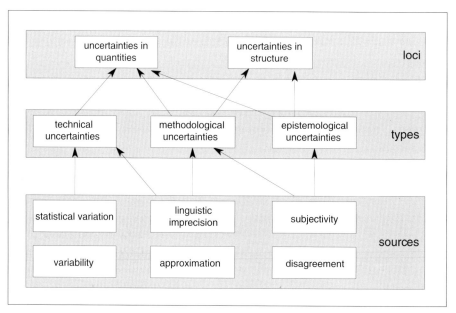

Figure 10.1 Taxonomy of uncertainty.

the core of this chapter is the notion that subjectivity is the result of adopting a specific perspective.

Furthermore, classical methods suffer from the fact that they only address uncertainties in model inputs and neglect the structure of the model itself. Estimates of minimum, maximum and best guess values used in classical uncertainty analysis are often erroneous or misleading (Frey, 1992). The fact is that such estimates ignore the interactions among multiple, simultaneous uncertainties, precisely the interdepencies and relationships which cannot be ignored in integrated assessment. Current methods thus give decision-makers no indication, regarding the magnitude and the sources of the underlying uncertainties and fail to translate uncertainty into concepts such as risk, a notion which corresponds closely to the experiences, practices and needs of decision-makers.

It is evident, therefore, that integrated assessment requires new tools, if we want to be able: i) to incorporate a variety of perspectives in relation to uncertainty and ii) to make uncertainty more explicit by expressing it in terms of risk. Building upon Janssen and Rotmans (1994), we attempt to approach this challenge by employing social-scientific means, i.e. theoretical insights and empirical results with respect to differences in perspectives. For this purpose we introduce the concept of multiple 'model routes' (van Asselt *et al.*, 1996; van Asselt and Rotmans, 1995; 1996), which will be clarified here by the extended metaphor of different walks through a landscape.

10.2 Model routes

Integrated assessment models designed to describe (aspects of) the functioning of the real world can be thought of as 'maps' of a landscape in which natural and human-induced pressures on the dynamics result in certain impacts. Both modellers and users have to deal with the fact that the landscape cannot be completely known in advance and that interpretations of what can actually be seen may differ significantly. Multiple model routes can then be thought of as representations of the different pathways followed by a group of hikers, reflecting different assessments of the landscape's uncertain features.

According to social psychology (Tversky and Kahneman, 1974; 1980; 1981), biased interpretations are seldom totally arbitrary, but rather tend to relate to value-systems or so-called perspectives. We propose to use social scientific insights and reasoning with respect to various perspectives as an organising framework to address subjective judgement in modelling efforts. The perspectives are then used to invest the alternative model routes with coherence. The hikers can be thought of as representatives of different perspectives. A pathway followed by a single hiker then corresponds to a perspective-based route. Translated into model terms this implies that a perspective-based model route is a chain of biased interpretations of the crucial uncertainties in an integrated assessment model. We intend to use a typology of perspectives as a heuristic device in order to describe consistent choices at the crucial points in the model where subjective judgement is required. Such points can be compared to man-made 'crossings' in a landscape. Model routes signify different choices at the various crossings and perspective-based model routes are then coherent chains of uncertainties coloured with the bias and preferences of a certain perspective.

It is the modellers' role to erect signposts in the landscape and to indicate them on the map to provide routes which lead reliably to a specific destination. This requires the modellers to wander through the landscape on successive reconnaissances, during which they successively adopt another perspective. In other words, modellers are obliged to depart from their own preferred route, and have to acknowledge that other routes are equally possible. Instead of providing merely one description of the functioning of (aspects of) the real world, thereby adopting a cavalier attitude to the uncertainties and ignoring the wide variety of alternative perspectives, modellers

A group of hikers embark on a walk from the same point in a landscape. They look at the landscape surrounding them, but they interpret what they see in different ways, so that they do not all march off in the same direction (*Figure 10.2*). For example, if they see a mountain ahead of them, one hiker will see it as a challenge and want to climb it, while a companion prefers a route that avoids the mountain for fear of falling off. A third hiker would like to enjoy the panorama at the top, but thinks climbing is too risky, and thus ascends by established footpaths. At every juncture the hikers are obliged to choose how they will cross the landscape. Furthermore, they cannot always assess the situation in advance. The hikers' assessments of the landscape behind the mountain may range from envisaging another, even higher mountain to a ravine or a plateau. Their different conceptions of the unknown landscape make them inclined to choose different pathways.

Moreover, the landscape might change as a result of human actions. As an example we might consider the building of a campfire, the impact of which may range from scorching the surface beneath the fire itself, to unexpectedly great and probably irreversible damage caused by a forest-fire. The risk assessments and thereby the attitude towards any specific action will certainly differ among the group of hikers: from abandoning the idea of a fire, or taking preventive measures such as building trenches filled with water round the site, to simply building a large campfire. Apart from being affected by human activities, the landscape can also be changed by natural factors like the weather, or events like earthquakes, which may create unexpected obstacles for the hikers.

Our hikers start out without much knowledge of the landscape and the processes within it, and similarly lack understanding of the consequences of both natural and human-induced changes on the features of the landscape. Nevertheless, they learn from experience how to survive. If, for example, they had once been washed out in a storm, because they had erected their tent on a hill, the hikers would subsequently pay more attention to their choice of campsite.

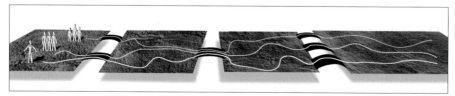

Figure 10.2 Metaphor of multiple routes.

need to illuminate the major uncertainties and to provide reasonable alternative interpretations. 'Erecting signposts' is thus equivalent to identifying crucial uncertainties in state-of-the-art knowledge, indicating them in the model, and then performing mathematical calculations consistent with the bias and preferences of a particular perspective. Less experienced users of the model can then choose to follow one of these routes to explore the landscape, instead of blindly hoping to stumble across a suitable path.

10.3 Framework of perspectives

Such an approach clearly requires a general framework of perspectives. Unfortunately, the social sciences do not provide a convenient, generally accepted typology of perspectives which is independent of time and scale. Notwithstanding the framework of perspectives we explore below, we wish to emphasise that the model route strategy makes sense independently of the choice of a certain typology. For the development of TARGETS, a top-down approach has been chosen, in which the analysis starts at the highest aggregation level, i.e. the global level, considering the Earth as a whole (Chapter 3). We have therefore chosen to adopt a top-down approach in our elaboration on perspectives. In the top-down approach certain social scientific insights and arguments are used to arrive at an aggregated typology. Conversely, the bottom-up approach takes as its point of departure empirical research in which individuals or groups are asked to express their interpretation of particular uncertainties. Such answers might then be clustered to arrive at a limited set of perspectives. The bottom-up approach might be promising for dealing with theme-specific or region-specific integrated assessment models.

Perspectives may be considered as aggregations of the different points of view humans have, and can be defined as consistent hybrid descriptions of how the world functions and how decision-makers should act. Perspectives can thus be characterised by two dimensions: (i) a world view, which entails a coherent view of how the world functions, and (ii) a management style, i.e. policy preferences and strategies. Following the Pressure-State-Impact-Response approach (PSIR, Chapter 2), the distinction between world view and management style is then reflected in the distinction between the Pressure-State-Impact-system (PSI) on the one hand, and the Response system on the other. Model routes in the PSI system can be taken to represent various world views, while model routes in the Response system signify different scenarios and therefore reflect various management styles.

In the sociology of science (Latour, 1987, 1993; Woolgar, 1981, 1988; Wynne, 1982) a strong and mutual interdependency between perception of the world and socio-cultural position is recognised. This means that the two dimensions 'world view' and 'management style' cannot be separated. Nevertheless, we use the distinction between them to structure various insights derived from a variety of disciplines with respect to perspectives. We analyse both world view and management style consecutively, and connect them using Cultural Theory as introduced by Thompson *et al.*, (1990) in order to provide a limited set of coherent perspectives.

World views

A 'world view' can be defined as a coherent description of: (i) a way of structuring reality, and (ii) an accompanying vision of the relationship between human beings and the environment (Zweers and Boersema, 1994). Such a view on the structure of reality comprises a myth of nature, which addresses issues such as ecosystem vulnerability, and a view on humanity. Philosophical anthropology, the philosophy of culture and ecology also address perceptions of ecosystem vulnerability and human nature (Zweers and Boersema, 1994), while environmental ethics is concerned with humanity's relationship to the environment and its understanding of, and responsibility towards, nature (Pojman, 1994; Zweers and Boersema, 1994). Insights from these disciplines enable us to describe some stereotypical conceptions.

Myths of nature

The extreme conceptions of ecosystem vulnerability can be described according to the various myths of nature identified by ecologists (Holling, 1979; 1986; Timmerman, 1986; 1989). There are four possible myths of nature: Nature Benign, Nature Ephemeral, Nature Perverse/Tolerant and Nature Capricious. Each myth can be represented graphically by reference to the metaphor of a sphere rolling in a landscape (*Figure 10.3*). Nature Benign has it that nature is very robust. The world is wonderfully forgiving. The most extreme perturbations likely to stem from any human activity will do no more than result in mild disruptions. Nature Ephemeral shows nature in a delicate balance. Small disturbances may have catastrophic results. Any man-made change in a natural system is likely to be detrimental to that system. Nature is thus characterised as fragile. Holling (1994) shows that the image of 'an unforgiving nature that can be overwhelmed by change' results from three axioms:

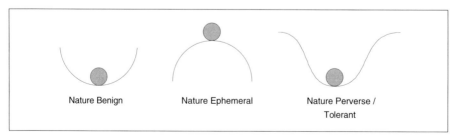

Figure 10.3 Myths of nature.

- all populations grow exponentially,
- physical population limits exist, and
- variability is a necessary condition for Earth's survival.

According to the myth Nature Perverse/Tolerant, nature is robust within certain limits, i.e. its rich differentiation ensures that it is forthcoming when approached in the right way, but retributive when pushed beyond certain bounds. If the disturbances are small, nature will return to equilibrium. However, as soon as a certain threshold is passed, disturbances pose a threat to the functioning of nature. In other words, nature can to a certain extent be controlled. Nature Capricious tells us that nature is random and unpredictable. It is unknown what will happen if there are disturbances.

As will be clear, these myths are the simplest models on the functioning of ecosystems. Kwa (1987) critizes the typology of myths of nature in which nature is understood as a 'productive machine that runs out of fuel, that can be overworked or in some other ways faces the possibility of being destabilised so that it cannot survive'. He argues that this one-dimensional view ignore the fact that stability in the narrow sense is not very important for most known ecosystems. Instead persistence, i.e. the ability of an ecological system to function outside equilibrium or in many metastable forms while developing new ones all the time, is much more important. His criticism reveals the need for an at least two-dimensional typology of images of nature comprising both stability and complexity. For the time being, in the absence of a better alternative we use the myths of nature as described above to indicate the range of conceptions of ecosystem vulnerability.

Views on humanity

Our image of the structure of reality indicates a view on humanity: who are we and what special qualities do we possess? It is argued that the images of humanity and nature are closely related, mutually influencing one another (Thompson *et al.*, 1990; Zweers and Boersema, 1994). Four views on human nature can be distinguished which can be represented by the same diagrams that describe the myths of nature (*Figure 10.3*). In the view associated with Nature Benign, human nature is extraordinarily stable. Human nature is described as essentially self-seeking, i.e. human beings are considered to be rationally self-conscious agents seeking to fulfil their ever-increasing material needs. Such a conception of human nature denies that individuals can be motivated by pursuing the collective good. The view on human nature tied to Nature Ephemeral holds that humans are born good, but are highly malleable. Just as human nature can be corrupted by 'evil' institutions (e.g. markets and hierarchies), so can it be rendered virtuous by an intimate and equal relationship with nature and other humans. This view of human nature is nicely expressed by Reich (1970) and dated back to Rousseau, the famous French philosopher who lived in the 18th century. Human satisfaction or self-realisation lies in spiritual growth and maturity, rather than in the consumption of goods. Concomitant with Nature Perverse/Tolerant is the idea of human nature which considers humans to be born sinful, but capable of redemption by good institutions for the sake of man and the environment. This conception of human nature is described by, for example, Howe (1974). Finally, Nature Caprious implies a view of human nature as being so unpredictable that it is essentially random in character.

Ethical attitudes

With regard to the variety of ethical attitudes towards the relationship between man and nature, two main streams of thought can be discerned: anthropocentrism and ecocentrism. Admitting that such a dichotomy is far too simple to account for the observed variation in ethical standpoints, we would like to add 'partnership' as a third alternative. A schematic representation of the core of each stream of thought is set out in *Figure 10.4*.

Anthropocentrism is the view in which the development of humankind is considered as the ultimate goal of the universe, while nature is seen merely as providing resources for humans to exploit. Humans are the only beings that have value in themselves, and therefore everything else is seen in terms of benefit to them, an attitude which Zweers and Broersma (1994) describe as 'humanity's monopoly on values'. Nature is seen as a purely factual, material entity. The fundamental attitude towards nature associated with anthropocentrism is 'supremacy' (Zweers and Boersema, 1994). Bacon (1628) and Descartes (1637) are considered to be among the first anthropocentrists (Achterhuis, 1995). The anthropocentric world view is held by the so-called 'technocrats', who have an unlimited confidence in technological possibilities, and for whom there are no limits to growth.

In discussions on environmental issues, ecocentrism is generally considered to be the opposite of anthropocentrism. Ecocentrism imputes intrinsic value to nature: which means that it can have its own 'aims'. Aims which will continue to be present even if humans are not involved in any way. Instead of applying human criteria to nature, each animal and every organism should be regarded as valid on its own terms. Nature is defined as a complex whole which organises and maintain itself and which includes everything, including humans. In other words, humans are 'plain citizens'

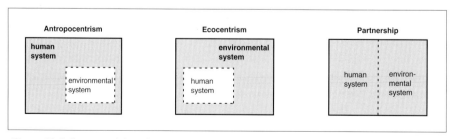

Figure 10.4 Streams of thought with regard to the relationship between man and nature.

(Leopold, 1949) of the biotic community. This approach can be described as 'participation' (Zweers and Boersema, 1994), i.e. subordination of humanity to nature. The ecocentric world view is, for example, found among radical environmentalists, pure ecologists and advocates of the Gaia hypothesis (Lovelock, 1988).

While positions that can be characterised as anthropocentrism and ecocentrism are widely held, describing the relationship between people and the environment as a partnership is less common. Following Zweers and Broersma (1994) we define partnership as the ethical attitude which views the Earth as a totality, with humans and nature being valued equally. In this view the mutual dependency between humans and nature is stressed. For example, humans are dependent for their survival on nature because of their need for natural resources, while nature's diversity and quality has to be protected against overexploitation. Thus, the goal is a balance between environmental and human values.

Cultural Theory
Admitting that no causal relationship can be established between a conception of physical and human nature on the one hand and an ethical attitude on the other, plausible combinations of both can be derived which enable us to describe coherent world views. Such views involve a certain attitude towards risk. This enables us to couple our efforts in developing a framework of perspectives to the so-called cultural theory of risk. Cultural Theory (Douglas, 1982; Douglas and Wildavsky, 1982; O' Riordan and Rayner, 1991; Rayner, 1984; 1991; 1992; Schwarz and Thompson, 1990; Thompson *et al.*, 1990) claims that distinctive sets of values, beliefs and habits with regard to risk are reducible to a small number of ideal types. Cultural Theory generally distinguishes five perspectives from which people perceive the world and behave in it, namely: the hierarchist, the egalitarian, the individualist, the fatalist and the hermit. Hierarchism, egalitarianism, indivdiualism and fatalism are not novel concepts. In some sense, Cultural Theory parallels conventional science. Hierarchism and individualism, for example, reflect a conventional duality expressed variously as 'collectivism versus individualism' and 'state versus market' (Grendstad, 1994).

We realise that Cultural Theory does not represent social science as a whole. Furthermore, we recognise that its schematism is rigid and cannot fully take into account the real-world variety of perspectives. The typology associated with the cultural theory of risk is nothing more but also nothing less than an attempt to systematically address the complex issue of different perspectives at a high level of aggregation, in which the analysts have attempted to take into account the whole body of theoretical and empirical work done in this field (Thompson *et al.*, 1990). Like any model it is merely a limited reflection of reality. However, in spite of all the gaps and inconsistencies, it seems legitimate and reasonable to use the types put forward in Cultural Theory to characterise the broad spectrum of world views in order to assess the boundaries of plausible interpretations. As far as our knowledge of social science extends, no other social scientific theory offers a comparable heuristic on such an aggregate level. In the following we indicate how views of nature and humanity as well as ethical attitudes might be associated with the perspectives distinguished in Cultural Theory.

Anthropocentrism does not see nature as a constraint on satisfying human needs. This cornucopian school of thought (Simon, 1980; 1981) presumes either that people have an unlimited capacity to find substitutes for scarce materials and to develop successful remedial strategies incrementally once the need is apparent, or that natural resources are abundantly available. Anthropocentrism in its more extreme manifestations can therefore plausibly be associated with the conception of reality as being robust (i.e. Nature Benign). This world view is generally associated with a risk-seeking attitude. Cultural Theory characterises it as 'individualist'. Grendstad and Selle (1995) demonstrate, on the basis of survey research, a correlation between anthropocentrism and the individualistic perspective.

Ecocentrism holds that current human activities are a driving force behind environmental change (Zweers and Boersema, 1994), implying an image of nature that is vulnerable to human activities. Thus, it is plausible to connect ecocentrism in its extreme manifestations with the idea of fragile nature (i.e. Nature Ephemeral). Adherents to ecocentrism tend to be risk-averse. This attitude is referred to as 'egalitarian' in Cultural Theory. Furthermore, Grendstad and Selle (1995) identify an empirical correlation between ecocentrism and the egalitarian perspective.

Partnership implies the need for a balance between human and environmental values, which seems to suggest that nature is forthcoming when approached in the right way, but retributive when pushed beyond certain limits. A connection between partnership and the perception of nature as being controllable within limits (i.e. Nature Perverse/Tolerant) therefore seems plausible. In terms of risk, partnership might be characterised as risk-accepting. Cultural Theory refers to this world view as 'hierarchist'. Fatalism in Cultural Theory terms is associated with the view of reality as being capricious. The fatalist 'survivalists', for whom everything is a lottery, are excluded from our framework of perspectives, because they cannot systematically be described by any characteristic function and are frequently excluded from, or uninterested in, active participation in debates (Rayner, 1984; 1994). For the same reason, we left out the hermit.

We take the so-called 'active perspectives', i.e. hierarchist, egalitarian and individualist, as stereotypes. The spectrum that lies in between, comprises a variety of less extreme, or rather hybrid, perspectives on the structure of reality and the relationship between people and nature. We argue that evaluation of coherent extreme perspectives is the best approach, given the current state of knowledge.

Management styles

Management styles are defined in terms of approaches towards response strategies, and thus include preferences regarding policy instruments. The present version of the

TARGETS model assumes a hypothetical world governor, who has the means to implement global policies (Chapter 3). In some sense the TARGETS model in itself can be considered as a hierarchist framework. We consider the effort of developing multiple model routes within such a biased framework as a first step towards a more pluralistic modelling approach, and we recognise that other frameworks – and not just modelling frameworks – are ultimately necessary to accommodate the full range of perspectives.

Accepting the notion of a world governor restricts the spectrum of management styles. Cultural Theory enables us to ascribe stereotypes of management styles to the various perspectives. Hierarchism is associated with a preference for bureaucratic management, while the egalitarian preference can be described as communal anti-managerialism. Individualism in its extreme manifestations advocates an anti-intervention laissez-faire attitude. In other words, both egalitarianism and individualism in their extreme manifestations would reject a decision-making process that assumes a world governor. Advocating a pragmatic approach, we restrict ourselves to a small subset of the true spectrum (*Figure 10.5*). We use the features of Cultural Theory to assess the management style that would be preferred by a hypothetical hierarchist, egalitarian and individualist world governor. Following Cultural Theory (Rayner, 1991; Schwarz and Thompson, 1990), the hierarchist management style can be characterised as 'control', while the egalitarian management style is seen as 'preventive'. The management style associated with the individualist world view is described as 'adaptative'.

Hierarchists maintain that it is possible to prevent serious global problems by careful stewardship of the opportunities that nature provides for controlled economic growth. Assessment of the limits to human perturbation of the environment is an

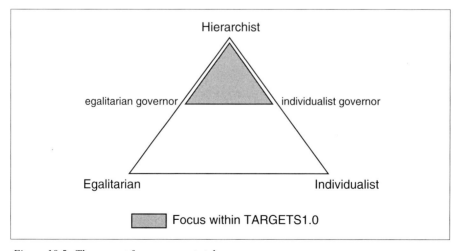

Figure 10.5 The scope of management styles.

essential feature of this approach. The 'control' management style can be associated with the risk-accepting attitude shown by hierarchists. The salient value here is system maintenance. Therefore, rational allocation of resources is the principal mandate given to managing institutions. Regulation and financial incentives are the preferred policy instruments.

The egalitarian management style implies prudence and prevention, and can therefore be characterised as risk-averse. The managing institutions are obliged to approach the environmental system with great care. Anticipation of the negative consequences of human activities is a central element in this management style. Activities which are likely to harm the environment should be abandoned. With regard to the capitalist economic system, drastic social, cultural and institutional changes are advocated in order to arrive at a sustainable socio-economic structure, regardless of short-term economic cost. For example, policy measures should be designed to radically reduce western consumption styles. The preventive management style stresses communication programmes, enhancement of education, and research and development, and demonstration incentives oriented towards appropriate technology.

Individualists hold that change provides new opportunities for human ingenuity that will be revealed through the workings of the market-place. Negative consequences of human activities will be resolved by technological solutions. Economic development should not be curtailed by policy measures, so as to ensure a high level of technological progress and increasing welfare. The adaptive management style can easily be related to the risk-seeking attitude associated with individualism. The managing institutions must allow the market system to operate freely, and should

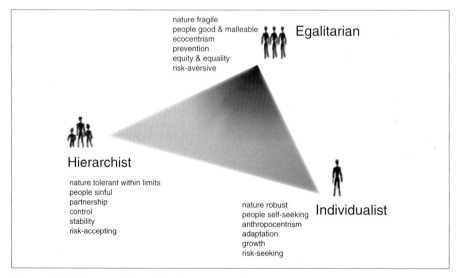

Figure 10.6 Characteristics of the three perspectives.

therefore refrain from any regulation or prohibition. Preferred policy instruments are financial incentives, and research and development programmes.

The main characteristics of the hierarchist, egalitarian and individualist perspectives including both world view and management style, are summarised in *Figure 10.6*.

10.4 Methodology

We use multiple model routes as a tool to communicate the wide range of views held by scientists. Notwithstanding such a pluralistic attitude towards science, we still believe that scientific claims have a special status. In this so-called 'realist view' of science (Mazur, 1996) in which both objectivity and subjectivity are seen as being relative, scientific knowledge is considered to be more 'objective' than other systems of belief about the natural and social world. The framework of perspectives described above is then used as a heuristic device to cluster wide-ranging *scientific* opinions.

Implementation of model routes
Model routes consist of chains of alternative formulations of model relationships and model quantities. How do we identify the quantities and relationships which express a subjective interpretation? Our first step is to analyse the relevant scientific debates in order to identify the major controversies associated with global change. Value differences are at the core of such disputes (Colglazier, 1991). Analysis of the controversies enables us to identify which model choices are influenced by value judgement. Such model choices comprise both the model formulations, i.e. the structure of the model, and the estimations of model inputs, i.e. the quantification of parameters. In a continuous dialogue between modeller(s) and analyst(s), we select the model parts which will be interpreted in line with the various perspectives. Sensitivity analysis is useful to arrive at a reasonable selection. Another methodology for selecting the main sources of uncertainty is based on the criteria magnitude, degree and time-variability is described in van Asselt and Rotmans (1995), and this is used in Chapter 16. We proceed by describing multiple interpretations of the uncertainties, using our framework of perspectives. The description of interpretations of model quantities and relationships in relation to the qualitative descriptions of the perspectives is extremely difficult. This means that selective compromises have to be made. In other words, we emphasise that such a translation exercise is not free of subjectivity. It is not clear whether different people would make the same assignments of a set of model characteristics to the different perspectives. Experience in applying the methodology has taught us that, in about 80% of cases, it is possible to arrive at rather unambiguous descriptions of the interpretations of the perspectives on a specific issue. With regard to the ambiguities, describing multiple perspective-based interpretations is an iterative effort involving

modellers, disciplinary experts, cultural theorists and other social-scientific experts. Notwithstanding such problems, the advantage of using the proposed framework of perspectives lies in the fact that we arrive at our descriptions in a systematic and reproducible manner, which thereby facilitates peer criticism.

The next step involves translation of the qualitative descriptions of the perspectives into model terms. In the case of model quantities, this means that alternative values have to be determined. In this initial phase, this is achieved by simply changing parameter values. When seeking more sophisticated implementation, we may consider subjective probability functions (Morgan and Henrion, 1990; Morgan and Keith, 1995) or alternatively the concept of 'fuzzy sets'.

In the case of uncertain relationships, functional forms need to be reformulated. In the simplest manifestation, this means either changing the function, or deleting, adding or changing the function's arguments. Notwithstanding their mathematical simplicity, such changes are fundamental in a conceptual sense.

We do not claim that the model route strategy enables to capture all types and sources of uncertainty in integrated assessment models. In order to evaluate technical uncertainties, classical uncertainty analysing techniques are used to provide uncertainty ranges associated with each perspective-based projection. If sampling methods are employed, a specific set of parameters (model inputs) is varied in a significant number of model simulation runs leading to 95% confidence intervals. Major difficulties in applying such sampling methods include determining the set of parameters. We used the alternative model routes as heuristic tools in the selection process: the parameters which serve as accessories for the model routes are those which constitute the sampling set (Chapter 11).

Risk assessment

We have indicated that the methodology based on multiple model routes allows for the inclusion of a set of perspectives. Moreover, this approach might also be valuable for communicating about uncertainty in order to narrow the gap between the policy-making community and the scientific world. Matching each perspective's management style to its respective world view is a technique we used to assess the utopias. A utopia is a world in which a certain perspective is dominant and the world functions according to the reigning world view. Dystopias describe either what would happen to the world if reality proved not to resemble the adopted world view following adoption of the favoured strategy, or vice versa, i.e. where reality functions in line with one's favoured world view but opposite strategies are applied. Thus dystopias are mismatches between world views and management styles (*Figure 10.7*). Dystopias are most useful with respect to communicating the role of uncertainty and its consequences for decision-making. If we examine a row in *Figure 10.7*, we see that the dystopias indicate the risks of decision-making in uncertain conditions by showing to what kind of future the chosen strategies may lead, in the event that the adopted world view fails to describe reality adequately. Uncertainty

	WORLD VIEW		
MANAGEMENT STYLE	egalitarian	hierarchist	individualist
egalitarian	*utopia*	dystopia	dystopia
hierarchist	dystopia	*utopia*	dystopia
individualist	dystopia	dystopia	*utopia*

Figure 10.7 Utopias and dystopias.

here is no longer a theoretical scientific concept, but a notion which might be usefully deployed by decision-makers in arriving at their decisions.

The world views associated with the different perspectives are used to establish perspective-based model routes in the PSI system, while the favoured management styles are used to develop multiple model routes in the Response system. Experiments involving juxtaposing combinations of model routes in the Pressure-State-Impact system with those in the Response system enable analysis of utopias and dystopias. The challenge is then to go beyond the multiple perspectives, and to arrive at general insights relevant for decision-makers which are valid independent of the preference for a certain perspective. We aim to indicate the risks associated with a specific global change issue by evaluating the whole range of utopian and dystopian outcomes and to assess whether the majority of the resulting images of the future show undesirable trends. The difficulty is then to define criteria to discriminate between desirable and undesirable. For the time being, we have taken two key indicators and we have then tried to define low risk, moderate risk and high risk areas for the resulting two-dimensional space. The chapters in Part Two discuss how this can be done for each of the global change issues identified. We recognise that this is merely a first attempt to communicate results obtained with integrated assessment models in terms of risk. We consider it as a promising avenue to be explored. Future research efforts will be devoted to more systematic elaborations of uncertainty, risk and integrated assessment models. The methodology based on multiple model routes is summarised in a flow chart (*Figure 10.8*).

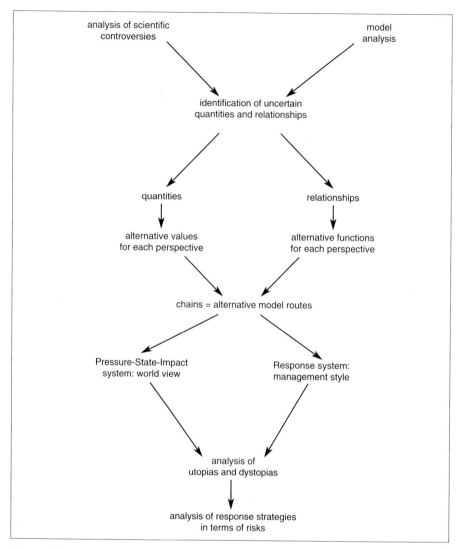

Figure 10.8 Flow chart of the methodology in Part Two.

Part Two Exploring images of the future

Towards integrated assessment of global change

11

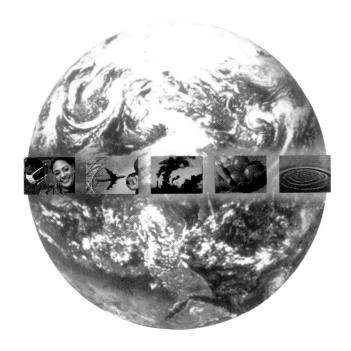

"Two dangers threaten the world: order and disorder."
Paul Valéry

11 TOWARDS AN INTEGRATED ASSESMENT OF GLOBAL CHANGE

Jan Rotmans, Bert J.M. de Vries, Marjolein B.A. van Asselt, Arthur H.W. Beusen, Michel G.J. den Elzen, Henk B.M. Hilderink, Arjen Y. Hoekstra and Bart J. Strengers

This chapter introduces Part Two of the book, which reports on experiments with submodels of TARGETS1.0 carried out to assess a number of global change controversies. The experiments include a range of utopias and dystopias to discover whether the problem at the core of each controversy is likely to occur, and if so, under what conditions. The hierarchist utopia, which reflects the assumptions behind many reputable scenario studies, is used as a reference case to explore issues such as population growth, demand for food, water and energy, the environmental consequences of these pressures, and a range of societal responses. The results of integrated experiments with the TARGETS1.0 model are presented at the end of this part of the book.

11.1 Introduction

Part One of this book described tools for performing integrated assessments of global change. The aim of Part Two is to gain insights by using these tools, both withthe separate submodels and with the integrated TARGETS1.0 model. The main goal of TARGETS is to put possible developments within the subsystems of the world into perspective in an integrated way. In this way we hope to provide a context for discussing global change and sustainable development. The quantitative modelling framework is used to support the qualitative framing of important issues. Though they are partial and limited in scope, the resulting images of possible global futures enable us to localise areas of tension and directions for sustainable development strategies.

One of the purposes of the simulation experiments with the TARGETS1.0 model is to develop scenarios. Scenarios are hypothetical sequences of events, constructed for the purpose of focusing attention on causal processes and decision points (Kahn and Wiener, 1967). They are descriptions of alternative futures, based on different views of how the world works and how it should be managed. Formal models can be used to give such images of the future a quantitative basis, so that way, they can serve as tools for developing strategies for sustainable development. Often scenarios are constructed from a single perspective on how people behave and how nature functions. We use three of the five perspectives of Cultural Theory: the hierarchist, the egalitarian and the individualist, to construct a variety of images of the future (see Chapter 10). This allows us to explore what is desirable, possible, and feasible.

This chapter describes the step-wise approach adopted to gain insights from the five submodels presented in Part One. It introduces the presentation of results and the experimental set-up, and three exogenous economic scenarios. We also briefly describe the procedure followed to deal with uncertainty. Next, we describe the hierarchist utopia to provide a background for the model analyses and to serve as a reference scenario for the submodel experiments presented in Chapters 12 to 16.

These chapters each start by summarising the present uncertainties and controversies in relation to population, health, energy, food, water, and global biochemical cycles. Then, the corresponding part of the hierarchist utopia is presented in more detail. Sensitivity analyses are used to indicate whether the controversy and the associated uncertainties significantly affect model outputs. Next, the egalitarian and individualist utopias are presented and differences in the various projections are explained. Analysis of the results allows a first crude evaluation of policy strategies, given the uncertainties in terms of risks. Finally, the insights from the model experiments are used to assess the major controversies. These chapters pave the way for the fully integrated experiments with the TARGETS1.0 model as a whole, described in Chapters 17 and 18. The concluding chapter, Chapter 19, summarises some of the lessons learned in building and using the TARGETS1.0 model.

11.2 Experimental set-up and uncertainty analysis

In the next five chapters, simulation experiments for the period 1900-2100 are presented for each of the five submodels discussed in Part One of this book. The aim of these simulation experiments with the submodels is twofold:

- to gain insight into the dynamics of the subsystems which are modelled and described in Part One, and
- to understand the consequences of the major uncertainties and controversies in terms of subsystem dynamics.

To this end the experiments presented in Chapters 12 to 16 are carried out with the various stand-alone submodels. Input scenarios for economic development, population, food, water, energy and environmental change provide a 'background' in the form of fixed inputs which are drawn from the reference scenario, that is, the hierarchist utopia. Because such a stand-alone analysis does not include the feedbacks from other submodels, the integrated experiments presented in Chapter 17 and 18 may provide significantly different outcomes.

Presentation of results
In the presentation of the simulation results in this and subsequent chapters, we

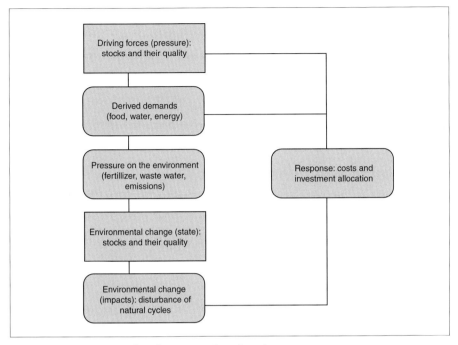

Figure 11.1 PSIR approach to the presentation of results.

follow the Pressure-State-Impact-Response (PSIR) approach introduced in Chapter 2. The major pressures come from population and economic activity. These two important driving forces lead to a series of derived demands for food, water and energy. These, in turn, can only be provided from natural resources such as productive soils, water resources and fossil fuel deposits. These processes of exploitation exert pressure on various parts of the natural environment. The resulting impacts interfere in many ways with the functioning of natural processes and are considered unsustainable in as far as they negatively affect adequate provision of both present and future human needs. *Figure 11.1* gives the framework for the sequence in which the submodels and the simulation results are presented.

Experimental set-up

It is necessary to choose a strategy for the experimental set-up because there are a huge number of possible experiments. The set-up of the simulation experiments in Part Two of this book is presented in *Table 11.1*. First, we introduce, in the next section of this chapter, three exogenous scenarios for Gross World Product (GWP). In the next step, the Population and Health submodel is run against a hierarchist background, that is, a hierarchist world view and management style for all other submodels. Hence, these experiments use the values for food intake per capita and access to safe water in the integrated hierarchist utopia as fixed, exogenous inputs.

This leads to three population projections, from which three welfare scenarios in terms of GWP per capita – or GWP/cap – are derived.

The submodels are run with world view W and management style M against a background B and with W, M, B either H (hierarchist), E (Egalitarian) or I (Individualist). Throughout this part of the book, we use a colour code for the utopian projections: blue for the hierarchist, green for the egalitarian and red for the individualist.

Applying three different world views and three different management styles within the Population and Health submodel produces nine scenarios, three of which are the previously mentioned utopian futures and six are semi-utopian futures (*Table 11.1*). In combination with three different backgrounds, these nine population projections give a total of 27 experiments. We use an UNCSAM uncertainty analysis (see Section 11.4) to explore the domain of model outcomes generated by these 27 experiments. Some of these experiments suggest a distinctly dystopian future, as will be discussed in the following chapters.

Chapter	(Sub-) model	Type of experiments	(Semi-) utopia	World view	Management style	Economy	Pop& Health	Background		
								Energy	Water	Land	Cycles
11	TARGETS	Integrated	Hierar	H	H	H	na	na	na	na	na
12	Pop& Health	Stand-alone	Hierar	H	H	H	na	H	H	H	H
			Egal	E	E	E	na	H	H	H	H
			Indiv	I	I	I	na	H	H	H	H
13	TIME	Stand-alone	Hierar	H	H	H	H	na	H	H	H
			Egal	E	E	E	E	na	H	H	H
			Indiv	I	I	I	I	na	H	H	H
14	AQUA	Stand-alone	Hierar	H	H	H	H	H	na	H	H
			Egal	E	E	E	E	E	na	H	H
			Indiv	I	I	I	I	I	na	H	H
15	TERRA	Stand-alone	Hierar	H	H	H	H	H	H	na	H
			Egal	E	E	E	E	E	E	na	H
			Indiv	I	I	I	I	I	I	na	H
16	CYCLES	Stand-alone	Hierar	H	H	H	H	H	H	H	na
			Egal	E	E	E	E	E	E	E	na
			Indiv	I	I	I	I	I	I	I	na
17	TARGETS	Integrated	Hierar	H	H	H	na	na	na	na	na
			Egal	E	E	E	na	na	na	na	na
			Indiv	I	I	I	na	na	na	na	na

H = Hierarchist, E = Egalitarian, I = Individualist, na = not applicable

Table 11.1 Set-up of the simulation experiments in Part Two.

In similar fashion, energy futures are explored with the Energy submodel TIME for three different population and economic developments but with the land, water and cycles submodels set according to the integrated hierarchist utopia. Then, developments in the water system are discussed on the basis of the AQUA submodel for a world in which the population, economy and energy futures can be hierarchist, egalitarian or individualist. Similarly, the TERRA submodel is used to analyse food and land-use developments; for these simulations the CYCLES submodel is run with hierarchist assumptions. Finally, the CYCLES model is used to simulate element fluxes with all other submodels being run according to the hierarchist, egalitarian and individualist assumptions. These experiments are, however, not the integrated utopias presented in Chapter 17 and 18, because the feedbacks from the CYCLES submodel to the other submodels are *not* taken into account. In this way we build on outcomes from previous simulation experiments in the presentation of the various submodel experiments. The integrated experiments in which the feedback interactions are operating are presented in Chapter 17 as integrated or 'full' utopias.

Uncertainty analysis
Chapter 10 discusses the importance of uncertainty in dealing with global futures and proposes multiple perspective-based model routes. In the following chapters we use the classical techniques of uncertainty analysis in addition to the perspective-oriented approach. The latter primarily addresses uncertainties in model structure while the former mainly deal with assumptions in model parameters. Complementary use of both methods leads to a more comprehensive uncertainty analysis. Here we briefly describe how we practically implement both uncertainty methods for the different submodels of TARGETS.

The implementation of multiple model routes in the TARGETS submodels means the incorporation of various model structures and parameter values. It is a matter of choice and definition whether one model structure differs from another. Strictly, a different model structure can be created by changing just one of its parameter values. Alternatively, one can construct a model in such a way that using a different model structure is merely a matter of parameterisation. Here, we use the strict definition: multiple model routes capture *any* change in model parameters or relationships. We distinguish three situations:
- changes refer to model parameters only, i.e. the mathematical expressions as such are not changed,
- changes refer to mathematical expressions for a specific relationship between two model variables, and
- changes refer to the use of an alternative set of equations to represent a specific part of the system.

The classical uncertainty methods use Monte Carlo based methods in conjunction with regression and correlation analysis. Here we use the software package

Uncertainty: some examples

An example of an uncertainty in a parameter which is largely based on ignorance about the climate system, and hence on scientific controversy, is the climate sensitivity parameter in the CYCLES submodel (Chapter16).

Assessments of the long-term costs at which fossil fuels can be supplied are intrinsically uncertain: only when a field is fully exploited, can its geological and economic history be fully understood. In the Energy submodel model we use different supply cost curves for gas and use the difference between the two areas as the parameter domain. *Figure 11.2* gives, as example, the multipliers for the three perspectives with which the capital-output ratio of gas production is multiplied as resource depletion proceeds. The area between the two perspective-based limits is used for the uncertainty analysis.

In the fertility module the effectiveness of population policies on the use of contraceptives is extremely uncertain. Hence, we use different relations to model the three perspectives. The hierarchist thinks the use of contraceptives is mainly determined by population policy options such as family planning. Considering the other two perspectives, the course of human development expressed by the HDI has a strong influence on the use of contraceptives. Another example of different relationships is the determinants of health within the health model. The egalitarian and hierarchist see the process of being exposed to e.g. malnutrition as causing one or more diseases. The individualist views health as being a consumer good with the only determinant being the level of health services.

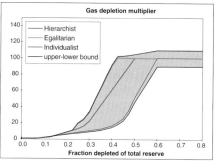

Figure 11.2 Assumed rise of the capital – output ratio for natural gas with depletion, used for uncertainty analysis.

UNCSAM (UNCertainty analysis by monte carlo SAMpling techniques) as developed at RIVM and specially designed for performing sensitivity and uncertainty analysis on a wide range of mathematical models (Janssen *et al.*, 1992). Major difficulties in applying such methods include determining the set of parameters, the uncertainty domain and the type of probability distribution.

An additional difficulty arises since we want to analyse the same set of uncertainty sources here (model parameters, mathematical relationships) as in the perspective-oriented approach and classical techniques do not usually address uncertainties in different mathematical relationships. We solved this problem by reducing mathematical relationships to model parameters and applying the following basic rules to determine the uncertainty domain for the different types of uncertainty:

- for model parameters it is given by the range of values used for the implementation of the three perspectives,
- if alternative mathematical expressions are used, it is given by the space spanned up by these expressions, and
- if alternative sets of equations are used, an independent estimate of the uncertainty domain of the parameters is used for each of them.

This procedure is in line with our assumption that the different perspectives cover the extreme values. In most cases the values used for the implementation of the perspective-based model routes are themselves often single-point and biased estimates. Hence, we have extended the parameter domain in the following way:

$$uncertainty\ domain = [\ a - MIN(\varepsilon \times a, \varepsilon \times (b-a)),\ b + MIN(\varepsilon \times b, \varepsilon \times (b-a))] \quad (11.1)$$

with a, b the lowest and highest value comprised in any model route, and ε the uncertainty factor set at 0.1. A uniform distribution has been used in all of our experiments.

11.3 Economic scenarios

The TARGETS1.0 model does not take into account any endogenous formulation of economic growth on the basis of capital, labour and technology dynamics. Instead, it is driven by three exogenous scenarios for Gross World Output (GWP), as provided by an economic scenario generator (Chapter 3). This generator models the economy as a capital stock which produces industrial output with a fixed capital-output ratio. Part of it is re-invested and it is this re-investment rate which is used to generate growth trajectories for industrial output. The remainder of industrial output is allocated as investments among the food, water and energy systems, service capital, and consumption-related capital. Investments in service capital are derived from the indicated service output per capita, which is determined by per capita industrial output (Meadows *et al.*, 1974). Health services and expenditures for water and energy are considered part of the service sector output; the remainder of the service output is called 'other services'. The capital-output ratio for the service sector excluding the capital stocks for water and energy provision is assumed to be constant. The remainder of GWP minus food expenditures is available for what in the present context is defined as consumption.

We use the growth rate in other services and consumption as an indicator of the system's potential to sustain growth in welfare while maintaining the basic life support systems. Of course, it is only a first crude approximation. In the future, we hope to link these calculations to a macro-economic model which brings in the rigour and consistency of an input-output framework. More accurate estimates of the costs of supplying food, water and energy are needed. Besides, other costs which may be associated with sustaining the life support systems are not accounted for, e.g. the costs of adapting to rising sea levels. Nevertheless, we look upon this exercise as an interesting approach to identifying the trade-offs between maintaining certain rather well-defined subsystems (health, food, water, energy) and expanding other economic activities.

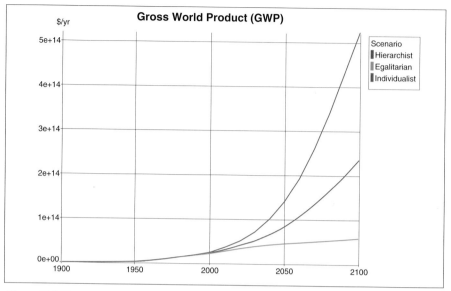

Figure 11.3 Exogenous scenarios for Gross World Product (GWP) for the hierarchist, egalitarian and individualist utopias.

For the hierarchist utopia assumptions about economic growth reflect values which are comparable with the IS92a scenario of the IPCC, i.e. medium economic growth (Leggett *et al.*, 1992). To arrive at an egalitarian utopia, we have assumed that the desired economic growth rate is much lower (comparable to the IPCC-IS92c scenario), and that a larger proportion of services reflect a society in which human values are more important. With regard to the individualist utopia, it is assumed that the economic growth rate is high (comparable to the IPCC-IS92e scenario). The economic scenario generator then provides three GWP trajectories (*Figure 11.3*), which are used as inputs for the simulation experiments.

11.4 The hierarchist utopia: a reference future

In the hierarchist utopia both the world view and the management style correspond to the hierarchist perspective; this generally reflects the opinions put forward by a number of prominent institutions and scientists (Rayner and Malone, 1996). It is also the perspective that, by its very nature, is expressed most explicitly in terms of projections, scenarios and assumptions. Hence, we have relied on scenario studies by institutions such as the UN (United Nations), the World Bank, FAO (Food and Agriculture Organisation), IPCC (Intergovernmental Panel on Climate Change) and WHO (World Health Organisation) to construct the preferred hierarchist world view and management style. The resulting utopia is considered to be a reference future. It is discussed here in order to provide a background for the experiments with submodels of TARGETS1.0 in the following chapters.

Such reference scenarios are given various names: 'Conventional Wisdom', 'Business-as-Usual' or simply 'medium'. Usually, this means that the underlying models are fed with parameter values and assumptions on which there is more or less consensus within the scientific and policy communities[1]. The future is often only an extrapolation of past and current trends. At a deeper level, reference scenarios also reflect tacit agreements on how the world should be interpreted and interacted with and how to deal with ignorance. For example, prices often play the conventional role in relation to supply and demand whereas innovations are introduced as exogenous parameter changes. Such implicit assumptions are not easy to uncover and they often change slowly because they incorporate perceptions of problems and explanatory frameworks. This also holds for the TARGETS1.0 model and its submodels. As far as the TARGETS submodels are concerned, we have attempted to be as explicit and transparent as possible.

More specifically, we have used the assumptions in the IS92a scenario as presented by the IPCC (Leggett *et al.*, 1992) to represent a hierarchist outlook on the future of the world. This scenario has been constructed as a reference scenario in the context of the climate change debate and it reflects a Business-as-Usual or Conventional Wisdom view on global population, economy, energy and emission developments. As such it is a warning against a 'do-nothing' policy and it is only partly justified to use it to represent the hierarchist utopia. Many government environment departments will not see it as a utopia at all. Indeed, it is currently being used as a yardstick against which governments can assess the need for climate policy and thus formulate mitigation measures. However, many institutions not related to environmental policy may still view its major features as realistic, desirable and acceptable.

In the hierarchist utopia there is a population of 11.6×10^9 in 2100 (*Figure 11.4a*), comparable to the medium projection of the UN (1993), with a life expectancy of 81 years, which is higher than current levels (*Figure 11.4b*). Water gets scarcer and water quality continues to deteriorate. Although total water withdrawals continues to increase, consumptive use slowly stabilises and pollution stops increasing. Average food intake is almost 25% above the 1990 level. A transition towards renewable energy resources has started, but fossil fuels still account for the bulk of energy supply. The resulting emissions cause environmental change: a 2.6 °C temperature increase and a sea-level rise of 0.6 m between 1900 and 2100. These changes are comparable to recent IPCC estimates (Houghton *et al.*, 1995).

What processes and developments explain this image of the future? A combination of family planning practices allows a further reduction in fertility rates

---

1   This type of scenario can be associated with 'surprise-free' futures (Kahn and Wiener, 1967) and also with 'muddling-through' (Lindblom, 1975).

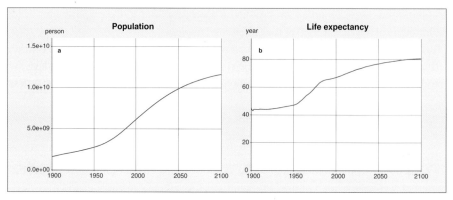

Figure 11.4a-b Population and life expectancy for the integrated hierarchist utopia.

to levels prevalent in countries like the USA (World Bank, 1993) today. Population is still growing in 2100, but there are clear signs of stabilisation. Health expenditures are rising to almost 10% of GWP and in combination with better and more widespread provision of food and safe water, this causes a significant reduction in disease incidence and a shift towards increased longevity.

It is assumed that all demands are met by the supply systems. Food consumption per capita continues to rise to an average of about 3,500 kcal/day per person, which is comparable to the present level in Western-European countries. The fraction of non-vegetable products in the diet increases slightly. Vegetable production, as a result, increases rapidly, but growth levels off in the second half of the next century (*Figure 11.5a*). This increase is largely made possible by an extension of irrigated land and an increase in the application of fertiliser – usually referred to as 'intensification of agriculture'. Irrigation is mainly responsible for the increase in water use. Arable land and logging in the developing countries expand at the expense of tropical forests. The

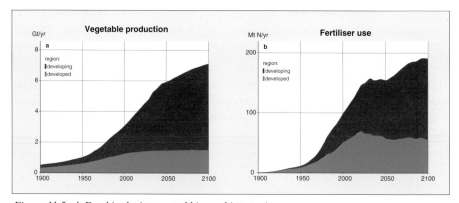

Figure 11.5a-b Food in the integrated hierarchist utopia.

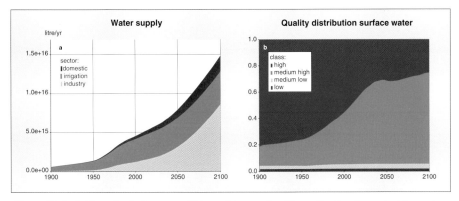

Figure 11.6a-b Water in the integrated hierarchist utopia. Water Quality is ranked from class A (suitable for all functions: high) to class D (low).

50% increase in total fertiliser use causes significant increases in anthropogenic nitrogen fluxes (*Figure 11.5b*). Higher nitrogen concentrations contribute to a deterioration of water quality. This in turn affects public water supply, which may negatively influence human health. The number of people suffering from malnutrition declines because on average food becomes more available. Chapter 15 provides a more detailed discussion of the hierarchist utopia with respect to food, while Chapter 12 discusses the impact on human health in quantitative terms.

Global water use can be expected to increase, mainly because of continuing growth in industrial water use and despite an increase in water use efficiency (*Figure 11.6a*). However, water consumption, i.e. the percentage of water withdrawal that gets lost through evaporation, stabilises. The increased use of water, especially by industry, results in a shift by water bodies to lower quality classes. Waste-water treatment, however, curbs the growth in the pollution load on the environmental

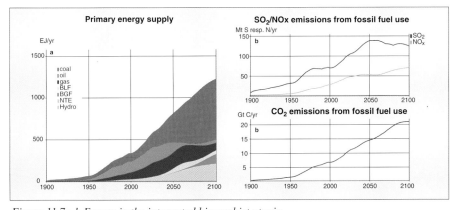

Figure 11.7a-b Energy in the integrated hierarchist utopia.

235

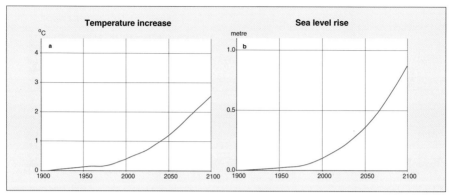

Figure 11.8a-b Increase in global average surface temperature and sea level in the integrated hierarchist utopia.

system. The net result is that water quality deteriorates significantly, with the fraction of water of the highest quality declining from 80% to about 30% (*Figure 11.6b*).

The exponential growth in levels of economic activity leads to increased demand for energy – three times the present level by the year 2100. Energy efficiency continues to reduce the use of energy per unit of economic output, partly as the result of rising prices (*Figure 11.9b*). In response to this, the energy system expands to supply the world with large amounts of oil and gas in the first half of the next century. Due to depletion of these resources, a transition to coal and renewable resources begins (*Figure 11.7a*). The resulting energy-supply pattern causes a rise in CO_2 emissions and a stabilisation and later on a decline in SO_2 and NO_x emissions (*Figure 11.7b*). Chapter 13 gives a more detailed account of these results.

Increasing anthropogenic CO_2 emissions result in an accumulation of CO_2 in the atmosphere of up to about twice the present level in 2100. This results in a

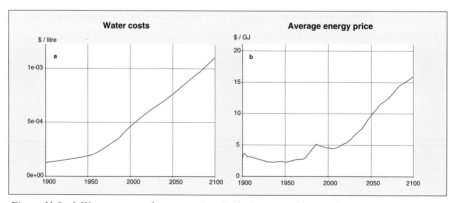

Figure 11.9a-b Water costs and energy prices in the integrated hierarchist utopia.

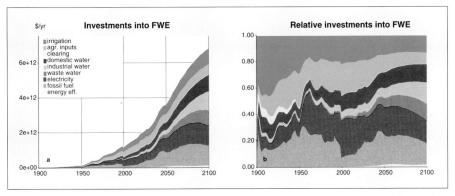

Figure 11.10a-b Investment flows for energy, water and food in the integrated hierarchist utopia.

temperature increase of about 2.6 °C in 2100, which is comparable to the medium range in IPCC estimates (*Figure 11.8a*). This estimate takes the cooling effect of aerosols related to SO_2 emissions into account. Chapter 16 provides a more detailed discussion of this part of the hierarchist utopia. The sea-level rise of about 0.9 m between 1900 and 2100, associated with this scenario, is the result of several processes such as thermal expansion, groundwater withdrawal and melting of glaciers and ice sheets (see *Figure 11.8b*).

One of the responses to increasing pressures and the resulting resource scarcity is a rise in the cost of supplying water and energy as shown in *Figure 11.9a and b*. The absolute and relative amount of investments for the food-water-energy supply systems associated with the production levels indicated above are shown in *Figure 11.10*. It can be seen that a declining fraction goes to food supply, whereas the fraction needed for the energy system slowly rises to 55%, after which it declines to 40%, which is comparable to the present share of investments in energy. Water supply requires an increasing proportion, mostly for industrial water supply and waste-water treatment. The investments required for the food, water and energy-supply systems are assumed to be part of industrial output (Chapters 3 and 17).

These results give an impression of a global future in which the hierarchist world view turns out to be correct and a corresponding set of policies is being implemented. The TARGETS1.0 modelling framework has enabled us to bring some quantitative consistency into such explorations of the future. In a way, such experiments serve as a kind of 'existence proof' in the sense that the resulting image of the next 100 years does not violate basic insights about population and resource dynamics. However, there are numerous other images of the future that also do not violate those insights. Moreover, important insights may have been left out because of ignorance or model-related restrictions, e.g. aggregation level. Hence, it is important to find criteria for selecting interesting scenarios. As explained in Chapters 3 and 10, we do this by constructing divergent sets of model parameters and equations on the basis of particular world views and management styles. The choice of these sets is derived from the existing scientific and political controversies as they emerge in the literature.

Hierarchist 'muddling through' – a narrative

One can use a narrative to put flesh on the above snapshots of a hierarchist utopia. As the world is increasingly confronted with nationalism, fundamentalism and anarchy, there is a growing awareness of the need for international collaboration. Effective policies for regulating trade and capital flows are implemented. Less industrialised nations get widespread support to combat poverty and their governments acknowledge the need for more effective management. Investments in education and infrastructure, among them access to safe water and sanitation, get high priority; so do family planning programmes. Multinational companies agree to comply with world-wide labour, safety and environmental regulations.

Economic growth, although modest, is sufficient to keep the health and resource transitions going. This is partly possible because people in the rich countries are willing to restrain their consumption of material goods as new ways are found to cope with unemployment. International co-operation in the fields of agriculture, energy and environment leads to the transfer of, for instance, solar technology and regulation of the use of pesticides and other toxic chemicals. Farmers in less developed regions get sufficient access to fertiliser and machinery to increase yields and meet food demand. In rich regions of the world, afforestation of the former agricultural lands leads to an expansion of forested areas. Agreements on the conservation of tropical forests are partly successful: about one quarter of the tropical forests in the developing regions remain in 2100, due to more sustainable forestry practices and regulated food trade.

The dematerialisation of the economy continues. Carbon emissions keep rising but the resulting climate change turns out to be rather mild, partly because timely adaptive actions are undertaken in endangered regions. This is also the reason why the world community takes no action when India and China greatly expand their coal production. Non-carbon sources of energy penetrate the market in the second half of next century.

Mainstream institutions are attacked from two sides. Egalitarians accuse them of 'muddling through'. They reproach them for letting large areas of tropical forests disappear and taking unacceptable risks with regard to climate change. From the other side, individualists fight the environmental regulations and bureaucracies which are introduced. They argue that such an approach hampers economic growth in the interests of avoiding long-term risks for which no firm evidence is available.

Population and health in perspective

"If we can anticipate at all, it can only be done by considering demographic trends in a broader socio-economic, cultural, and biological context."
W. Lutz, Future demographic trends in Europe and North America (1993)

12 POPULATION AND HEALTH IN PERSPECTIVE

Henk B. M. Hilderink and Marjolein B. A. van Asselt

The controversy at the core of this chapter is whether current and anticipated trends in economic and social development, combined with projections for population growth, are compatible with maintaining good health for all. Whereas there has been a doubling of life expectancy over the past century, there are still considerable uncertainties about how the health transition will continue. The way in which such trends are interpreted depends to a large extent on the perspectives held by scientists and policy-makers. In this chapter, developments in population that are related to policy and health dynamics are interpreted according to the three perspectives outlined in Chapter 10 and are compared to UN projections. The consequences for population size and the health risks of stagnating economic growth and of food and water crises are assessed in four cases.

12.1 Controversies related to population and health

World population has more than tripled between 1900 and 1996, rising from about 1.6×10^9 (Kuznets, 1966) to 5.8×10^9 (UNFPA, 1996), i.e. an average annual growth of 1.3%. Such a rate of growth is extremely rapid by any historical perspective. Population growth and its potential consequences for people and the environment have long been an issue of concern. Since the time of Malthus (1789) the population issue has been the subject of a furious debate, both in the scientific and the policy community (Ehrlich, 1968; Cohen, 1995). Specific concerns about population growth have varied from one generation to the next. Shortly after World War II, for example, the issue was stated in terms of 'overpopulation', while in the mid-1960s it was framed as 'the world food problem' (see also Chapter 15). More recently, the population issue has been addressed in the context of sustainable development (Eberstadt, 1995). The fundamental controversy still centres on the size of the future world population and its consequences, namely: "*Will future population developments result in overpopulation in environmental and/or socio-economic terms?*"

In addition, it has become clear that in some of the world regions the environmental assets essential for human health are scarce and becoming scarcer (Haines, 1991; McMichael, 1993; McMichael and Martens, 1995; WCED, 1987). At the same time it can be said that with regard to human health, the 20th century has been a tremendous success, despite rapid population growth (Eberstadt, 1995).

Roughly speaking, global life expectancy has more than doubled, from about 30 in 1900 (Preston, 1976) to 66 in 1996 (UNFPA, 1996). These different points of view are reflected in the following controversy on the future of human health: *"Will a healthy life for all be possible, given current and anticipated socio-economic and environmental trends?"*. While the health level of a population is an important factor underlying both morbidity and mortality patterns, both fertility and health are crucial determinants of population size and structure. The two issues are thus clearly related and can be combined into one single controversy reflecting a wide variety in points of view on the health transition (see Chapter 4), namely: *"Is continuing population growth irreconcilable with a healthy life for all, given current and anticipated socio-economic and environmental developments?"*. There are different interpretations of the uncertainties with respect to transitions in demography and health (Cohen, 1995). This chapter describes various perspectives and the associated uncertainties. Such a pluralistic approach accepts the inevitable existence of uncertainty and might thus enable us to address the controversies in an adequate and balanced manner. This section starts with an overview of the uncertainties concerning interpretation and projections with regard to population and health issues.

12.2. Population and health uncertainties

The demographic data do not allow the health transition to be described unambiguously: demographic and epidemiologic transitions are open to a variety of interpretations concerning the underlying processes. Considering the controversy as aggregated chains of uncertainties offers a framework for structuring the related uncertainties. Using this framework allows analysis of each controversy at the level of a specific type of uncertainty. In interpreting a chain of uncertainties, consistency can be reached by relating uncertainties to three perspectives. For population and health, the uncertainties are clustered into scientific and policy uncertainties. Major scientific uncertainties relate to ethical aspects: *"Are there any limits to population growth?"*, to demographic aspects: *"At what level of society does fertility behaviour change?"*, to public health determinants: *"What are the main determinants of disease?"* and to socio-cultural aspects: *"Is the demand for health an important societal issue?"*. The policy uncertainties taken into consideration relate to disagreements within health-policy: *"What are appropriate measures for controlling fertility, which are the major health determinants, and what is the relevance of health to policy?"*. For the sake of brevity, we have summarised the various types of uncertainties, their characteristics and the associated viewpoints in *Table 12.1*. Much of the literature we have used for analysing population and health uncertainties are included in this table. For a more comprehensive and elaborate description of these uncertainties, we refer to the article by van Asselt *et. al* (1996).

Uncertainties	Description	Points of view	Explanation
Scientific uncertainties			
Fertility			
* ethical uncertainties	Are there limits to population growth?	1. no limits	1. for example (Eberstadt, 1995; Kasun, 1988; Petty, 1899; Simon, 1980; 1981)
		2. physical limits	2. for example, (Brown et al., 1996; Ehrlich, 1968; Malthus, 1789; Meadows et al., 1991; 1972)
		3. limits in terms of environmental quality and quality of life	3. for example, (Coffin, 1991; Gore, 1992; Hardin, 1968; Keyfitz, 1993; King, 1990; King and Elliott, 1994; Mitchell, 1991)
* demographic uncertainties	Which factors trigger structural change in fertility behaviour?	1. micro view	1. individual characteristics (Murdoch and Oaten, 1975)
		2. macro view	2. global socio-cultural characteristics (UNDP, 1995)
Health			
* socio-cultural uncertainties	How is health perceived?	1. health as consumption good	1. market-oriented
		2. health as human capital	2. cost-benefit approach
		3. health as human asset	3. social approach
* public health uncertainties	What are the main determinants of disease?	1. social-economic and environmental	1. broad health determinants. (Hurowitz, 1993; King, 1990)
		2. health services	2. (Scrabanek and McCormick, 1992)
		3. ageing as health determinant	3. (Olshansky et al., 1990; Olshansky et al., 1991; Verbrugge, 1989)
Policy uncertainties			
Fertility	What are appropriate means to control fertility?	1. family planning	1. (Robey et al., 1992)
		2. human development	2. (Becker, 1991; Pritchett, 1994)
		3. anti-abortion	3. abortion should be prohibited on religious and cultural grounds
Health	How can health be improved adequately?	1. selective medical care	1. (Walsh and Warren, 1980; World Bank,
		2. comprehensive health care	2. (WHO, 1978)
		3. curation	
		4. prevention	

Table 12.1 Uncertainties associated with the population and health controversy.

12.3 Perspectives on population and health

To be able to address the controversy on population growth and human health adequately, we need to identify coherent clusters of interpretations of fertility and health uncertainties. We use the three perspectives of the hierarchist, the egalitarian, and the individualist heuristically to construct coherent descriptions of the demographic and epidemiological dynamics. This section gives descriptions of elements of these three perspectives on fertility and health. Viewpoints expressed in actual debates usually represent combined positions associated with these stereotypical perspectives.

The hierarchist perspective

Three elements in the description of the hierarchist have guided our effort in arriving at an interpretation of the scientific uncertainties, namely: (i) the myth of nature, i.e. robust within limits, (ii) the perception of human nature as being sensitive to authoritative incentives, and (iii) the driving force, i.e. stability. Furthermore, hierarchists are generally associated with the viewpoints of authoritative institutions, which in the field of population and health are, for example, the WHO and the UNDP. Applying the hierarchist myth of the world to the ethical uncertainties leads us to the interpretation that vigorous population growth will eventually be disastrous if the Earth's physical limits and thus its carrying capacity are exceeded. Demographic uncertainties are seen from a 'macro' view, in which fertility behaviour is said to be primarily determined by the attitude of authorities, such as the State and the Church. In line with the hierarchist ideal of stability, we associate the hierarchist interpretation of the socio-cultural uncertainties with the 'health as human capital' view, in which health is conditional to economic growth. Improvement in health makes an important contribution to economic development; health is thus narrowly defined as the absence of disease. Efforts in health are primarily evaluated in economic terms, with the focus on the working population, and the hierarchist thus advocates that human health as deserving of policy interests and action. Because organisations such as the WHO stress the importance of the quality and accessibility of medical care in determining human health, the hierarchist can be associated with the 'health-services' point of view. Here, access to and a high quality of medical care are seen as major determinants of health status. Adherents to this view underpin their argument with studies showing that increased expenditure on medical care is a determinant of lower mortality and morbidity levels, especially in the later stages of the health transition.

The need for a population policy is based on the comparison of the actual population size with the maximum size derived from the carrying capacity. The preference for control helps us to associate the hierarchist attitude towards population policy with the 'family planning' view, in which high fertility is said to be predominantly a consequence of inaccessibility or prohibitive costs of contraceptive services. The hierarchist attitude towards health policy is associated with the 'cost-benefit' approach. Health services should be oriented towards cure and primarily directed to those groups in society which are the most relevant for the functioning of the socio-economic system. We have attributed the 'anti-abortion' attitude to the hierarchist but fully admit to having had serious problems in defining the hierarchist attitude towards abortion (van Asselt and Rotmans, 1996).

The egalitarian perspective

For the egalitarian interpretation, the following elements are of considerable importance: (i) nature is fragile, (ii) human nature is malleable, (iii) the ecocentric attitude, and (iv) equity is a driving force. In this context we argue that the egalitarian associates rapid and large population growth with societal and environmental problems such as poverty, hunger, environmental degradation and resource depletion. In the egalitarian view the tolerable limit of population levels is thus coupled to ecological and social criteria. As to public health uncertainties, the egalitarian myth on nature supports the interpretation that biological limits operate on life in general, including human life. For egalitarians it is thus highly unlikely that life expectancy at birth will exceed the currently estimated biological upper limit. Considering this aspect combined with the perception of people as part of a complex whole, we associate the egalitarian with the 'broad health determinants' point of view that human development is the main determinant of the health conditions for the population. Adherents of the broad health determinants view refer to the burden of studies in Western European countries revealing a correlation between higher income and higher social status, and longevity and good health. Furthermore, this view maintains that environmental developments, such as climate change, ozone depletion and land degradation pose problems for human health. For example, changes in global climate may alter the distribution of vectorborne diseases or the productivity of agriculture (Martens *et al.*, 1995a, 1995b). With regard to the interpretation of the demographic uncertainties, the egalitarian can be associated with the meso approach, in which both micro- and macro-processes influence the circumstances in which people live and make their choices. Concomitant with the egalitarian ideal of equity, we associate the egalitarian interpretation of socio-cultural uncertainties with the point of view characterised as 'health as human asset'. Here, an unhealthy life is considered inhuman and the primary goal of governing institutions should be to promote health as a general asset of society. The main objectives of (public) health services are the prevention of premature death and the improvement of the quality of life. Cost-effectiveness alone cannot be a criterion in evaluating efforts into health improvement.

The egalitarian management style can be characterised as preventive. Socio-cultural prospects and the environmental outlook have to be taken into consideration when deciding whether population policy is necessary. The preference for preventive measures together with the egalitarian perception of human choices conditioned by their circumstances makes it plausible to associate the egalitarian with the 'human development' point of view as population policy. This kind of policy aims at improving the status of women, e.g. stimulating better education, participation in the employment market and empowerment. In health policy, the egalitarian is associated with 'comprehensive health care', which permeates many, also non-medical, aspects of daily life, such as good housing and sanitation, pollution control, road improvements, public transport, a safe work environment, stable interpersonal relationships, sufficient income and education. This kind of health policy centres on the prevention of illness. Referring to groups which are in general characterised as egalitarian, such as women's rights groups, we advocate that besides a belief in legalisation, safe and reliable abortion facilities are provided. It should, however, be noted that the egalitarian expects people to abandon the use of abortion and take preventive birth-control measures (van Asselt and Rotmans, 1996).

The individualist perspective
Scientific uncertainties of population and health are interpreted from the following assumptions: (i) nature is robust, (ii) human nature is self-seeking, (iii) economic growth and a dominant role of the market are the driving forces, and (iv) human ingenuity, especially in technology is important. In this view there are neither limits to population growth nor to the length of the human lifespan. People are looked upon as a resource. Furthermore, individualists argue that notwithstanding the so-called 'population boom', the human population today lives longer on the average, eats better, and produces and consumes more than at any other time in the past. At the same time air and water quality in developed countries have improved remarkably, signifying environmental progress. Associated with this point of view are scientists (e.g., Murdoch and Oaten, 1975) who, referring to the observed propensity for

	Hierarchist	**Egalitarian**	**Individualist**
Scientific uncertainties			
Fertility			
* ethical uncertainties	physical limits	socio-economic and ecological limits	no limits
* demographic uncertainties	social, economic and cultural environment (macro view)	both environmental and individual circumstances (meso view)	individual characteristics (micro view)
Health			
* socio-economic uncertainties	health as human capital	health as human asset	health as consumer good
* public health uncertainties	health services	broad health determinants	ageing
Policy uncertainties			
Fertility	family planning anti-abortion	human development	legalisation of abortion
Health	selective health care, cure	comprehensive health care; prevention	market-oriented

Table 12.2 Overview of the population and health uncertainties associated with the three perspectives.

families in prosperous societies to have fewer children, argue that continued economic growth increases affluence in the world sufficiently to stabilise population growth. In relation to the demographic uncertainties, the individualist is associated with the 'micro' view, which holds that if individuals prefer a small family, they will find the means by which to plan their families. Reference is made to France, where total fertility was already at low levels when modern contraception methods had not yet become available and where family planning was condemned by public authorities. Changes in either the fertility levels or health status are induced by the socio-economic situations of individuals. The market mechanism will then ensure the availability of contraceptives, abortion facilities and private health services if needed. The individualist can thus be associated with the 'health as consumer good' view, stating that individuals ascribing value to good health will, if affordable, determine their own level of health by buying their health services. Individualists advocate a *laissez-faire* attitude, allowing optimal functioning of the market mechanism. They can be associated with the view of considering population and public health policy as simply redundant. Individuals should be given full freedom to make their own choices. Hence, individualists advocate full legalisation of abortion, the facilities for which, however, should be provided by the market system.

12.4. Three images of the future

The three perspectives within the population and health controversy are used to create multiple perspective-based model routes in the population and health submodel (see Chapters 4 and 10). The uncertainties associated with the population and health controversy are great enough to provide room for fundamentally different model descriptions. For example, while morbidity and mortality are described in the egalitarian and hierarchist routes as causal chains from exposure to disease and mortality, the individualist route is determined primarily by economic conditions and incentives. Obviously, multiple model routes are only valid reflections of reality if they represent the historical observations and available data.

The three perspectives on the population and health dynamics and policy options yield significantly differing images of the future, which we evaluate in terms of the major indicators, namely: total fertility rate, population size and life expectancy, expressed in life expectancy and associated Disability-Adjusted Life-Expectancy or DALE (Chapter 4). Besides these indicators, another important outcome is GWP per capita; the three GWP-trajectories of the *economic scenario generator* are used as exogenous economic inputs. It should be noted that in this chapter the assumptions with respect to food and water availability and environmental change do not vary among the various perspectives (see Chapter 11). The implications of this will be discussed at the end of this chapter. *Figure 12.1a-e* gives a general overview of the three utopian projections.

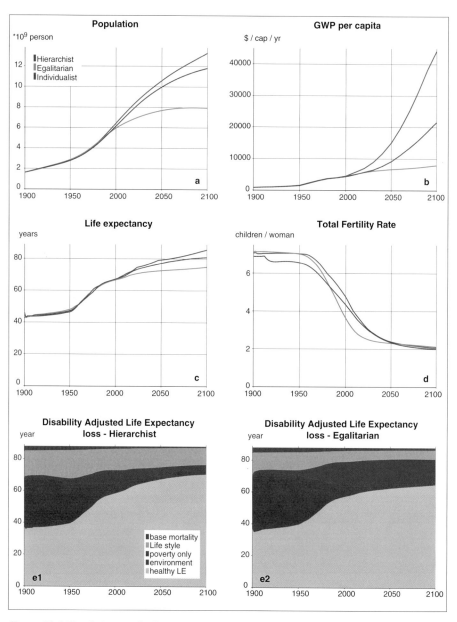

Figure 12.1 Simulation results for the three utopias.

Hierarchist utopia

The hierarchist utopia presents a future in which the population increases, but at a declining rate. By 2100 the population is 11.8×10^9, about a doubling of the present size at an average GWP of 22,000 dollar per capita. This is the result of two major developments. On the one hand, fertility is declining due to an increasing use of contraceptives, a diffusion process stimulated by family planning programmes and increased socio-economic developments. On the other, there is an increase in life expectancy, which stabilises at 81 years at the end of the next century, comparable to the present female life expectancy in Japan (UN, 1995). This latter development is due to a better health status, improvements in food availability and access to safe drinking water for the population as a whole. The general advance in human development causes a substitution of low SES-related health risks, i.e. poverty, malnutrition, safe drinking water, by high SES-related health risks and brings about a relatively high increase in smoking and blood pressure as disease determinants (*Figure 12.1e*). Mortality levels are stable after 2025, namely a crude death rate of 9 deaths per 1000 persons, comparable to the present situation in the Netherlands (World Bank, 1995). With fertility falling to around the replacement level by the end of the 21st century, a stable population level in the 22nd century can be anticipated. Due to a successful decline in fertility and epidemiological transitions, degreening and longevity result in an ageing population in the hierarchist future. For 2100 the hierarchist utopia shows a decrease in the percentage of young people from the present 35% to 20% (a level comparable to the present situation in the United Kingdom), and an increase in elderly people from the present 13% to an unprecedented high level of 25%. Degreening is evoked by the fast decline in fertility levels at the beginning of the next century: 0.6 children per woman per 10 years in the 1995-2040 period. With reference to the population and health controversy, the simulation results associated with the hierarchist utopia presents a future with a population comparable to the medium UN projection (1995) and a health status beyond levels presently observed.

Egalitarian utopia
In the egalitarian utopia the future population stabilises at a rather low level of 7.9×10^9 in 2100 and a GWP level of $ 8000 per capita, due to a very fast fall in the number of births and a moderate increase in the average lifespan from 66 years in 1995 to 76 years in 2100. The health situation in 2100 in the egalitarian future is comparable to the situation presently observed in countries such as the United Kingdom and the United States (UN, 1995). The improvement in health results from an upward trend in health services. However, in absolute terms the investments in health in the course of the next century are relatively small due to a moderate economic growth during the whole 21st century. The relatively minor improvement in life expectancy, compared to the hierarchist and the individualist utopia, can be partly explained by the fact that the egalitarian perspective considers people to be fragile creatures. Notwithstanding the improvement in water and food availability, the total health risks associated with low socio-economic status remain high, also reflected in the morbidity and mortality components. The decline in fertility results from a fast and early diffusion of the use of modern contraceptives resulting in a

reproductive behaviour comparable to the United States today (World Bank, 1993). Egalitarian policy incentives towards better education for all, and women in particular, enhance the collective awareness and the associated willingness to involve global concerns in individual decisions. The population structure is significantly affected by the changes in fertility and mortality patterns. The degreening process appears early in the next century due to the a rapid decline in fertility at a rate of 1.0 child per decade in the next 40 years. Health services are equally distributed among the population, as a result of which the elderly population is in a good health. By the end of the next century, only 18% of the people older than 65 years suffer from illness. The egalitarian utopia shows that a healthy life for all, including the elderly, is possible in a situation of low economic growth and a small population.

Individualist utopia
Notwithstanding a successful transition in fertility – total fertility falls to the replacement level in the course of the 21st century – the population in the individualist utopia continues to grow to 13.3×10^9 people in 2100. This growth is primarily due to an enormous extension of the life expectancy, from 66 years at present to 86 years in 2100. Such an expectation is unprecedented. Mortality levels for each age group decline very fast as a result of the high health-services level. The health investments associated with these developments show a large increase from the present $ 240 to $ 2000 per person per year in 2100. These large investments in health services are feasible due to the high economic growth leading to a GWP level of almost $ 45,000 per capita. The individualist utopia is a future in which the human development index, comprising literacy, life expectancy and income (see Chapter 4), reaches the maximum level before the end of the next century. The increased socio-economic development is the major driving force underlying the changes in fertility and mortality patterns. While the individualist utopia features successful fertility and health transitions, the fertility and mortality patterns evoke a continuous trend towards degreening and ageing. By the end of the 21st century the percentage of young people fluctuates around 18% and the elderly form 28% of the total population. It is worth noting that the population structure in percentages closely resembles the egalitarian utopian population structure. However, the transitional pathways and the ultimate population in 2100 differ significantly: the individualist population size is, roughly speaking, almost twice the egalitarian one. In this case, the simulation results indicate a large population in good health, in a situation of high and continuous economic growth.

12.5. The plausibility of the projections

The spectrum emanating from the three utopias runs from 7.9 to 13.3×10^9 people in 2100, while the life expectancy projected for 2100 varies between 76 and 86 years

(*Figure 12.1c*). Our projections when compared with the UN projections (UN, 1993, 1995), of which the 2100 values range from 6.4×10^9 as a low projection to 17.6×10^9 as a high projection, shows a smaller range. Why do our projections span a smaller range than the ones provided by the UN? First of all, the utopian projections all assume that the fertility transition will succeed; it is only the onset and the transition rate that differ among the three perspectives. In designing their scenarios, the UN assumes fertility levels to be constant for the next 50 to 100 years and vary between 1.6 in the low-fertility case to 2.6 in the high-fertility case for the second half of the next century. The high UN projection comprises futures in which the fertility level is far above replacement level, and thus yields a higher population projection then our utopian projections. Secondly, our population and health submodel accounts for the recognised mutual relationships between fertility and mortality. The utopian population projections presented in this chapter presume a secure food supply, access to clean fresh water for everyone and moderate environmental changes. The result is that the major health determinants such as food security, safe drinking-water supply and health services, evoke in all utopias an epidemiological transition towards low mortality levels. This implies a high life expectancy, which, in turn, increases the level of human development, stimulating a decrease in fertility level. On the other hand, a lower fertility level causes a smaller increase in the population, thereby enhancing the availability of resources per capita, which in turn results in better health conditions. These causal relationships imply that a high life expectancy excludes a high fertility level for the world at large. Our projections therefore, given the assumption of fixed GWP trajectories, do not comprise a scenario which describes a large excess of births over deaths for a healthy population. The high UN projection seems to presuppose such an implausible development.

The low UN projection provides a picture in which the population declines very fast: a decline of 1.4×10^9 in 50 years during the second half of the 21st century. None of our utopian projections shows such a fast decline. The low UN projection presupposes an excess of deaths over births. None of the stages of the demographic transition features a situation in which the crude death rate exceeds the crude birth rate. In other words, the low UN projection assumes a very low fertility level (1.6 children per woman as early as 2050), and/or an extraordinary situation featuring very high mortality levels, possibly caused by food and severe water shortages. The utopian projections in this chapter assume a hierarchical background, which does not account for serious lack of food and water.

Within our knowledge, hardly any projections on global life expectancy are available. One is provided by the UN (1995) in which they associate an expectation on longevity with the medium population projection. They foresee an increase in life expectancy up to 74.4 years in 2050, which is within the range provided by the three utopias (*Figure 12.1c*). Even less information is available on global health and morbidity levels. The WHO (1996) merely provides disease-specific mortality data describing the present state-of-the-art, but does not describe the current global health

If we expose our utopias to statistical uncertainty methods (*Figure 12.2*), which implies that the main parameters used to differentiate the three perspectives have been varied randomly (Chapter 11), an uncertainty range varying from 5 to 15×10^9 persons in 2100 is obtained; this now constitute the low UN projection. The low perspective-based projections can be associated with very low fertility rates, which result from the uncertainties in the diffusion process with respect to contraceptive use. It is worth noting that the hierarchist route provides the lowest projections; this indicates that if the use of contraceptives diffuses very fast, up to a level comparable with the present use in the United States, in a society which is hierarchically organised, the fertility level may fall dramatically and remain very low during the next century. Such a society will eventually face an enormous greying of the population, far beyond the reach of any present expectations.

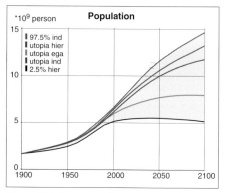

Figure 12.2 Outcomes of uncertainty analysis.

situation in terms of morbidity and healthy life expectancy. Data is lacking with regard to the past century and no projections for the next century have been given up to now. Given these circumstances, it is rather difficult to assess the plausibility of our projections on human health.

12.6 Risk assessment

Dystopias

Dystopian experiments indicate what the consequences of a particular population and health policy strategy might be if the demographic and epidemiological dynamics function according to another perspective. For example, a hierarchist management style applied to a world in which demographic and epidemiological dynamics function according to the individualist rules (*Figure 12.3*) results in an increasing population size, reaching 13.5×10^9 people in 2100. The associated total fertility rate of 2.2 children per woman can be compared to present fertility levels in middle-income countries such as Poland (UN, 1995). Due to individualist behaviour of the people, fertility levels are less influenced by hierarchist policy measures than in the utopian case, which explains the higher number of births. The life expectancy in this dystopia is about 80, comparable to the hierarchist utopia, but much lower than in the individualist utopia. The individualist world view implies a strong relationship between the economic situation and the health status; these dynamics will hardly be influenced by the policy measures as employed in the hierarchist management style.

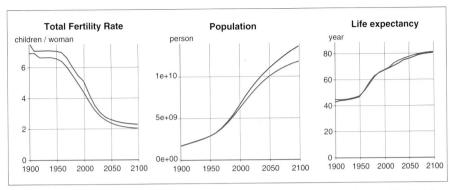

Figure 12.3 Dystopia of hierarchist management style within an individualist world. The blue curves are for the hierarchist utopia (compare Figure 11.4).

Another striking example of a dystopia is an egalitarian management style in a hierarchist world (*Figure 12.4*): the population is larger than in the egalitarian utopia: 8.4×10^9 people in 2100 compared to 7.9×10^9. However, the number of children per woman per year is significantly lower, i.e. 1.8 versus 2.1 in 2100. The average of fertility levels in high-income economies is presently 1.8 (UN, 1995) where the

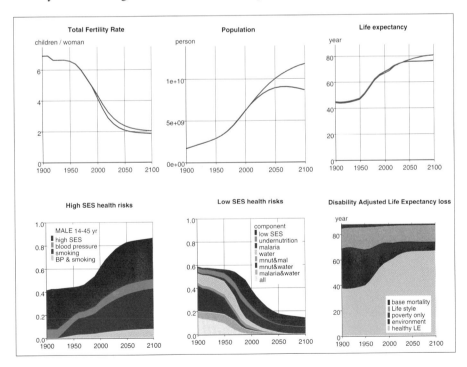

Figure 12.4 Dystopia of egalitarian management style within a hierarchist world. The blue curves are for the hierarchist utopia (compare Figure 11.4).

utopian level of 2.1 is considered to be at replacement level. Therefore we would expect a lower population instead of the higher one in this model experiment. Analysis of the life expectancy trends, however, shows that the life expectancy in the dystopian situation improves faster than in the utopian case, which results in a higher population increase. The better health status is mainly due to a higher economic growth rate and a corresponding higher health services level in the dystopian case: the hierarchist expects a much higher economic growth than the egalitarian. In this case a prolonged longevity more than outweighs a decline in births.

Governance in terms of risk

The utopian and dystopian experiments enable us to characterise the three management styles presented and the associated policy strategies in terms of risk. We do not pretend to carry out a full, statistics-oriented risk analysis. However, we do aim at characterising the different types of governance with a more qualitative interpretation of risk. Realising that the notion of risk is value-loaden and notwithstanding the differences among perspectives, we have defined various classes of risk: low (that is, safe), medium and high risk. Regarding the population and health controversy, population numbers and health status (expressed in life expectancy), are considered to be the crucial indicators. Risk intervals are determined for both indicators. We used Cohen's comprehensive survey of estimates of the Earth's carrying capacity to estimate the limiting values. Over the last four centuries he found more than 65 estimates of how many people the Earth can support. One striking feature is that no convergence of the estimates occurred over time, which allows us to take all estimates as being equally important. Considering only the highest number given (when an author states a range), half the estimates fall below 12×10^9 (*Figure 12.5a*). If the lowest number is used, half the estimates fall below 7.7×10^9 (*Figure 12.5b*). We took these medians to establish the risk classes (i) low-risk comprising population numbers below 7.7×10^9, (ii) moderate-risk comprising population numbers between 7.7×10^9 and 12×10^9, and (iii) high-risk comprising population numbers above 12×10^9.

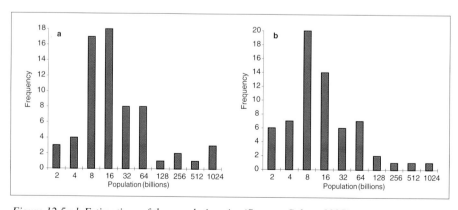

Figure 12.5a-b Estimations of the population size (Source: Cohen, 1995).

We took current observations as the starting point for risk intervals for life expectancy. The current global life expectancy is 66 years (UNFPA, 1996). We consider health levels below 66 years as havinh a high risk. The current average life expectancy in the developed countries of 77 years (UNFPA, 1996) is used to determine the low-risk class. The risk intervals with respect to health are defined as: (i) low risk: health levels above a life expectancy of 77 years, (ii) moderate risk: health levels of which life expectancy is between 66 and 77 years, and (iii) high risk: health levels below a life expectancy of 66 years. By implication, a future characterised by a population number below 7.7×10^9 with an average life expectancy exceeding 77 years is considered a highly desirable situation in demographic and epidemiological terms. Comparison of the utopias and dystopias with this desirable state (*Figure 12.6*) enables the different types of governance in terms of risk to be characterised.

Hierarchist governance turns out to be rather robust. The population numbers associated with the dystopias are within the range of 11×10^9 to 13.5×10^9, with a state of health which is in any case free of risk. The population continues to increase by the end of the 21st century in both dystopian cases due to fertility rates above replacement level (varying from 2.2 to 2.5) and low mortality levels. Hierarchist governance can therefore be characterised as moderately risky, where the risks solely pertain to the population controversy.

Under the condition that government incentives induce some structural changes, egalitarian governance results in a combination with a hierarchist and egalitarian world view in a stabilisation of the population at levels between 7.9 and 8.8×10^9 in

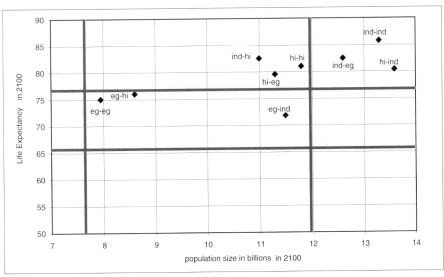

Figure 12.6 Representation of the simulation results in 2100 compared to risk areas for the utopias and dystopias.

2100. Life expectancy is in these situations robust to egalitarian governance, and increases to a level of 76 years in 2100. If society functions according to the 'laws of the market', the population continues to increase. By the end of the next century this dystopian case features a population of 11.5×10^9 people. Health prospects increase, but the improvement lags behind the previous cases due to the high population. We conclude that this type of management can be characterised as being moderately risky, with regard to both population and human health. Individualist governance is especially promising with regard to human health. Both dystopian and utopian situations show upward trends with respect to life expectancy. However, with regard to population numbers and structure, the consequences of individualist governance differ significantly. Ageing due to a decrease in fertility rate below replacement level, and a continuous growth due to a global fertility comparable to levels presently observed in East Asian and Pacific countries, seem equally plausible. For a society built upon equity and solidarity, the market-oriented management style of the individualist is likely to result in population numbers falling in the high-risk interval. Individualist governance is regarded as moderately to highly risky in population terms, while it is classified as safe and low risk when emphasising the health prospects.

Our pluralistic approach enables us to address the population and health controversy in terms of risk. Analysis of *Figure 12.6* shows that none of the utopian and dystopian futures is considered to be highly risky, although some of them are likely to enter the high risk area in the course of the 22nd century. On the other hand, none of the management styles yields a future which is in any case safe. The analysis of dystopian and utopian experiments suggests furthermore, that moderate population sizes and fairly good health are most likely where society is to a certain extent collectively creatable, so that governing incentives might create favourable conditions. However, high population numbers are also reconcilable with an extremely good health in a market-oriented society. Referring to the core controversy of this chapter, our experiments suggest that an improvement of the global life expectancy to a level comparable to the present situation in developed countries is quite possible and that this seems reconcilable with significantly differing future population trajectories.

12.7 Population, health and global change

The assessments presented in the previous sections are built upon the assumption that in the next century food will be secure, fresh and clean water available to all, environmental change moderate and the desired investments in health services provided (Chapter 11). Anticipating the integrated experiments performed with the TARGETS model as a whole (Chapter 17 and 18), various sensitivity analyses on the assumptions pertaining to food and water availability and health services in particular have been carried out. This section discusses some of these experiments in order to indicate the mutual dependency of the population size and structure, and its health status on socio-economic and environmental circumstances. We will analyse population and health prospects for the following cases: (i) food and water availability remains on the present levels; (ii) the desired investments in health services cannot be realised; (iii) both socio-economic and environmental circumstances do not improve in the course of the 21st century and (iv) there are food and water crises and a lack of the desired health investments.

Case 1: Lack of improvement in food and water availability
Both the hierarchist and the egalitarian model routes take account of environmental factors as determinants of fertility and mortality. *Figure 12.7* shows the results for the hierarchist and egalitarian projections if food intake remains at the present global average of 2660 Kcal/person per day and public water supply coverage is 75%. This results in a population stabilising at the level of 10.8×10^9 in 2100, while the life expectancy improves more slowly, reaching a level of 78 years in 2100, which can be compared to the present situation in Australia and the Netherlands (UNFPA, 1996). The difference with the utopian case can be fully attributed to the environmental determinants. Stagnation of food and water availability at current levels will bridle the improvement in life expectancy, and thereby curb the population growth. In the egalitarian utopia such a stagnation evokes a decline in population down to the level of 7.5×10^9 in 2100. In both the hierarchist and egalitarian projections, health risks for people with a low socio-economic status due to environmental threats are not eliminated, as is the case in the utopian projection. The environmental-related health

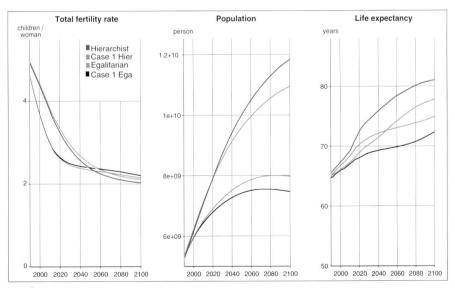

Figure 12.7 Results of Case 1: developments in the period 1990-2100 if there is no improvement in food intake and public water supply.

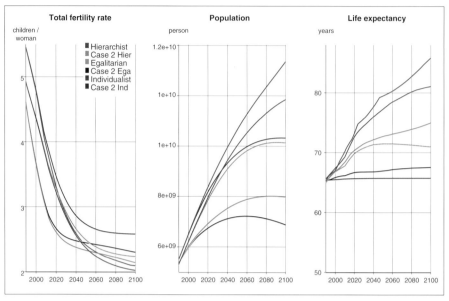

Figure 12.8 Results of Case 2: developments in the period 1990-2100 if health service levels remain constant.

risks almost remain at the 1990 level: 38% in 1995 versus 33% in 2100. Those developments together result in a health pattern in which 50% of the loss of healthy life-years can be attributed to environmental threats.

Case 2: Stagnation of health investments

The health services level is a determining factor in all perspective-based interpretations. Let us assume a situation in which investments in health stagnate at the present level, i.e. a global average of $ 240 per cap per yr. Projections for the year 2100 (*Figure 12.8*) then run from 6.9×10^9 in the egalitarian to 10.1×10^9 in the hierarchist and to 10.3×10^9 in the individualist perspective. Compared to the utopian images of the future, lower health services induce population sizes which are about 20% less than the utopian sizes. Life expectancy varies over the different perspectives from 66 years in the individualist case to 71 years in the hierarchist case. Comparison with the utopian projections shows these levels to be significantly beyond the utopian cases. The individualist projection yields the most worrisome projections: average life expectancy remains at the current level. Also the healthy life expectancy does not improve.

Case 3: Lack of socio-economic and environmental improvements

Figure 12.9 shows the projections associated with a future in which socio-economic and environmental circumstances do not improve. If food, water and health services

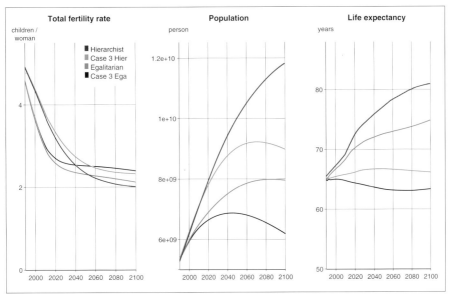

Figure 12.9 Results of Case 3: developments in the period 1990-2100 if there are no improvements in food, water and health services.

remain at current levels, the hierarchist model route features a future in which the population size reaches a level of 9.0×10^9 people in 2100, while the health level, measured either in life expectancy or in healthy life expectancy, does not deteriorate compared to the present global average. However, in the egalitarian assessment an 'overshoot and collapse' situation is foreseen: a decline to 6.1×10^9 by 2100 and a worsening of the life expectancy to the level of the 1980s, i.e. 63 years. The population numbers associated with this image of the future resemble the low UN projection and reveal the conditions under which the low UN projection for 2100 is plausible: (i) completion of the fertility transition in the first half of the next century, (ii) environmental factors constituting the major health determinants, and (iii) continuation of the present situation with regard to food, water and health services. Notwithstanding the fact that both the low UN projection and the egalitarian assessment of the socio-economic and environmental trends described show an overshoot-collapse situation, the rate of decline in the low UN projection is much larger, so that the low UN projection can still not be fully explained.

Case 4: Food and water crises

Imagine a future in which the food availability drops to the level of 1950 (2000 Kcal/person per day) and water coverage declines to 40%, which is comparable to the global average of the mid-1960s. We further assume that the health investments remain at present levels. Under these conditions, all projections (*Figure 12.10*) reveal

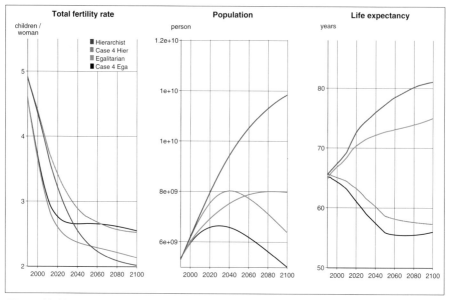

Figure 12.10 Results of Case 4: developments in the period 1990-2100 if food and water levels fall to 1950-1960 levels and health services remain constant.

'overshoot-and-collapse' scenarios. The population projection shows an increase in the first half of the next century and a fast decrease afterwards. The individualist health projection shows a continuation of the present health status (comparable with case 2), while the hierarchist and egalitarian projections show a significant deterioration of human health. In the course of the next century, the life expectancy drops to a level comparable to the global average of 20 years ago (56 to 57 years), thereby compressing the healthy life expectancy to less than 50 years. Can the low UN projection be explained by such crises? The egalitarian perspective expects a future in which the population declines to 5.1×10^9 in 2100, and in which the global life expectancy in the 21st century drops to the level presently observed in Ghana. In this projection, the fertility rate is slightly above the replacement level, so that the decline in population numbers can be fully ascribed to the deterioration of the quality of life, leading to higher mortality rates. However, the rate of decline, although fast is still lower then in the low UN projection. In other words, while the 2100 population numbers associated with the low UN projection can be reproduced, the trajectory featured in the low UN scenario cannot be explained by even one of our experiments.

The four cases indicate the extent to which population and health projections as described in the previous sections are susceptible to trends in food, water and health services. The experiments teach us that socio-economic and environmental developments significantly affect population size and structure, and health status. If

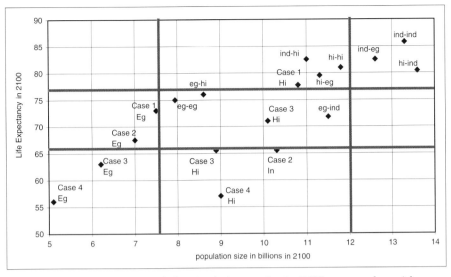

Figure 12.11 Representation of the simulation results in 2100 compared to risk areas including the four cases.

we add those experiments to our image featuring health versus population (*Figure 12.11*), we notice that the area signifying low to moderate population numbers combined with a health status at or below present levels comprises more points than the ones solely based on utopian and dystopian projections (*Figure 12.6*). In other words, addressing the health and population controversy with the statement that a high life expectancy is most likely, is highly conditioned by the assumptions concerning trends with regard to food, water and health services. While a high life expectancy seems to be reconcilable with a wide variety of population trajectories, the cases as performed in this section tell us that wretched health conditions can generally be associated with overshoot-collapse situations. In case of stabilisation or increase of the population the health level is likely to improve, even in the case of less favourable environmental and socio-economic circumstances.

12.8 Conclusions

What do the utopian projections tell us with respect to plausible futures in the light of state-of-the-art knowledge? An integrated approach towards fertility and mortality tells us that population projections exceeding 15×10^9 inhabitants on Earth by 2100 are implausible, even if we take into account the variety of perspectives on demographic and epidemiological dynamics. The three images of the future suggest that an improvement in life expectancy in the course of the next century is probable.

Such an extension of longevity is generally associated with a better health status.

With regard to the population and health controversy the utopian and dystopian experiments tell us, that a future featuring both low population numbers and a high life expectancy is not to be expected. Risks for the future are primarily located in the population controversy. Notwithstanding these risks, our experiments reveal that a doomsday scenario featuring excessive population numbers in miserable health conditions is not very plausible.

The experiments addressing the case where there is no improvement in the availability of food and water suggest that under such circumstances population growth will slow down in the course of the next century due to the fact that mortality levels remain high. Such images of the future are further associated with unhealthy conditions, which can be attributed primarily to environmental threats.

If health services remain at current levels, the health prospects for those living in the next century are likely to be similar to ours. Taking into account that the current global average reflects severe regional health problems, for example, in Africa, it is likely that in the case of a stagnation of local health services such regional circumstances will persist.

With regard to the demographic and epidemiological pathways into the next century, these experiments suggest that a continuation of current socio-economic and environmental circumstances might evoke an overshoot-collapse situation. Our experiments thus suggest that overshoot-collapse scenarios are to be expected in the case of a food and water crises, combined with a lack of sufficient health services. From these experiments we conclude that the low UN population projection is also highly implausible.

Building upon both the utopian and dystopian projections, and the sensitivity experiments, we therefore conclude that population growth and a healthy life seem to be reconcilable under improving socio-economic and environmental circumstances. However, given deteriorating socio-economic and environmental conditions, population growth is likely to be curbed by increasing morbidity and mortality levels. Recognising the significant dependency of population and health projections on socio-economic and environmental developments, an integrated assessment is needed to address the controversy more adequately. As feedbacks between socio-economic, environmental and demographic developments are not considered in the experiments presented in this chapter, an integrated approach is likely to yield more insights. Furthermore, integrated assessment allows us to evaluate the socio-economic and environmental impacts of various population trajectories, and thereby enables us to assess potential limits to population growth. A more integrated assessment of the population and health controversy is presented in Chapters 17 and 18.

13
Energy systems in transition

"Man, using his muscles alone, is not a very powerful engine."
F. Braudel, The Structures of Everyday Life (1981)

13 ENERGY SYSTEMS IN TRANSITION

Bert J.M. de Vries, Arthur H.W. Beusen and Marco A. Janssen

In this chapter we present simulation experiments and outcomes of the energy submodel TIME. First, the major controversies and uncertainties are discussed. Next, the cultural perspectives are introduced with reference to world energy, after which we clarify the way in which these are linked to assumptions and model routes. Some results of sensitivity and uncertainty analyses are also given. We discuss a few energy dystopias which could emerge if, for a given population-economy scenario, the world view and the management style within the energy system are discordant. Some conclusions are presented about the plausibility of and risks related to the utopian energy futures. The impacts of the emissions from fossil fuel combustion on water, land, and element cycles are discussed in the next three chapters.

13.1 Introduction

In 1886 Jevons warned in his book 'The coal question' about the rapid depletion of British coal fields threatening the British Empire. Numerous appraisals of coal, oil and gas availability have been made since then, many of them for strategic reasons. Environmental issues and the two oil crises in the 1970s have intensified the debate on fossil fuel use. Later on, it has been broadened by incorporating demand side management and renewable supply options and by including macro-economic aspects. Controversies and uncertainties about the future development of the world energy system abound. Can energy demand really be influenced and to what extent are price changes the right instrument for this? How important are changes in life-style and in the nature of economic activities, and what is the role of technical innovation? Is the world really facing depletion of its high-quality oil and gas resources and will it show up in the form of sudden price increases and supply disruptions or will it be a smooth transition towards alternative fuels? Are the new technologies which supply energy from non-fossil sources really as promising and competitive as their advocates claim?

The major controversies relate to the question: *How can energy demand be met in an adequate and reliable way within the constraints set by socio-economic developments and goals, available energy-supply options and environmental integrity?* This formulation emphasises a few key characteristics of energy in society. First, energy is, in a variety of ways, a necessity of life which should be satisfied at such levels of cost and reliability that do not constrain human activities. In rural areas of non-industrialised regions, the emphasis is on activities like cooking, food preservation and water supply. In urbanised regions of industrialised countries, energy is also used to operate factories, heat dwellings and offices, transporting people and goods, etc. Secondly, energy has to be supplied from a variety of resources which involve a whole spectrum of technologies that require capital and skilled labour for there operation. As such, the energy-supply system is a major part of an industrialised nation's economy. Its dynamics are governed by a complex interplay between resource endowments, prices, technologies and strategic aspects. Thirdly, energy supply, conversion and use as we know it today has numerous impacts on the natural environment. Some of these have led to serious environmental damage but can be dealt with by a combination of technology, capital and political will. Other impacts, first and foremost the contribution to the enhanced greenhouse effect due to CO_2 emissions from fossil fuel combustion, are likely to be more serious and are probably less easily mitigated. The above question can be split into more specific questions:

- how will energy demand – in whatever form – develop in relation to population, economic production and consumption patterns ?
- how will technical innovations in combination with changing fuel prices affect the relation between end-use energy demand and secondary fuel use?
- how much energy from fossil fuels will be available and at what costs?
- which alternatives – for all energy forms – will be available and at what costs, and at what rate can they be expected to penetrate the market?
- should the combustion of fossil fuels be constrained because of the enhanced greenhouse effect?

In view of the transition concept introduced in Chapter 2, of special interest is the question as to which transition pathways can be envisaged from finite fossil fuel resources to non-fossil resources and technologies. In this chapter we focus on these questions and the associated controversies and uncertainties by constructing perspective-based sets of assumptions which are then explored within the framework of the energy model described in Chapter 5. First, we discuss the major issues. Although we follow a different approach for energy from the one used for water (Chapter 14), there is quite an overlap in the issues and controversies – for instance, whether the energy problem is primarily seen as a supply or a demand problem. The debate on energy, however, has a longer history and is somewhat more crystallised.

13.2 Major controversies and uncertainties

Declining energy intensity
The first item in the above list is about what will happen with the energy intensity in MJ per unit of economic activity. It has been declining in the industrialised regions but as yet it is unclear whether the underlying trends will persist. Even more

uncertain is how energy intensity will change in the less industrialised regions of the world. There are strong upward pressures: the industrialisation process itself and the introduction of the 'modern' consumerist life-style. On the other hand, the availability of more energy-efficient technologies offers large opportunities to these regions for catching-up (Grübler and Nowotny, 1990).

Recent scenarios show a rather surprising agreement on the possibility to reduce energy intensity significantly. A study done for Greenpeace (Lazarus, 1993) claims that energy intensity can be reduced between 1990 and 2100 to 4.6 MJ/$, a more than threefold reduction. A possible future sketched by Kassler (1994) called 'dematerialisation', considers a similar drop to 4.5 MJ/$. A recent IIASA/WEC study (1995) gives a range between 4.6 and 7.7 MJ/$ for the year 2050. Four recently published energy scenarios for the European Community assume energy intensity to fall by 1.1 to 1.7 % per yr (EC, 1995) over the next 20 years; an inventory of analyses for the USA gives a range of 0.8-1.3 % per yr (EMF, 1996)[1]. Although one should be aware of the different backgrounds of these studies and the probability of wishful thinking and collective bias[2], it should be noted that agreement on such drastic reductions was completely absent in the early 1980s.

In a recent and fairly comprehensive overview of scenario studies made for the IPCC (Alcamo, 1995), it appears that almost all analyses assume a significant decrease in the overall energy intensity, 0.45 to 1.45 % per yr between 1990 and 2100, as a result of the three factors mentioned in Chapter 5: structural change, autonomous energy efficiency improvements (AEEI) and price-induced energy efficiency improvements (PIEEI). An overview of AEEI-values (see Equation 5.2) used in recent energy models range from 0 to 1.1% per yr in global energy models and from 1.12 to 2.85% per yr in energy efficiency scenarios (Matsuoka *et al.*, 1995). It is partly a matter of focus: "Where there is no great attention paid to energy conservation, the annual rate is between 0 and 0.5%, whereas if large energy savings are assumed, this rises to 1.0%". According to Matsuoka *et al.* (1995) the feasible range is between 0 and 1.5% per yr for the long term. One of the major controversies has to do with the effects of rising energy prices as expressed in the PIEEI-factor. Most experts agree that rising energy prices will induce energy conservation but estimates of the price-elasticity suggest great uncertainties in the rate and degree. The price elasticity is difficult to measure and differs for different sectors and countries partly because of varying substitution possibilities. It may be time-dependent, becoming smaller once more profitable measures have been taken (Dargay and Gately, 1994). Moreover, energy prices relative to interest rates and wages may actually be the relevant variable. An important role is played by government-supported research and development (R&D) programmes.

1 The larger part of this decline in energy intensity is from shifts in activity compositions and the replacement of older by newer, available and more efficient equipment; further increases in equipment efficiency and price-induced effects are minor in almost all model studies (EMF, 1996).

2 See, for example, Sterman and Richardson (1983) on the evolution of estimates of ultimate recoverable oil in the USA.

Depletion of fossil fuel resources
The second item in the above list is about the long-term supply-cost curve for coal, oil and gas. Estimates of fossil fuel resources and reserves abound in the literature (de Vries and van den Wijngaart, 1995). There is general agreement that the coal resource base is large enough to sustain present levels of production throughout the next century without major cost increases. Most researchers expect rising costs to find and produce the as yet undiscovered deposits of oil and gas but there are large uncertainties and controversies on when and how much. Estimates during the 1980-1995 period of ultimately recoverable oil and gas range from 8000 to 40000 EJ, respectively. Most estimates lack specific information on costs or probability. Nevertheless, the general attitude nowadays is that resource depletion is not an important issue anymore. One should be cautious about this because the apparent consensus may simply reflect an unwillingness to acknowledge the fact that for most countries the era of cheap and nearby oil and gas deposits to fuel industrial development is either over or will never arise.

It is known that oil shales and tar sands can provide large additional amounts of oil, possibly up to three times the conventional oil resource base (Edmonds and Reilly, 1985). For gas there is the hypothesis of huge reservoirs of pressurised gas and clathrates (Lee, 1988). Another controversial option is the liquefaction and/or gasification of coal, which could supply the world with oil and gas substitutes for a long time to come. The prospects for such conversion processes have diminished since the initial euphoria of the 1970s, and now only electricity generation through combustion of coal-based synthetic natural gas in combined-cycle plants is considered a promising option.

Alternatives to fossil fuel
Until the early 1980s the prevailing view on future energy supply was that fossil fuels and nuclear power would dominate the scene in the 21st century, although renewable energy options might play a role too. More recently, the trend towards more flexible, convenient and clean forms of energy appears to favour natural gas and new fuels like methanol and hydrogen which could be derived from a mix of nuclear and renewable sources. Nuclear energy still offers the prospect of a non-carbon energy source, but new major options for further decarbonisation are electricity from solar cells and from wind turbines. Another option to reduce net anthropogenic CO_2 emissions is the production of liquid and gaseous fuels from biomass. This could be as an expansion of present usage forms such as agricultural residues, but most of it will have to be in the form of 'commercial' or 'modern' – as opposed to 'traditional' – biofuels, in which case biomass can become a substitute for petrol in the transport sector or for coal in electric power generation.

There are still major controversies on the rate at which the costs of fuels or electricity from these supply technologies can be brought down, and hence about their penetration rate (Johansson *et al.*, 1993; Lenssen and Flavin, 1996; Statoil and

Energy Studies Programme, 1995; Trainer, 1995; Williams, 1995). First, there is the worldwide controversy on the acceptability of nuclear fission technology, which depends, to some extent, on the prospects for safer reactor designs and acceptable solutions to radioactive waste disposal. Even more uncertain is the possibility of breeder and nuclear fusion reactors. Second, most analysts agree that the large cost reductions of solar photovoltaics in the last few decades will continue, but how much the reduction for large-scale market penetration is to be is controversial and uncertain too. Third, the option of deriving liquid or gaseous fuels from biomass has rapidly gained prominence in long-term energy scenarios, but there are large uncertainties on costs and land requirements, and on the interference with food production and climate change (IIASA/WEC, 1995). There are similar controversies about the cost and acceptability of energy carriers like hydrogen and promising technologies like fuel cells, about how new supply technologies will fit in the energy system and about the efectiveness of R&D programmes. Whereas on issues of energy efficiency and fossil fuel resources a convergence in expert views may have occurred, this is less true for the role of non-fossil fuel options. In the Business-as-Usual future of the IPCC-IS92a scenario, the fossil fuel share is still 56% by 2100. The FFES scenario for Greenpeace claims that a complete phase-out of fossil fuels is feasible at an almost threefold increase in GWP per capita level (Lazarus, 1993). The IIASA/WEC study (1995) suggests that the fossil fuel share can be reduced to a maximum of 20% by 2100 in an 'ecologically driven' scenario .

Emissions from fuel combustion

Fossil fuel (product) combustion accounted in 1990 for over two third of anthropogenic emissions of CO_2, NO_x, and SO_2 (Alcamo, 1994). Future CO_2 emissions will be largely determined by the level of population and economic activity, the energy intensity and the share of non-carbon fuels. Given the previously discussed controversies and the resulting uncertainties, it is not surprising that projections of the CO_2 intensity in 2100 ranges from 105% to 10% of its 1990 level (Morita *et al.*, 1995). In view of the relative scarcity of low-cost oil and gas *vis-à-vis* coal, many official forward projections indicate an increase in coal use and in CO_2 emissions. If such a future unfolds, the need to take action beyond a 'no regrets' strategy will become more pressing. Removal of CO_2 from exhaust gases could become one of the necessary responses[3]. For NO_x, and SO_2 there is a variety of emission abatement options, but their introduction will often depend on regional circumstances.

Despite the controversies and uncertainties, there is a widely held conviction that the world energy system will undergo a transition over the next century. Most of the

---

3   CO_2 removal may become feasible in the future for large-scale combustion processes (Blok *et al.*, 1993). It is not considered in the present model.

above elements will be part of it, but it is difficult, if not impossible, to predict their relative importance. A final set of questions relates to the feasibility of such an energy transition from a macro-economic point of view. Future expansion of the energy system will require enormous investments, an increasing share of which will be needed in the presently less developed regions (Dunkerley, 1995). Capital and/or energy shortages may become a constraint to economic growth if the proportion of electricity increases and capital-intensive options like nuclear and solar power get a larger market share. On the other hand, important learning effects and an increasing share of capital-extensive biofuels may mitigate this problem. Evidently these issues are hard to resolve in the face of large uncertainties on capital markets and technological performance. Another constraint may be posed by land in case of large-scale introduction of biomass-based fuels : a sizeable part of presently cultivated land may be needed.

13.3 Perspectives on world energy

Given all these controversies and uncertainties, what is the use of making long-term (energy) projections? As has been set out in Chapter 10, we will attempt to address this problem by formulating coherent sets of assumptions which are considered representative for a particular perspective. The three perspectives are the hierarchist, the egalitarian and the individualist, which not only reflect a preferred way of interpreting the world, but also of managing it. Each set of world view and corresponding management styles makes up a utopia: a future in which the world behaves and is managed according to that view. In this section we will briefly – and necessarily somewhat caricaturally – describe these perspectives as far as the future world energy system is concerned (see, for example, Schwarz and Thompson (1990) and Thompson (1982)).

The hierarchist perspective
The hierarchist wishes to avoid disruptions to the smooth functioning of the energy system in view of its consequences for economic growth and voter behaviour. To this end the hierarchist institutions of society will anticipate and respond on the basis of scientific expert knowledge. The need for governance structures is emphasised. There is a preference for a risk-reducing control approach; decisions should be supported by the outcomes of cost minimisation, cost-benefit analysis etc. Technologies which can be planned and controlled are favoured and issues like oil dependence and public acceptance rank high[4]. Energy consumers can and should be guided towards 'rational energy use' – which is the justification for regulation, taxes,

---

[4] In the context of ambitious government plans for nuclear power expansion in the USA and the former USSR, the phrase 'nuclear priesthood' was coined; in France some spoke of 'Les nucleocrates' (Simonnot, 1978).

information campaigns and the like. Hierarchist institutions will tend to suppress egalitarian and individualist counterforces unless they become a threat to their power, in which case they will be accommodated (e.g. the Greens, markets).

With regard to the previously introduced controversies, the hierarchist will make a prudent assessment of the potential for energy conservation and have an institutional bias towards large-scale supply-side options. Resource estimates will be rather conservative and there will be a cautious approach to the issue of climate change. Hierarchists will support cost-effective 'no-regrets' measures which reduce the risk of climate change, but are keenly aware of the fact that fast and stringent cutbacks in CO_2 emissions may be socially disruptive and create competitive disadvantages. Hence, a carbon tax should be 'realistic' and only be introduced if an internationally negotiated consensus is reached to avoid windfall profits or free riders (see, for example, Hourcade *et al.* (1995) on carbon tax evaluation). R&D programmes for new energy supply and efficiency options should get government support, because they too stimulate economic growth and (national) status.

The egalitarian perspective

The egalitarian wishes to reduce inequity and stresses the rights of those without a voice: our children, the poor, and nature. High and rising CO_2 emissions are seen as one more sign of humans' maltreatment of the Earth which may lead to catastrophes. Mathematical tools and models can play only a minor role because many of the issues at stake cannot be expressed in numbers or money. Egalitarians will advocate a morally founded justification for government regulation and support programmes. The more general issue of (under)development is seen as part of the problem; the claims of less industrialised regions that they should bear only a small part of the burden are supported. Egalitarians will be suspicious of large multinational (energy) companies whose concentration of money and knowledge makes them as much a part of the problem as of the solution. From an egalitarian perspective, science and technology can certainly solve part of the problem but add as much to it as long as their course is governed by centralised and commercial interests, and market ideology.

The egalitarian will embrace the 'precautionary principle' as a way to express his/her risk-averse attitude. Energy futures will be judged not only in terms of costs, but also with regard to distributive aspects and ecological impacts. A modest economic growth will probably be necessary but it should narrow the present income gap between the rich and the poor. Energy taxes are promoted as means to change wasteful production and consumption practices. Energy demand projections are much lower than official ones and have to be met to a large extent with non-fossil sources (Lovins, 1991). There will be a preference for decentralised and clean technologies, and therefore a natural tendency to focus on energy end-use needs and efficiency (Johansson *et al.*, 1989). Estimates of fossil fuel resources are on the low side, whereas the prospects of renewable energy sources are usually on the high side if compared with the hierarchist perspective. Development of renewable sources should be strongly supported by government RD&D programmes and indirectly by taxing carbon fuels because the market is often inadequate due to existing barriers and distortions.

The individualist perspective

For the individualist, entrepreneurial freedom and unhampered working of market forces gives the best guarantee for increasing material wealth and at the same time solving resource and environment problems. If energy-supply companies can operate in a regime of free trade and with a minimum of government regulation and interference, price signals will steer the transition away from fossil fuels before they are depleted. CO_2 emissions are probably increasing less than official expectations suggest – a view which may give rise to a somewhat odd coalition with egalitarians. The Earth itself is also far more resilient than we tend to think, so climate change impacts are probably exaggerated by those advocating strict measures. Moreover, there are several and relatively cheap options for adaptation (Nordhaus, 1991). The key resource is human ingenuity: human skills generate science and technology, which will bring options one cannot even imagine at the present (Simon, 1980). Not much can be said about the distant future in any case – what further opportunities and progress will, for instance, emanate from information technology, biotechnology, space technology? Technology is also the major driving force behind economic growth, which will ultimately benefit the poor.

The individualist emphasises the opportunities which arise from the search for new resources and new technologies to supply and conserve energy. Energy resources turn out, over and again, to be more abundant and cheaper than expected. Policy measures like a carbon tax are unnecessary. First, there are still too many uncertainties about the enhanced greenhouse effect and possible climate change to accept drastic measures. Secondly, they are ineffective because industries will move to other countries and consumers will stick to certain life-styles whatever the costs.

Of course, in the real-world actors rarely express their views in such a caricatural way. They are in constant interaction and often have strategic and public relations in mind as well. Moreover, positions may be implausible or even inconsistent when stakeholder share only part of the underlying values and judgements. For example, the egalitarian concerns about nuclear reactor safety and climate change have increasingly been incorporated in hierarchist policy formulation in the form of regulatory and negotiation frameworks. Similarly, the energy business community – part of which is rather hierarchist – is advocating the need for more efficient and environmentally friendly resource use options, at the same time emphasising the virtues of the market and the limitations of command-and-control approaches (Schmidheiny *et al.*, 1992). There is also the paradox that the egalitarian expectation of fast innovation in energy efficiency and non-carbon energy supply

Perspectives on energy

In their report 'Our Global Neighbourhood', the Commission on Global Governance warns that "the measures required to avert risks must be put in place immediately and those already in place – the Framework Convention on Climate Change... – must be rapidly and substantially strengthened" (Commission on Global Governance, 1995, pp. 83). Elsewhere they argue that "Energy efficiencies are an economic imperative for developing countries faced with capital expenditures to satisfy growing energy needs... And it is clearly in the interest of the industrial world to ensure that these countries have the financial and technological support required to meet these needs...A contribution could be made to alleviating the global warming problem through energy or carbon taxation..." (Commission on Global Governance, 1995, pp. 84, 212).

Many have argued that the situation requires drastic policy interference. In a report to Greenpeace the Stockholm Environment Institute puts the issue of life-style in the forefront (Lazarus, 1993): "Achieving a fossil free energy future will require major changes in energy policy and life-styles. The wasteful high energy consumption path that the North has enjoyed has to end. Future energy use will have to be extremely efficient, and increasingly based on sustainable renewable energy sources such as solar, wind and biofuels. The basis of that wasteful life-style is of course the economic growth and development path that we have chosen."

Or could it be that policy measures are more harmful than beneficial to resolve the climate change problem ? One of the two scenarios presented by Kassler (1994) of Shell Planning is called New Frontiers. It pictures a world of high economic growth in the less developed regions in which environmental problems are solved by market instruments. Renewable energy sources mitigate the threat of climate change: "As they progress along their learning curve, first capturing niche markets and then gradually expanding, new energy sources may well become commercially competitive over the next decades and start to be visible around 2020. [ ] Technologies will compete but the market will decide. [ ] With this perspective in mind, the idea of 'saving hydrocarbons for future generations' is perhaps unduly conservative. [ ] ... this scenario ... would have powerful implications for the climate change debate... There is an exciting challenge lying ahead: reaching New Frontiers following a path which makes economic sense. The industry has the capability ... Policy makers must also create the market conditions allowing this to happen." Not surprisingly, the mirror image of this scenario, called Barricades, is more dystopian: "liberalisation is resisted and restricted because people fear they might lose what they value most. [ ] There is increasing divergence between rich and poor economies...[ ] In the developed world, a number of non-governmental organisations... cause energy to be regarded as something bad and to be used sparingly, leading to an unfavourable investment climate in this sector".

Some assess other options as well: "A radical technological opton would be geoengineering, which involves large-scale engineering to offset the warming effect of greenhouse gases... The advantage of geoengineering over other policies is enormous, although this result assumes the existence of an economical and environmentally benign geoengineering option" (Nordhaus, 1994, pp. 80, 96). Or can nuclear energy rescue us? "...the growth in world population... and in human aspirations will likely generate a large demand for end-use energy over the next three hundred years... Only two options for expansion appear viable : coal and nuclear. [ ] If undesirable global warming... results from carbon dioxide generated by coal burning, a tolerable level of fossil fuel use can be established and the remainder made up by the nuclear option... [ ] None of the options for supplying the needed extra energy presents any important risk to life or health. " (Nathwani *et al.*, 1992, pp. 256, 259).

and imminent depletion of cheap oil and gas, is at odds with their fear that the high CO_2 emissions of the Business-as-Usual scenarios become reality (see e.g. Lenssen, 1996). Evidently, our implementation of the three perspectives into the energy model is only a first attempt to introduce real-world divergence in interests and values into a quantitative modelling framework.

13.4 Simulation results for the three utopias

In the previous section we introduced the three 'active perspectives' on world energy futures. The qualitative characterisations of perspectives and management styles have been translated into a set of assumptions and model routes. In Chapter 11 we gave a brief description of the integrated hierarchist utopia. Here, it will be discussed in some more detail, including a sensitivity analysis, followed by the egalitarian and individualist utopias. These are actually semi-utopias because only the driving forces (population and GWP) and the energy model assumptions are changed, while the water, land and cycles submodels are run according to the hierarchist utopia (see Chapter 11). The population/health utopian scenarios which provide inputs for the energy submodel experiments are described in Chapter 12.

Within the energy submodel a number of parameters has been chosen the same for all three perspectives. As to structural change (see Equation 5.1) we assume a further decline in average end-use energy intensity for the residential, services and other sector. For transport and electricity, however, it is assumed to keep growing in the next few decades. The lower limit on the AEEI-factor is set at 0.2 for heat and 0.4 for electricity. For another set of parameters we have made perspective-based assumptions, which are a reflection of the controversies and uncertainties outlined above. Some of these are related to expectations on energy intensity, and on end-use and conversion technology: the AEEI factor, the energy conservation cost curve and

Parameter	Hierarchist	Egalitarian	Individualist
AEEI ('technology')	average 1%/yr, all sectors	faster	faster
PIEEI ('prices')	moderate	cheaper and long payback times accepted	much cheaper and short payback times
TE (thermal electric) efficiency	rising to an average 50% in 2100	rising to an average 52% in 2100	rising to an average 60% in 2100
NTE (non-thermal electric) cost	moderate improvement	moderate improvement	fast learning, hence cheaper
coal cost	slow increase	removal of subsidies, hence fast increase	removal of subsidies and no learning in SF (Surface Coal), hence fast increase
gas resource base and cost	medium estimate	less, at higher cost	more, at lower cost
BLF/BGF (Bio Liquid/Gaseous fuels) cost	moderate improvement	higher labour cost, more severe land constraint	lower labour cost, less severe land constraint
carbon tax	no	towards $ 500 per tC ($12.5 per GJ) in 2020, constant thereafter	no

Table 13.1 Perspective-based model routes : indication of assumptions.

its rate of decline, thermal electric conversion efficiency and the learning coefficients for non-thermal electric power generation (NTE). As to energy efficiency, desired payback times for energy conservation measures and premium factors for coal have been varied. For NTE, the base load factor has also been differentiated. A second group has to do with the fossil fuel resource base and its exploitation: the long-term supply cost curves for coal, oil and gas, labour costs in underground mining and the learning coefficient for surface coal. A third group is related to biofuels (BLF/BGF): learning coefficients, labour and land costs, and the influence of land scarcity on biofuel yields. The management style is implemented on the basis of three policy variables: a carbon tax on secondary fuels, an RD&D programme for NTE and RD&D programmes for biofuels. The assumptions made for the present simulation experiments are based on a mixture of simulation experiments and literature estimates, and summarised in *Table 13.1* (see also Chapter 5 and de Vries and Van den Wijngaart (1995) and de Vries and Janssen (1996)). We have endeavoured to implement three quite divergent views on the energy system into a single model structure. However, such an attempt can only be partially successful as the model itself is also biased, for example, because of the importance given to relative prices in driving substitution processes.

Reference case: the hierarchist utopia

In the hierarchist scenario the AEEI factor is on average 1% per yr. Coal for electric power generation remains relatively cheap because governments support their coal industries for strategic and employment reasons. NTE options experience moderate learning of 10% decrease in specific investments for every doubling of cumulated production but cost reductions are counteracted by a declining base load factor due to storage and transport costs. The ultimately recoverable oil and gas resource base is rather large (72,000 and 60,000 EJ, respectively) but only 60% and 30%, respectively, are recoverable at cost levels less than 20 times the 1900 level. This reflects the rather conservative attitude of hierarchist resource estimates. The learning rate for surface coal is kept at a moderate 10%. Labour costs rise for underground coal but this is partly offset by a doubling of capital-labour ratios. Commercial biofuels are also assumed to have a 10% learning rate, which brings costs down to the level of $ 10 to $ 15 per GJ. Only for BLF is a modest R&D programme assumed; no carbon or energy taxes are applied. The assumptions are chosen in such a way that they reproduce important parts of the IPCC-IS92a scenario (Leggett *et al.*, 1992).

Use of secondary fuels and electricity increases from the present 220 EJ/yr to over 800 EJ/yr by 2100 (*Figure 13.1*). The largest growth is in electricity and the industrial sector. The share of electricity in final demand climbs from the present 19% to over 40% – a level which has almost been reached now for the US residential sector. About 40-45% of demand reduction between 1990 and 2100 is from autonomous improvements (AEEI). There is an additional reduction in the energy intensity of 3%

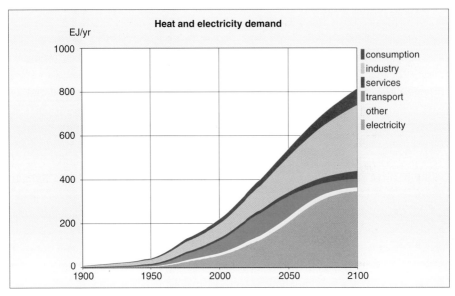

Figure 13.1 Sectoral non-electricity (heat) energy demand and electricity demand in the hierarchist scenario.

for electricity up to some 40% for the transport sector due to rising energy costs. By 2100 over 50% of the electricity is generated in non-thermal electric (NTE) power plants. Of the thermal electricity, 90% is generated by burning coal. The costs of coal-fired electricity rise, but penetration of NTE, stabilises the average electricity price. Coal production in the scenario increases almost fivefold to about 700 EJ/yr, near the level in the IPCC-IS92a scenario *(Figure 13.2)*. The proportion of coal decreases until 2010 after which it starts rising; oil and gas will be depleted by the end of next century and biofuel-based substitutes have partly taken over *(Figure 13.3a)*. In combination with medium economic and population growth, carbon emissions rise throughout the next century to over 20 GtC/yr by 2100 compared to about 6 GtC/yr in 1990 *(Figure 13.4)*. Such an emission trajectory would lead to a CO_2 concentration using the hierarchist route for the CYCLES submodel of about 550 ppmv by 2100. This is considered 'acceptable' in view of expected risks.

Simulated price paths for coal, crude oil and natural gas are shown in *Figure 13.5*. Coal prices show a smooth and small increase, partly because surface-mined coal emerges as a cost-stabilising option which counteracts the rather steep increase in the cost of underground coal caused by depletion and rising labour costs[5]. The rise in oil and gas prices induces the penetration of biomass-derived fuels. Penetration of liquid biofuels (BLF) leads to a decline and later on, when land constraints become

---

5   It should be noted that coal liquefaction and gasification are not explicitly taken into account. In the IPCC-IS92a scenario, coal use is assumed to take place in the form of liquid and gaseous coal-based fuels.

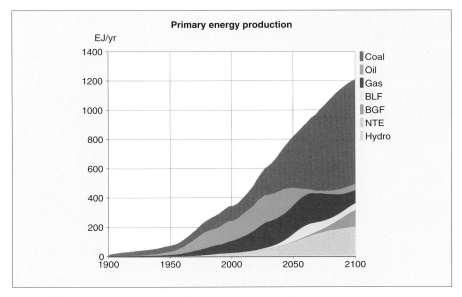

Figure 13.2 Primary energy production by fuel type in the hierarchist scenario.

more serious and learning ebbs away, a rise in the price of Light Liquid Fuel (LLF) at a level of about $15 per GJ or about $100 per barrel[6]. This price includes non-price barriers and in the present model formulation it is primarily an indication of the price differential needed to let commercial biofuels penetrate the market[7]. A similar pattern evolves for Gaseous Fuel (GF). The simulations suggest that biofuel technologies with the hierarchist characteristics would penetrate without the need for major demonstration projects. Of course, this hinges on the assumptions on the long-term supply – cost curves for conventional oil and natural gas and the biofuel production function.

Two important system characteristics are the over-all energy intensity and the average energy price. The latter gradually increases to about three times the 1990 level by 2100. Although energy use per capita doubles, there is a continuous decline in the energy intensity calculated as the ratio of primary fuel supply and GWP, from the present 14 to about 5 MJ/$. *Figure 13.6a* shows another system characteristic: the investments in the energy system. Almost half of these investments go into the electricity system, due to the capital-intensive nature of the NTE options. Because biofuel yields approach their limits, expensive oil re-enters the market and the investments in conventional oil exploitation remain significant in the second half of next century. Overall cumulative investments in the 1990-2020 period are in the

---

6   In the model biofuels only penetrate the markets for Light Liquid Fuel (gasoline, kerosene, etc.) and natural gas.

7   A better researched production function for biofuels and more insight into the substitution dynamics is needed to refine this analysis.

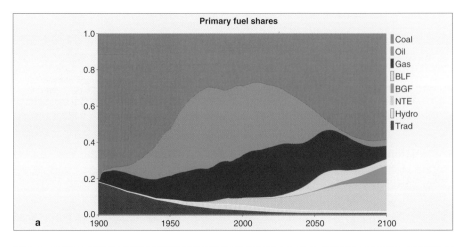
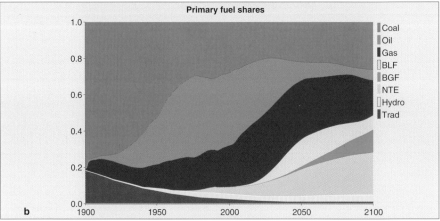
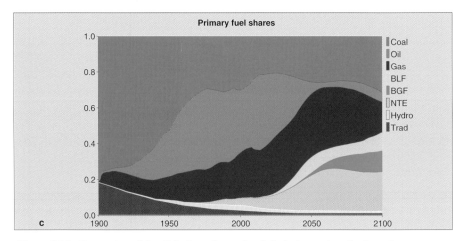

Figure 13.3 Proportion of fossil fuels and non-fossil fuel alternatives in the primary energy supply for the hierarchist (a), the egalitarian (b) and the individualist (c) utopia.

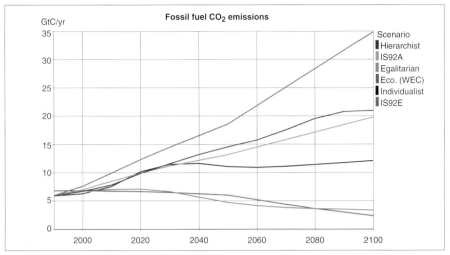

Figure 13.4 The CO_2 emission trajectories from fossil fuel burning in the three utopias for the period 1990-2100. Also indicated are three scenarios from other reports which can be associated with the three utopian perspectives. Note the difference between the individualist utopia and the high-growth IS92e scenario of the IPCC.

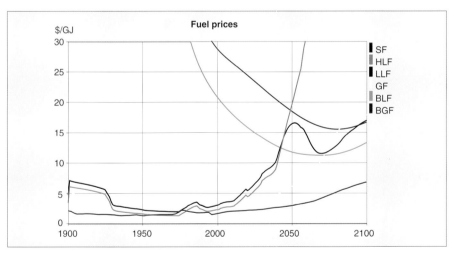

Figure 13.5 Changing prices of various fuels force fuel substitution into the hierarchist scenario. As fossil fuel resources are depleted, fuel costs rise but the rise in Light Liquid Fuel (LLF) and Gaseous Fuel (GF) costs are stabilised by the cost reductions in BioLiquidFuels (BLF) and BioGaseousFuels (BGF).

order of $ 18\times10^{12}$ (1990). This compares reasonably well with the recent estimates of cumulative capital requirements of $ 16\times10^{12}$ (1990) for a medium-growth scenario (IIASA/WEC, 1995). The overall energy expenditures, defined as the product of secondary fuel use and prices, rise as a percentage of GWP from about 6%

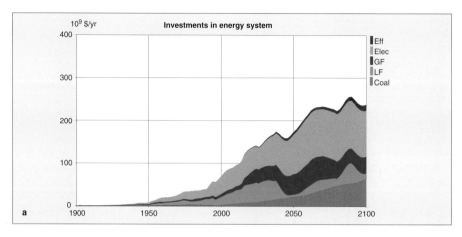

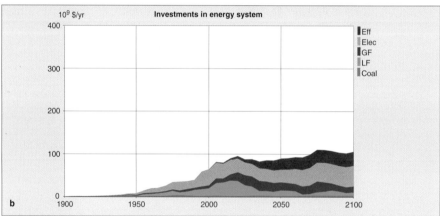

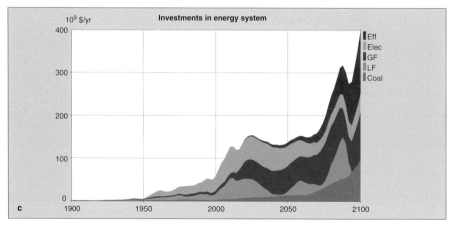

Figure 13.6 Energy system investments for the hierarchist (a), the egalitarian (b) and the individualist (c) utopia. Investments in enegy efficiency are underestimated because we assume replacements at no cost.

Sensitivity analysis

We performed a number of sensitivity experiments for the hierarchist scenario in which some key variables (energy demand, primary fuel production, CO_2 emission) have been calculated for a range of parameter values. The structural change parameter, i.e., the assumption on the relationship between sectoral activity and end-use energy demand (before AEEI and PIEEI, see Chapter 5) is kept constant for the three perspectives like another influential variable: the conversion efficiency from secondary fuel to end-use energy. Neither are included in the sensitivity analysis. The assumptions on the AEEI rate and the PIEEI cost curve induce changes in secondary fuel use in the order of about 15% for the domain of values found in the literature. A rather important assumption is the lower limit on the AEEI factor. The influence of the cost-curve decrease rate and the desired payback time only becomes pronounced (that is, more than ±10%) for values which are rather extreme, e.g. a 1% per yr decrease rate and payback times of more than eight years

Electricity use only drops significantly if more optimistic assumptions are used for the AEEI factor and the price elasticity. We also explored what would happen if the structural change parameter for electricity demand remains at the present high level and is reduced by half for non-electricity demand – a transition to an 'all-electric' society. The share of electricity in final demand rises from 19% in 1990 to 51% instead of 42% in 2100. The resulting CO_2 emissions in 2100 are reduced by 2.3 GtC/yr as compared with the reference case, giving an indication of what successful introduction of electric cars, for example, could mean.

On the supply side, the assumption that thermal electric conversion efficiency rises to 70% in 2100 instead of 45% causes a 30% reduction in fossil fuel use for electric power generation. However, equally large reductions occur, at least in the long term, when the NTE learning coefficient is doubled and/or NTE can be operated at a high (0.8) base load factor. Another sensitive parameter is the low cost of coal for electricity generation: a doubling can be expected to induce a major shift towards oil and gas which in turn will stimulate the introduction of biofuels. For the cost assumptions on coal, oil and gas, it is the shape of the long-term supply cost curve determining the oil and gas depletion cycle and hence fuel substitution dynamics, which really matters. The results of these sensitivity analyses have been used in the implementation of the perspective-based model routes.

at present to 10% for the second half of next century, which is comparable to the high level in the 1980s (*Figure 13.7*). The slow rise in the next 40 years reflects the increasing costs to produce oil.

These simulation results describe a medium-growth world in which the energy transition is only partly realised. Energy intensity decline is impressive; biofuels and non-thermal electricity generation do play a role, but abundant resources bring coal back to the forefront in the second half of the next century, when oil and gas resources become uncompetitive. This, and one of its consequences, rising CO_2 emissions, is one of the more controversial aspects of this scenario.

The egalitarian utopia

If the world is managed by egalitarians, there will be more incentive to develop energy efficiency-oriented technology and stimulate its penetration[8]. We assume that with active support from the NGOs the AEEI rate can be raised to 1.5% per yr and

---

8 We have not changed the structural change multiplier, as such changes would require more research. There are good arguments for an egalitarian world to have a lower growth elasticity because of changing life-styles, more public transport and the like. On the other hand, the lower GWP growth rate slows down in the model, at a rate at which structural change contributes to a lower energy intensity.

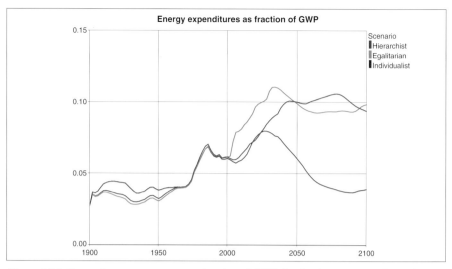

Figure 13.7 Expenditures on energy as a fraction of GWP for the three utopias. Expenditures are calculated as the product of secondary fuel and electricity use and their respective prices, excluding a carbon tax.

that the decline in the PIEEI cost curve is twice as fast as in the hierarchist scenario. Moreover, consumers are willing to use longer payback times because of information campaigns and concern about impending climate change. It is also assumed that coal use is actively discouraged in both the end-use and the electricity generation market due to its environmental disadvantages. The major policy instrument is a worldwide carbon tax increasing to $ 500 per tC (about $ 12,5 per GJ) in 2020 and constant thereafter. This would be accepted after successful negotiations during which regions like China and India are convinced to revise their coal expansion plans and to focus instead on oil and gas, the availability of which increases because of energy conservation efforts in the industrialised regions. Later on, their economies will be strong enough to introduce the renewables, by then significantly cheaper.

In the egalitarian utopia the population is only 8×10^9 at $ 7000 per cap in 2100 (see *Figure 12.2b*). Mainly as a result of this and the carbon tax, the trajectory of secondary fuel use is almost 70% below the hierarchist scenario. The proportion of electricity grows towards 50%. The AEEI factor runs about 10% point below the hierarchist scenario values. The price-induced energy conservation increases to 35% (services) – 55% (transport) by 2100 as compared to 5-10% in 1990; for electricity it is still a low 5%. Electricity generation in an egalitarian utopia will use less coal because it is more costly. Moreover, efficient combined-cycle and fuel-cell power plants lead to a higher average thermal electric conversion efficiency and NTE options are vigorously supported. As a result, fossil fuel use is down by a factor of almost 4 compared to the hierarchist utopia. There are some 18,000 large power plants less than in the hierarchist scenario.

With regard to fossil fuel supply, the more conservative estimate of low-cost natural gas availability – also reflecting the attitude that such valuable non-renewable resources should be saved for future generations – allows for an earlier and faster penetration of modern biomass-based fuels. The result is that primary fuel production peaks at 400 EJ/yr in 2025 and coal production remains at the 1990 level, while its proportion drops to 20-25%. Renewable sources increase their contribution to almost 50% by 2100 (*Figure 13.3b*). This is reflected in CO_2 emissions peaking at about 7 GtC/yr between 2000 and 2030, after which they decline to 3-4 GtC/yr (*Figure 13.4*). The carbon tax discourages the use of fossil fuels and especially coal: its price increases fourfold between 2000 and 2020. Investments flow into energy efficiency and non-fossil fuel-based electricity generation to the extent that in the second half of next century over two-thirds of total investments go to these two options[9]. The absolute investment level is modest, at most twice the present one (*Figure 13.6b*), but as a fraction of GWP, it rises to 10% around 2040 after which it slowly declines to about 8% (*Figure 13.7*). In the egalitarian utopia the present generation indeed makes a sacrifice for the next, but whether and how much this will benefit these future generations is the question. It will be discussed in Chapter 17 and 18.

The individualist utopia

For the individualist, a utopian world will be driven by markets and prices, and technological innovation. The differentiation with regard to the other perspectives has been introduced by higher rates of energy efficiency improvements and fast learning for non-fossil supply options once the prices signal their competitiveness. The consumer will tend to use a short-term horizon, hence short desired payback times. Like the egalitarian, the individualist supposes that the price of coal will go up because it is inconvenient and subsidies are removed (Kassler, 1994; IIASA/WEC, 1995). For surface-coal mining environmental impacts absorb the cost reductions from learning. The assessment of natural gas resources is optimistic: the same amount as for the hierarchist is available at half the cost. Options for high-efficiency thermal conversion will fulfill their promise: by 2100 thermal efficiency reaches an average 60%. NTE capacity can be operated at a high base-load factor. Biofuels become cheap because of fast learning and cheap labour.

All this technological optimism leads to an individualist utopia in which energy use does not exceed the hierarchist level of about 800 EJ/yr by 2100, 40% of which is in the form of electricity. This is possible with the high economic growth rate because the energy intensity declines to a very low 2.5 MJ/$ due to 50-70% autonomous efficiency improvements and 20-30% price-induced efficiency improvements with respect to 1990. NTE rapidly penetrates the electricity generation

---

9 Without the - high - carbon tax, CO_2 emissions are about 5.4 GtC/yr by 2100. The investments in energy efficiency are underestimated because we do not consider replacement costs.

market up to 50% by 2050 and 80% by 2100. Biofuels grow to a rather small 10% by 2050 as they have to compete with cheap natural gas. However, by 2100 they contribute in the order of 25%, when both oil and gas have become scarce and expensive (see *Figure 13.3c*). Coal use increases to some 250 EJ/yr by 2100 as compared to over 700 EJ/yr in the hierarchist scenario. These changes together lead to a stabilisation of CO_2 emissions at 10-12 GtC/yr from 2030 onwards (*Figure 13.4*). The investments in the energy system rise steeply after 2030, when fossil fuel depletion starts to play a role (*Figure 13.6c*). As a fraction of GWP, they are of the same order of magnitude as in the egalitarian utopia (*Figure 13.7*). This reflects the technological optimism of the individualist borne out in the form of cheap and abundant non-fossil fuel options to supply energy for a huge economy with highly efficient energy consumers. Prices of oil-based fuels increase and are successfully stabilised by cheap biofuels, but after 2060 biofuels start to face land-related constraints and prices go up. Coal prices go up only slightly faster than in the hierarchist scenario because the slower depletion rate partly compensates the cost increasing factors.

13.5 Uncertainties and dystopias: some more model experiments

The implementation of the three perspectives gives an indication of the uncertainties which surround any projection of energy-related variables. In this section we present an uncertainty analysis in which for a given population-economy scenario and energy perspective, input variables and parameters are varied across the uncertainty domain generated by the implementation of the three perspectives. Next, we discuss futures in which the dominant management style within the energy system is at odds with how the world really is. There are 24 such semi-dystopias (see Chapters 10 and 11). We choose to highlight only a few of them, on the basis of the plausibility and consistency of the related stories.

Uncertainty ranges for the three utopias
We performed an UNCSAM analysis to assess the uncertainty involved in the various scenarios. For each of the three utopias, all the input parameters and variables used for the differentiation of the three perspectives are varied uniformly throughout the domains. The population and GWP scenarios are given their utopian values and are not varied. *Table 13.2* shows the 2.5 and 97.5 percentile values for a few key output variables; the values for the utopias are given in parentheses. *Figure 13.8* presents the CO_2 emissions that correspond with these experiments.

For the hierarchist world 95% of the paths of secondary fuel and electricity use in 2100 fall between 610 and 790 EJ/yr. The uncertainty on primary energy supply is slightly greater; CO_2 emissions in 2100 vary between 11.5 and 17 GtC/yr. Major parameters and variables which contribute to the uncertainty band are: the AEEI and its lower limit for industry and electricity, the relative cost of labour in underground coal mining, the learning coefficient for NTE and biofuels, and the thermal efficiency of fossil-fired power plants. The rather narrow uncertainty band is partly due to the fact that the relationship between end-use energy demand and activity levels is the same for all three perspectives. The divergence in the assumptions on NTE options and biofuels is reflected in the factor 2 between the upper ($ 20 per GJ) and the lower ($ 9.5 per GJ) probability paths for the average energy price by 2100. This creates a rather large difference in the incentive for energy conservation but the final impact is relatively minor due to the effect of the assumed rise of the marginal cost of energy conservation. Also, a number of uncertainties may cancel each other out.

In the egalitarian utopia, uncertainty bands for energy demand and CO_2 emissions are in relative terms similar to the ones in the hierarchist world, but are quite small in absolute terms, despite a rather large uncertainty in the average energy price ($ 8.5 - $ 16 per GJ by 2100). This indicates that the assumptions on population and economic growth dominate. In the individualist utopia, the average energy price in the coming decades is much below the uncertainty bands for the hierarchist and the egalitarian. However, after 2025 it starts to exceed the hierarchist and after 2055 the egalitarian value spectrum. By then the resource constraints become more severe because cumulated production is the highest of all three utopias. From a resource-depletion point of view, the individualist world view indeed favours the short-term benefits. For energy demand and CO_2 emissions the relative

Variable value in 2100	Hierarchist		Egalitarian		Individualist	
	<2.5%	>97.5%	<2.5%	>97.5%	<2.5%	>97.5%
Secondary energy use (EJ/yr)	610	790 (810)	190	220 (250)	980	1220 (800)
Primary energy production (EJ/yr)	830	1110 (1230)	270	330 (300)	1280	1630 (1070)
CO_2 emission (GtC/yr)	11.5[a]	17[a] (20)	2.5[b]	3.6[b] (3-4)	14.5	21.5 (12)
Average energy price ($/GJ)	9.5	20 (16)	8.5[c]	16[c] (15)	13	24 (11)

[a] Peaking in 2080 at 13 GtC/yr and 18 GtC/yr, respectively.
[b] Peaking in 2000 at 6.5 GtC/yr and in 2025 at 7.6 GtC/yr.
[c] Peaking around 2040-2060 at $ 13.5 per GJ and $ 18 per GJ, respectively.

Table 13.2 Uncertainty ranges for the three utopias (utopia values in parentheses).

uncertainties are comparable with those in the other two utopias. In absolute terms they are large, 14.5-21.5 GtC/yr for the period 2080-2100, which emphasises the risk aspect of such a future if the climate turns out to be sensitive.

It is interesting to analyse the position of the utopian values relative to the uncertainty ranges. It turns out that the hierarchist projection of energy use and CO_2 emissions in the utopia is outside its uncertainty bands (see *Table 13.1* and *Figure 13.8*). The reason is that the technological optimism of both the egalitarian and the individualist world view weigh heavily, and making the hierarchist estimates implausibly high, given its medium population and economic growth projection. For the same reason, energy use and CO_2 emissions in the egalitarian utopia fall above or at the upper end of the uncertainty range. For the individualist utopia the opposite happens. The rather conservative estimate of size and cost of oil and gas resources and the diverging views on coal prices in the hierarchist and egalitarian scenario pull the individualist utopia down to the extreme low end of its uncertainty range. In 2100, for instance, utopian CO_2 emissions are between 2050 and 2100 10-12 GtC/yr, whereas 97.5% of all uncertainty experiments show an emission path above 13 GtC/yr, indicating that the estimate of CO_2 emissions in the individualist utopia have a fair chance of being exceeded.

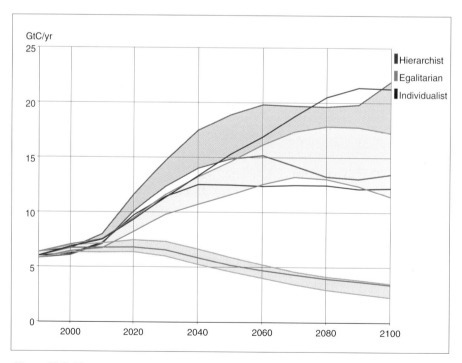

Figure 13.8 The uncertainty ranges in the CO_2 emission paths for the three scenarios for the period 1990-2100. The shaded areas indicate the 95% percentile for the three utopias.

Dystopian futures

Non-utopian futures are, in the present context, scenarios in which the world view and/or management style with regard to the energy system differ from those applied for population/health and economy. Although such discordance does not always have disruptive consequences, we will refer to such scenarios as dystopias (see Chapters 10 and 11). We confine the discussion here to two variables. *Figure 13.9* shows the energy expenditures along the *y*-axis as a fraction of GWP in 2100 and the CO_2 emissions in 2100 along the *x*-axis. Energy expenditures, defined as the product of secondary fuel and electricity use and their respective prices but excluding any carbon tax, are used as a proxy for the economic cost. The triangles within the solid lines are drawn around the three points with the population-economy perspective, i.e. the 'background', individualist (I) and egalitarian (E). The experiments with a hierarchist background give results in-between.

With the hierarchist or egalitarian world view on the energy system, the high-growth individualist background would raise emissions far above the IPCC-IS92a level of 20 GtC/yr (upper points in triangle I: HII, HHI, EII, EHI[10]). This would be disastrous in a world with a sensitive climate system (see Chapter 16). The technological optimism of the individualist would more than halve these emissions at substantially lower relative expenditures (lower points in triangle I: III, IHI). The dashed triangle (I*) shows how the situation changes with an egalitarian management style, i.e. with a $ 500 per tC carbon tax: relative expenditures rise by more than 2% of GWP and emissions are 40-55% lower for the hierarchist and egalitarian world view (upper points in triangle I*: HEI, EEI). With an individualist view on the energy system, the carbon tax is equally as effective in reducing CO_2 emissions as an egalitarian view.

The low-growth egalitarian background would imply emissions in 2100 below 8 GtC/yr, irrespective of the way in which the energy system is seen. However, costs differ significantly: with a hierarchist or egalitarian world view, energy expenditures as a fraction of GWP are 2-3 times higher (upper points in triangle E: HEE, EEE). The individualist assumptions on technology roughly halve emissions to less than 4 GtC/yr (lower point in triangle E: IEE). If in this situation a carbon tax is applied, emissions are reduced even further but at very high relative costs for a hierarchist or an egalitarian view on the energy system (upper points in dashed triangle E*: HEE, EEE). Even if the individualist technological and resource optimism were only partly justified, a carbon tax in a world with a low population and economic growth of the egalitarian utopia would be an unnecessary and expensive venture.

Of course, management styles which are obviously in disagreement with real-world observations will not be maintained for a whole century (see Janssen, 1996).

---

10 The index WMB means world view W, management style M and background B. Note that the points HII and HHI coincide because the management style of the hierarchist and the individualist hardly differ.

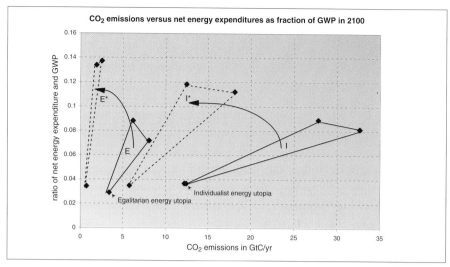

Figure 13.9 CO_2 emissions versus net energy expenditures as a fraction of GWP in the year 2100: the egalitarian and individualist utopias and dystopias, and the consequences of an egalitarian management style ($E \rightarrow E^*$; $I \rightarrow I^*$). Assessment of risks associated with high CO_2 emissions are given in Chapter 16.

No human response to dystopian tendencies is modelled. Another shortcoming is that the introduction of a carbon tax is not supposed to influence the economic growth scenario. If such a tax were to slow down economic growth, its (non-) implementation could have much larger, though not necessarily more dystopian repercussions. Yet, it is clear from these simulation experiments that the aspiration of a high-growth world has to rely on a combination of technological optimism and

Dystopian narratives

One possible dystopia is indicated by the upper right points in triangle I (Figure 13.9). High economic growth is successfully pursued and the consumer society is too seductive for most world citizens to resist. The expected decline in energy intensity and decarbonisation does not take place. Energy conservation and renewable energy turn out to be expensive. Vested interests ranging from reluctant oil-exporting countries to China and India opposing any curtailment of their coal expansion plans obstruct the implementation of regulatory policies. Attempts to introduce an egalitarian management style are confronted with bureaucratic opposition and system inertia. In such a high-growth world (individualist) with disappointing results from technology and emission-reduction policies, CO_2 emissions would soar to the high levels feared for by some egalitarian groups.

The upper part of the dashed-line triangle, E^*, is a different story. If public opinion is swayed towards an egalitarian management style in a low-growth future, emissions will drop because of stringent emission reduction policies. If the climate system then turns out to be insensitive, an unnecessarily large burden has been placed upon the economy – most of which, however, cannot be explicitly modelled[11]. The large energy expenditures may aggravate poverty and inequity, and add to social unrest.

11 Here the feedback parameter for economic growth can be experimented with.

energy taxing if the world wants to avoid emissions exceeding the 10 GtC/yr level. Also, if the emissions of 5-10 GtC/yr or an equivalent 500-1000 GtC cumulative emissions during the next century are considered an acceptable range, the imposition of a high carbon tax would be a heavy and unnecessary burden in a low-growth world.

13.6 Conclusions

We have used the Energy submodel TIME to investigate possible energy futures. Implementing divergent estimates of important model parameters and performing experiments with the resulting perspective-based model routes produces a wide spectrum of energy futures. The hierarchist utopia closely resembles the IPCC IS92a scenario which is often used as a yardstick in climate policy research. The egalitarian utopia resembles several published scenarios, as does the individualist utopia. There may be important differences, for example, due to the omission of model experiments which account for coal liquefaction and gasification or an 'all-electric' or hydrogen-based energy economy. Nevertheless, we feel confident that the three utopian scenarios form good basis for the integrated TARGETS1.0 simulations discussed in subsequent chapters.

The uncertainty analysis performed with the Energy submodel confirms the view that the IPCC IS92a scenario – which is similar to the hierarchist and can best be called a coal scenario – is rather implausible. For the chosen population and GWP scenarios an integrated approach such as that used in the Energy submodel indicates a CO_2 emission range of 2.5-21.5 GtC/yr by the end of next century. Both the hierarchist and the individualist utopia imply rather high environmental risks. The egalitarian utopia poses a much smaller threat to the environment, but it may jeopardize the material welfare of large groups of people. It seems both utopias have a rather low probability of occurrence because of the many counterbalancing forces wthin the system.

Most probably the next decades will see the effects of all three utopias. In different ways, they will all contribute to a transition away from an energy-inefficient and carbon-based energy economy. But only events which can neither be anticipated nor modelled at present can bring global CO_2 emissions below 1990 levels within 30-40 years. Some fear a sudden change in climate across large parts of the world or a severe disruption in oil trade. Others look forward to such events as the discovery of a huge new gas province or a technological breakthrough in the ways in which we use and produce heat and power. Such differences in perspective will always colour the outlooks on the future. This chapter provides answers to some of the questions posed in section 13.1; the rest are in Chapter 16.

14
Water in crisis?

"The water that holds the ship, is the same water that engulfs it."
Chinese saying

14 WATER IN CRISIS?

Arjen Y. Hoekstra, Arthur H.W. Beusen, Henk B.M. Hilderink and Marjolein B.A. van Asselt

The central question in this chapter is: can the world population be provided with an adequate and sustainable supply of clean water? We report on experiments with the AQUA submodel and attempt to offer insight into the role of population and economic growth, intensive agriculture, technology and pricing. Again three cultural perspectives are applied to establish model routes which reflect different assumptions and societal responses. The model experiments provide a number of water projections for the next century. We review low, medium and high risk developments and the effects of various water policy strategies. The chapter concludes with a proposal for future policy priorities to safeguard a sustainable supply of clean fresh water.

14.1 Introduction

There are too many uncertainties to give a simple answer to the question whether we will have enough clean water for the next century. However, we can do some exploratory work if we make certain assumptions. In this chapter, we present such an exploration based on the AQUA submodel (Chapter 6). As a heuristic device for composing coherent sets of assumptions, different perspectives are used (see Chapter 10). First, we discuss some of the major questions, uncertainties and controversies related to water and sustainable development. We then look for coherence in the different points of view by considering the controversies from three perspectives: the hierarchist, the egalitarian and the individualist. Next, we use the three perspectives to implement 'model routes' within the AQUA submodel and to present model-based projections of water in the next century. Finally, we distinguish low, moderate and high risk futures, and analyse the effects of different water policy strategies. More detailed background information can be found in Hoekstra (1997).

14.2 Questions related to water

Will water play a key role in development and does it therefore warrant a major focus in studies on global change? Or is water secondary to issues such as economic development, human health, energy supply, food supply, biodiversity and global warming? Some researchers and institutions assign a great deal of importance to hydrology, water pollution, irrigation and public water supply and sanitation. Such interest in water has been expressed in, for example, the *Global 2000 report* (Barney, 1980), the *International Drinking Water Supply and Sanitation Decade* (WHO, 1984; 1991), *Agenda 21* (UN, 1992), the *International Geosphere-Biosphere Programme* (IGBP, 1990; 1993) and various reports of the Worldwatch Institute (Brown *et al.*, 1993; 1996) and the World Resources Institute (WRI, 1992; 1994; 1996). As noted by McCaffrey (1993), some observers predict that disputes over fresh water may even escalate into 'water wars'. However, other researchers of global change only lightly touch upon the issue of fresh water. One can find this, for example, in *Limits to growth* and its successor *Beyond the limits* (Meadows *et al.*, 1972; 1991), *Our common future* (WCED, 1987) and in Shaw *et al.* (1992).

The central question for our analyses here is: *Can the world population be provided in the long term with sufficient, clean fresh water without threatening the ecological functions of water?* The answer depends on assumptions on, for example, the physical or economic limits to fresh-water supply, the development of water-saving technology, our ability to adapt our life-style to water scarcity and the extent to which we can shape our environment. Given the expected population growth and supposing that the intensity of water use will not decrease, authors like Ambroggi (1980), Postel (1992) and Kulshreshtha (1993) show that on a global scale the difference between actual and potential supply will decline considerably. Besides, they point out that water scarcity problems already exist due to the regional disproportion between water demand and potential supply. However, there are also researchers who are more optimistic about water scarcity. Engineers indicate that using water more efficiently and applying re-use techniques can lower demands significantly. Through reservoir construction and desalination of salt or brackish water, humans are also able to increase potential water supply. In countries with a water scarcity, new techniques are indeed being developed and used[1]. Many economists do not expect serious water problems, because, as they argue, considering water as an economic asset will result in far more efficient water supply (Anderson, 1995b).

The central question posed above can be broken down into more specific questions: What makes clean fresh water scarce and how can scarcity be overcome? How much fresh water can be supplied? What determines potential water supply and how can it be increased? What determines water demand and how can demand be managed? These questions along with a gamut of opinions will be discussed in the next section. We will limit ourselves there to uncertainties and controversies that relate to water demand and supply. However, in section 14.4 where we discuss the perspective-based model routes in AQUA, we also show how we deal with uncertainties on the nature of the water system (how robust is it, how it will respond to human disturbances, etc.).

---

1 It is not a coincidence that 60% of the world's desalination capacity is found in the water-poor but energy-rich nations of the Persian Gulf and that Israel has cut its water use per irrigated hectare substantially by applying more efficient, self-developed techniques (Postel, 1992).

14.3 Major controversies and uncertainties

14.3.1 What makes water scarce?

Due to uncertainties on the nature of scarcity, its causes and possible solutions, this simple question has a variety of answers. To better understand different perceptions of water scarcity, three extreme points of view will be discussed (*Table 14.1*). The perceptions in the scientific literature actually form a spectrum of these extremes.

Water scarcity as a supply problem
Many plans for water resources management conceive water demand as something 'given' and water supply as something 'to meet demand'. Water demand is then not considered a policy issue, but a fact emanating from economic developments, population growth and an increased demand for irrigated cropland. The actual issue is understood to be the provision of enough water with a quality sufficient for the relevant sectors in society, leaving enough to fulfil ecological requirements. Water policy should then aim at proper management of the physical water system, an approach found all over the world, e.g. the Second Water Master Plan of the Netherlands (Ministry of Public Works, 1985) and the extensive water resources development study for northern West Java (PPPP and Delft Hydraulics, 1989). A characteristic feature of this type of study is that most of the attention is given to analysing available water quantities and qualities, and water supply infrastructure. If relevant, this may include the effects of erosion, flooding,

Issue	Points of view	Key words
Water scarcity	(1) supply problem	shortage of supply, making supply meet demand, water supply infrastructure
	(2) demand problem	growing demands, potential supply limited, demand policy, human behaviour
	(3) market problem	supply costs, water price, market pricing
Potential water supply	(1) total runoff	fresh-water renewal rate
	(2) stable runoff in inhabited areas	losses due to flooding, pollution, etc.
	(3) clean fresh-water stock	non-renewability, depletion, pollution
	(4) no limits	new technologies (desalination, reuse, etc.)
Water demand	(1) a given need	demand independent of supply
	(2) a manageable desire	politics, social preferences, customs
	(3) in equilibrium with supply	price mechanism, technology

Table 14.1 Summary of the controversies.

human-induced evaporation and climate change. Perceived in this way, water scarcity is expressed in physical terms. Indicators to express water scarcity are, for example, various aridity indices (Shiklomanov, 1993; Leemans and Born, 1994; Deursen and Kwadijk, 1994) or an index like 'length of the growing period' (Oldeman and Velthuyzen, 1991). Water quality is described in terms of physical, chemical and biological properties. Problems are stated in terms of violation of standards. It is perhaps not surprising that this perception of water scarcity is often found among natural scientists (hydrologists, climatologists) and engineers. Potential water supply can be increased by constructing dams and reservoirs that stabilize runoff.

Water scarcity as a demand problem
Another point of view is that potential supply is limited and that demand cannot continue to increase. The actual driving force behind growing water scarcity is the augmentative demand. Falkenmark *et al.* (1987), for example, regard the increase of water demand as the major cause of the present water problems. An important force behind growing demands is population growth. In nearly all parts of the world, the fraction of available fresh water being used increases, which is a sign for action in the regions that have reached critical levels. Solutions for water scarcity should somehow manage demand and thus human behaviour. The increasing water shortages have led to interest in minimum water requirements, distinguishing between primary and secondary purposes of water. Especially primary uses like water for drinking deserve attention, making a social measure like public water supply coverage another indicator that can be associated with this view (WHO, 1984).

Water scarcity from an economic point of view
Simon (1980) states that the only meaningful measure of scarcity in peacetime is the cost of the asset in question. A substantial group of scientists holds this view for water as well (e.g. Turner and Dubourg, 1994; Anderson, 1995b). The view has also caught on in politics, since one of the 'guiding principles for action', embodied in the Dublin Statement on Water and Sustainable Development (ICWE, 1992) is that "water has an economic value in all its competing uses and should be recognised as an economic good". Within this view, it is not hydrological or water-quality data but the costs of water supply that are the correct indicators for water scarcity: limited potential supply and effects of water pollution are or should be accounted for in the water supply costs. Solutions for water scarcity are being sought in options like charging full costs to water users and privatising water supply companies.

14.3.2 What determines potential water supply?

The total fresh-water stock on Earth is finite, about 35×10^6 km^3, and comprises only 2.5% of the total water stock (Shiklomanov, 1993). About two-thirds of the fresh water is stored as ice and snow in the Antarctic and Arctic regions and is not available for human use. Fresh groundwater forms nearly one-third of the total fresh-water stock; fresh-water lakes and rivers comprise only 0.27%. Although stocks are finite, the cycling process and thus the renewal of fresh water is infinite. This, coupled with the high variation of fresh-water flows in time and space, might be the reason for different estimations of potential water supply. Below we discuss four main streams of thought, each having something of a stereotype character (*Table 14.1*).

Total runoff as a measure of potential water supply
The most common approach is to take the total runoff in a river basin as a measure of the fresh-water availability or potential water supply in that basin (WRI, 1992; 1994; 1996; Kulshreshtha, 1993; Shiklomanov, 1995), the basic idea behind this being that fresh water is a renewable resource. This implies that it makes more sense to look at the renewal rate than at the size of the stock. The fresh-water renewal rate of a river basin, expressed in km^3/yr, is defined as total precipitation minus total evaporation (which amounts to the total runoff). An advantage of this approach is the unambiguity of the definition, which leaves no room for dissension other than that on the runoff data, which are generally highly inaccurate (Speidel and Agnew, 1988). A major drawback is that it does not account for losses due to flood runoff, runoff in remote areas and pollution, thus giving a profound *overestimation*. Criticism from another side is that the approach is limited to assessing fresh-water recharge, ignoring, for example, the possibility of desalinating sea water. From this point of view, the approach will yield a *conservative* measure of potential water supply.

Stable runoff in inhabited areas as a measure of potential water supply
Some authors regard the total runoff in a river basin as an upper limit to potential water supply and propose reductions for losses due to flooding and runoff in uninhabited areas (e.g. Ambroggi, 1977; 1980; Postel *et al.*, 1996). This approach results in a much lower assessment of potential water supply than if one were to consider total runoff. On the global scale, for example, Ambroggi (1977; 1980) arrives at a figure of 23% of the total runoff; Postel *et al.* (1996) give a figure of 31%. Another reduction could be used to account for pollution and to guarantee a certain minimum runoff for maintaining aquatic and riverine ecosystems. Besides, it would be better to reckon with a dry year instead of an average year, to ensure that the measure for potential water supply also holds during dry years. The 'reduced runoff approach' has the advantage of carrying rather precise and prudent information on potential water supply, but the definitions used may give rise to many different interpretations and calculations.

The clean fresh-water stock as a measure of potential water supply
A third view on potential water supply is through water stocks instead of flows, considering water as a non-renewable resource. Two key words in this view are depletion and pollution, both processes that reduce the clean fresh-water stock.

According to this view, the remaining clean fresh water is a better measure for the effect of intensive water withdrawals and pollution, and thus for the capacity for more withdrawals than the ratio between water use and total (or stable) runoff. This view is uncommon among scientists but popular among interest groups who actually observe that water is getting scarcer: water tables decline and pollution affects large quantities of water. Fresh-water lakes and rivers form only 0.27% of the total fresh-water stock, about 93×10^3 km^3. However, lakes and rivers are the major source of water: about 71% of the world water supply is derived from fresh surface water (Kulshreshtha, 1993). Therefore the size and quality of the fresh surface water stock are considered important measures for potential water supply. The fresh groundwater stock, nearly one-third of the total fresh-water stock, provides the remaining 29% of the global water supply. Since depletion of deep aquifers and declining groundwater tables are serious issues today, as well as salt-water intrusion in coastal areas and other types of contamination (Postel, 1992), groundwater quantity and quality are also considered important indicators for potential water supply.

No limits to potential water supply
Although the concepts of potential water supply mentioned above differ considerably, they correspond in their recognition of *some sort of limitation*. However, as Falkenmark (1989) observes, it is by no means generally accepted that a limit to potential water supply actually exists. Both engineers and economists have a certain opposition to the so-called 'water barrier' concept. They adhere to a technological optimism in which problems of scarcity are supposed to be solved with new technologies. A confirmation of this view is found in the growing desalination capacity in many water-poor regions. In Saudi Arabia, for example, desalination of salt or brackish water is already responsible for about 20% of the total water supply (Gleick, 1993). Another possibility is water re-use after treatment (Dean and Lund, 1981) for either the same or another purpose, thus creating a large new source of fresh water; only losses have to be added from outside the recycling system. Other unconventional technologies to extend our resource base, attracting attention in recent decades but still in an experimental and a conceptual stage, are cloud seeding and towing icebergs (Ambroggi, 1977; White, 1983).

14.3.3 What determines water demand?

In the 1900-1990 period global water demand grew by a factor of seven (Shiklomanov, 1993) due to both an extension of water-demanding activities and an intensified water demand per activity. Clearly, population and economic growth, and technological development, have been important determinants, but what mechanisms have been at work is less understood. We will proceed to discuss three different views on how water demand may develop.

Demand as a given need
The most common attitude towards water demand is probably to consider population growth and agricultural and industrial developments as given processes, and to reckon with certain water use intensities per sector (Kulshreshtha, 1993; Shiklomanov, 1995). "Will future water needs be met?" is a question in the latest WRI report (1996), which upholds the assumption that there are no complex dynamics between demand and supply, although there is a certain demand which simply has to be supplied. An advantage of this mechanistic view on water demand is that it enables us to make projections of future water demand in a relatively simple way, using scenarios mainly for population, economic growth, industrial activity and irrigated surface area. A major drawback is that the actual processes determining water demand are ignored: political choices, customs, individual preferences, the economic demand–supply mechanism and technological development.

Demand as a manageable desire
Another view on water demand is that only water use related to 'basic needs' is a necessity. Water demand above the minimum requirements is considered a luxury and largely subject to social and political desires. Political allocation priorities strongly influence the extent of water use in the different sectors. Large irrigation schemes have been planned by governments and are still being planned all over the world. A review of public water supply projects financed by the World Bank has shown that about 65% of the average supply costs are covered by public funds; this figure is even larger for irrigation costs (Serageldin, 1995). At community level, water demand is thought to be largely a function of customs and human behaviour, which may change through improvement of environmental awareness or for example by a water-tax.

Demand as an economic concept
From an economic point of view, water demand is constantly seeking an equilibrium with water supply. Through the price mechanism, water demand is closely related to water scarcity. Scarcity leads to higher prices which result in lower demands and incentives to develop more efficient technology. Estimates show for example that a 10% increase in price would decrease agricultural water consumption in California by 6.5% and cut overall consumption by 3.7% in the seventeen US western states (Anderson, 1995b). Criticism on the economic conceptualisation of water demand is that it reflects an economists' ideal of rather than the present world. Anderson (1995b) recognises this when he notes that despite the evidence that most water projects do not make economic sense, political pressure continues to allow these projects to proliferate because the interest groups benefitting from them constitute a formidable political force.

14.4 Perspectives on water

The uncertainties and controversies discussed in the previous section are now placed in the context of the cultural theory (Chapter 10). *Table 14.2* summarises how differently the hierarchist, the egalitarian and the individualist perceive issues of water demand and supply. The table also presents different views with respect to the nature of the water system. In the first half of this section we clarify the three perspectives; in the second half we show how we have translated the perspectives into different model routes in AQUA.

The hierarchist

A typical characteristic of hierarchists is to regard scarcity as a supply problem. Their management strategy is to look how they can manage their resources: it is a problem of *make-supply-meet demand*. They regard stable runoff as an appropriate measure for potential water supply and use total runoff as an upper limit to potential water supply. Water demand is seen as a given need, depending on a mix of factors, including economic development, water prices, technology and usage. Although there is willingness to strive for more efficient water use, institutional interests interfere. The goal to provide people with proper water supply and sanitation often justifies the use of subsidies. In line with their perception of nature as both perverse and tolerant, hierarchists assume that temperature changes will alter the hydrological cycle to some extent, but not dramatically. Also the effects of other disturbances such as water withdrawals and pollution are assimilated as long as they do not reach critical levels.

Issue	Hierarchist	Egalitarian	Individualist
Water scarcity	supply problem	demand problem	market problem
Potential water supply	stable runoff	stable runoff in inhabited areas or clean fresh-water stock	total runoff or no limits
Water demand	a given need	a manageable desire	in dynamic equilibrium with water supply
Response of hydrological cycle to temperature change	medium response	high response	low response
Response of hydrological cycle to increasing water use	medium response	high response	low response

Table 14.2 Uncertainties and controversies from three perspectives.

The egalitarian
Egalitarians are prudent in assessing water resources and account for temporal and spatial variabilities. Water scarcity is conceived as a problem caused by growing water demand and pollution; the solution is to manage human needs in the form of policy incentives and shifts in social customs and preferences. Applying small-scale water-saving and re-use technology can lower water-use intensities in all sectors. Environmental impacts should be reflected in the price of water as a tax. Since egalitarians attach great importance to equity, access to safe drinking water and sanitation facilities for everyone is a principal policy goal. According to the egalitarian, the fragile dynamic equilibrium of the water balance is easily disturbed by human activities. Intensive water use, human-induced temperature change and deforestation may considerably affect stable fresh-water availability and sea level. Fertilisers and wastes from households and industries will not only remain on some hot spots, but will also widely spread through the water system.

The individualist
In the water scarcity debate, the individualist point of view largely coincides with what was earlier described as the economic point of view: water is an economic good and should be managed as such. Extension of the stable runoff through reservoirs or of the resource base through desalination are considered as options to increase water use efficiency. In the debate on potential water supply, individualists hover somewhere between considering total runoff as a proper measure and water as an unlimited resource. In the latter case, they fully reckon on the human capability of using desalinated seawater for all our purposes. Total runoff is considered fully available, since remote or flood flows can also be made available if necessary. Water demand is – or should be – determined by the price mechanism: higher prices as a result of increased scarcity will lower demand and stimulate the development of more efficient technology. Subsidisation of water, at present done all over the world, is rejected. In the case of high water scarcity, high-tech options for water supply (e.g. desalination) could be stimulated by governmental institutions. An active policy in public water supply and sanitation is not needed because economic development will adequately increase public water supply and sanitation coverage. In line with their perception of nature as being robust, individualists tend to consider possible disturbances of the hydrological cycle as of minor importance. If intensive water use, deforestation or temperature change is still to have some effect on the hydrological cycle, the changes will be slow enough for us to adapt.

Perspective-based model routes in AQUA
In implementing perspective-based model routes in AQUA, we distinguish between a 'world view' and a 'management style'. The world view is characterised by specific model formulations (*Figure 14.1*) and parameter values (*Table 14.3*) and the management style by specific response strategies (*Table 14.4*). The hierarchist's

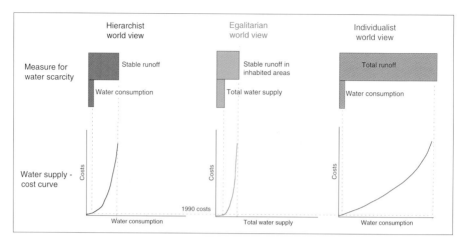

Figure 14.1 Schematic representation of perspective-based model formulations. Water consumption is defined as the part of the total water supply which is lost through evaporation and cannot be re-used. The figures are based on data for the year 1990.

position has been operationalised in the model by assuming medium estimates taken from the literature for growth and price elasticities of water demand. Low estimates are used for technological development and for the diffusion rate of new technology. Stable runoff is considered a measure for potential water supply; the ratio between consumptive water use and stable runoff is taken as a measure of water scarcity. Water-supply costs are supposed to increase as a function of this ratio. The increase of public water supply and sanitation coverage is supposed to increase with GWP per capita. Medium estimates from the literature are used for the response of the global

Perspectives in the literature

Shaw (1994) states that: "Once the needs of an area have been established, be it for a single town's domestic supply, the irrigation water for a commercial plantation, or the total demands of a whole country, and some continuing requirements for the future made, then the engineer must investigate the availability of the resources."

Postel (1996) responds to this point of view: "For all its impressive engineering, modern water development has adhered to a fairly simple formula: estimate the demand for water and then build new supply projects to meet it. It is an approach that ignores concerns about human equity, the health of ecosystems, other species, and the welfare of future generations. In a world of resource abundance, it may have served humanity adequately. But in the new world of scarcity, it is fueling conflict and degradation. [] Nearly the entire spectrum of conservation and efficiency options cost less than the development of new water sources".

This view is again contrasted with the perception of Anderson (1995b), who would rather believe in the market mechanism: "Experience around the world has demonstrated over and over again that the only successful way to avoid shortages is to rely on free market pricing and allocation. The same is true for water. [] If markets in water were permitted, demand would be reduced, supply would be increased, water would be reallocated, and the specter of water crisis would vanish."

Parameters	Unit	Hierarchist world view[1]	Egalitarian world view[1]	Individualist world view[1]
Growth elasticities of water demand (Eqn. 6.1)	-	medium	low	high
Price elasticities of water demand (Eqn. 6.1)	-	medium	low	high
Diffusion rate technology (Eqn. 6.2)	yr^{-1}	low	high	medium
Fractions consumptive water use	-	medium	high	low
Glacier melting parameter α (Eqn. 6.4)	1/°C per yr	medium	high	low
Glacier melting parameter β (Eqn. 6.4)	°C	medium	low	high
Climate sensitivities of ice sheets	mm/yr per °C	medium	high	low
Initial imbalances of ice sheets	mm/yr	medium	high	low
Standard deviations on water quality	mg/l	medium	high	low

[1] For documentation and justification of the exact figures, see Hoekstra (1997).

Table 14.3 Perspective-based parameter values.

hydrological balance to an increase in global temperature (e.g. medium sensitivities of glaciers and ice sheets from Warrick *et al.* (1995)). We also use medium estimates for standard deviations of water-quality distributions.

The egalitarian view is represented in AQUA by assuming low growth and price elasticities of water demand, a high increase in the maximum possible water-use efficiency through new appropriate water-saving technology, and a high diffusion rate for the spread of this new technology. We use a prudent estimation of potential water supply by reckoning with stable runoff in inhabited areas. Water supply costs

Response variables	Hierarchist management style	Egalitarian management style	Individualist management style
Water pricing (ratio of water price and actual cost)	Increasing charges (to 75% in 2025)	Water-taxing (to 110% in 2025)	Market pricing (to 100% in 2025)
Development of water-saving technology	No strong technological push	Strong push for small-scale technology	Stimulation of high-tech. if water gets scarce
Public water supply & sanitation coverage	Depending on economic growth	Active policy: 100% in 2025	Depending on economic growth
Relative use of the different water sources	No change	Decreased use of groundwater	Increased desalination

Table 14.4 Perspective-based responses.

greatly increase if water gets scarcer; water scarcity is calculated as the ratio between total water supply and stable runoff in inhabited areas. A strong water-pricing policy (water tax) is applied and there is an active policy to increase public water supply and sanitation coverages. To prevent groundwater depletion (and resulting sea-level rise), the egalitarian promotes the use of more surface water instead of groundwater. High estimates are used for temperature response parameters, consumptive fractions of water supply and standard deviations of water-quality distributions. The latter implies that – in comparison with the hierarchist case – more water is estimated to fall within the worst water-quality classes.

The individualist point of view has been operationalised in AQUA by assuming high growth and price elasticities of water demand, a sharp increase in maximum possible water-use efficiency through new high-tech water-saving technology, and a medium diffusion rate for the spread of new technology. As a measure for potential water supply we use total runoff; water-supply costs increase as a function of the ratio between water consumption and total runoff. We assume a market pricing policy. The increase in public water supply and sanitation coverages are assumed to depend on economic development in the same way as in the hierarchist world. In the future, the global desalination capacity will grow since there are more and more regions where supply from fresh-water sources becomes more expensive than desalination of salt or brackish water, especially in case of drinking water supply. There is a weak response of the global hydrological balance to an increase in global temperature. Water pollution is supposed to be concentrated in some areas (low standard deviations for water-quality distributions).

14.5 Water in the future: three utopias

We will discuss here three so-called 'water utopias': hierarchist, egalitarian and individualist, all representing a future that develops according to a coherent set of assumptions on 'how the world works', in combination with a management style that matches this world view. In other words, we take the same perspective for both the world view and the management style (Table 11.1).

Reference case: a hierarchist water utopia
The context for the hierarchist water utopia has already been presented in Chapter 11. For water, the most important characteristics are:
- the continued economic growth, resulting in a Gross World Product (GWP) of about $ 1990 250×10^{12} in the year 2100 (Chapter 11), comparable to the IS92a IPCC scenario (Houghton *et al.*, 1992);
- the decline in population growth, resulting in about 11.8×10^9 people in the year 2100 (Chapter 12), comparable to the medium scenarios of the World Bank (1991) and UN (1990);

- the continued deforestation in the tropics and some reforestation in the temperate zone, with a global forest area of about 3000 Mha in 2100;
- the continued but decreasing growth of irrigated cropland, approaching an area of about 350 Mha in 2100.

Under these circumstances, global water supply keeps increasing during the next century, with industry becoming the largest water-use sector and not agriculture (*Figure 14.2*). Domestic and agricultural water use keep growing, but the growth rates level off. Water-use intensities decrease for all sectors, especially the industrial one, where water saving by reuse and substitution is easiest. Nevertheless, industrial water use grows exponentially due to the increasing growth of industrial production. Since the *consumptive* part of industrial water withdrawals is relatively small – i.e. evaporation is small and most of the water withdrawal returns to the fresh-water reservoir – agriculture remains the largest water *consumer* and is thus responsible for the largest impact on the terrestrial water balance. The growth of global water consumption is considerably lower than the growth of total water supply and levels off well below the potential water supply (*Figure 14.3*). Although the level of water use strongly increases, water is used more efficiently and in a less consumptive way. This is what we earlier called the fresh-water transition (Hoekstra, 1995) and what is discussed in more detail in Chapter 17. Water scarcity expressed as the ratio between water consumption and potential water supply grows by a factor of about 2 during the next century. This will result in a significant increase in global average water supply costs. Expenditures in the water sector will grow accordingly. However, expressed as a fraction of GWP, expenditures will first decrease and then slowly increase again (*Figure 14.4*). The decrease during the first 25 years of the next century relates to the water-pricing policy assumed.

The total fresh-water availability on Earth has not significantly changed during this century and neither will it during the next century, but the stable part of it (and thus potential water supply) has and will continue to change due to a number of factors. *Figure 14.6* shows both the effect of some individual factors and the net effect. In the next century, new fresh-water reservoirs will increase the stable fresh-water availability, but not as fast as in the past 50 years, since the number of new reservoir projects will be limited. The future increase of stable fresh-water availability due to new reservoirs will compensate the loss due to land-use changes (especially deforestation). On a global scale, the effects of water use and climate change on stable fresh-water availability are small, since regional differences cancel each other out.

The sea level in the hierarchist utopia is expected to have risen about 105 cm in 2100 (as compared to 1900). About 25% is due to thermal expansion of the oceans, 17% to melting of glaciers and 6% to the melting away of Greenland (*Figure 14.7*). Antarctica has a small negative effect on sea level (-1.7%), likewise the construction of new reservoirs (-1.4%). The collective contribution of defores-

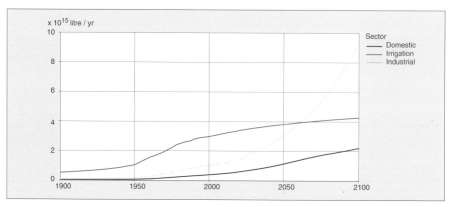

Figure 14.2 Sectoral water supplies in the hierarchist utopia.

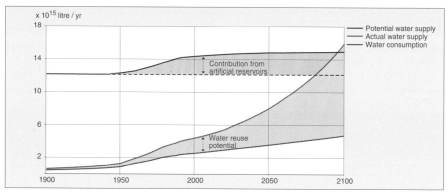

Figure 14.3 Potential and actual water supply and water consumption in the hierarchist utopia. Actual water supply exceeds potential supply in the second half of the next century, which is possible because of water re-use; water consumption does (and can) not exceed potential supply.

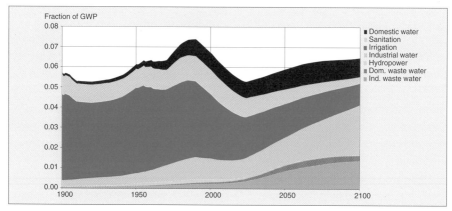

Figure 14.4 Expenditures in the water sector as a fraction of GWP in the hierarchist utopia. The temporary decline up to the year 2025 relates to the hierarchist water-pricing policy: the price increase lowers total water demand and thus total expenditures.

Comparison of the water supply projections with the literature

There have been a number of researchers who have made estimations of future water supply. *Figure 14.5* shows the results of some of the most important world-wide analyses since the seventies. Falkenmark and Lindh (1974) give two projections up to the year 2015. The figure only shows their high projection, the highest projection found in the literature: 10840 km^3/yr in 2015. Even their low projection which includes major reductions in industrial water use is high if compared to other projections. Shiklomanov (1995) estimates the global water use in 2025 at 5188 km^3/yr. This projection is based on a global irrigated cropland area in 2025 of 329 million hectare and a reduction of water-use intensities in the period 1990-2025 of 10-25%, depending on the region. L'vovich and White (1990) suggest that under a conservative growth of population and industrial production, the global water withdrawal in the year 2080 might be about 5300 km^3/yr. According to them, this might be considered a very optimistic estimate.

Raskin *et al.* (1995, 1996) project global water use in 2050 at 4300 km^3/yr, assuming middle-range estimates for population and economic growth and progressive improvements in water use efficiencies. For irrigation, they assume a global area of 311 million hectare by 2050 and a 10% reduction in water use intensity. For water use intensities in manufacturing, they assume an improvement of 60% for OECD-regions by 2050. For thermo-electric generation, they assume a 30% reduction. Not only the projected but also the present water use in Raskin *et al.* (1995, 1996) is much lower than for the example in Shiklomanov (1995) due to the use of another set of basic data. They used national data from WRI (1994), which refer to years between 1970 and 1992 but were taken as initial data for 1990. Finally, Margat (1994) gives a low and a high projection of future global water supply, considering the period 1990-2025. We show only Margat's low projection, the lowest projection of future water supply in the literature. This is mainly caused by three factors, which are all related to the estimation of industrial water use. First, industrial water use is supposed to link up with population growth and not with economic growth or the growth of industrial production. Second, Margat assumes an efficiency improvement for industrial water use in developed countries of 40% in 2010 and 50% in 2025. Third, Margat's estimates do not include water for cooling thermo-electric power plants.

None of the studies explicitly distinguishes the effect of prices, policies or the effect of a shift from private to public domestic water supply. The studies also do not apply economic or ecological limitation to actual water supply (supply is assumed equal to demand). The major drawback of the studies is therefore that they hardly give insight in the actual mechanisms that determine water demand and supply. Another drawback is that the studies – except Falkenmark and Lindh (1974) and Margat (1994) – provide only one scenario, which gives the wrong impression of a rather certain forecast. Nothing is less true if we compare the projections of the different authors. With our model experiments, we hope to give more insight into the effect of the actual processes and how we can interfere.

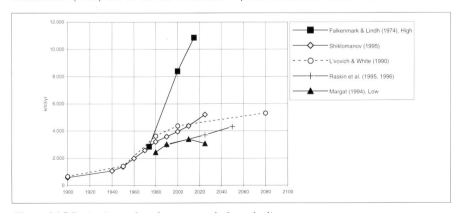

Figure 14.5 Projections of total water supply from the literature.

tation and loss of wetlands together is only 0.2%. A large contribution of about 55% is from loss of groundwater, largely due to groundwater withdrawals but also to decreased percolation as a result of land-use changes. As for the climate-related components of sea-level rise, the hierarchist projection largely corresponds with the central values of the IPCC (Warrick *et al.*, 1995). Assuming that the net direct anthropogenic contribution to sea-level rise (from groundwater, land-use changes and reservoirs) is zero, the IPCC has found a central estimate of sea-level rise in 2100 of 49 cm (if compared to 1990) or 57 cm (if compared to 1900). Without the contribution from groundwater, we arrive for the hierarchist utopia at a value of 40 cm and 47 cm respectively.

An egalitarian water utopia

In the egalitarian water utopia, we use the egalitarian world view and management style, and the hierarchist context ('background') of the reference case, but with an egalitarian (smaller) growth of the economy and population (Table 11.1). A striking difference from the hierarchist utopia is the relatively small increase in and then stabilisation of water supply simulated (*Figure 14.8*). This is not only due to a smaller growth in industrial water supply, which now does not exceed agricultural water supply, but also to lower water-use intensities for all water-use sectors. Water supply costs per litre increase more than in the hierarchist utopia due to the more prudent assessment of water scarcity by the egalitarian. However, as a result of the lower water use, the costs stabilize and become even lower than in the hierarchist utopia by the end of the next century. The annual expenditure in the water sector as fraction of GWP becomes less for the same reason.

The sea-level rise in 2100 in the egalitarian utopia is about the same as in the hierarchist utopia (*Figure 14.9*). A larger effect of climate change is compensated by a reduced contribution from groundwater, due to the lower water use (*Figure 14.8*) and reduction of groundwater withdrawals in the egalitarian utopia (*Table 14.4*). Although we assumed the same temperature increase as in the hierarchist utopia, the climate-related contributions to sea-level rise are larger than in the hierarchist utopia, due to the higher values for the response parameters.

An individualist water utopia

The projection considered here is based on an individualist world view and management style and the same 'background' as in the reference case, but with an individualist (larger) growth in the economy and population (Table 11.1). Total water supply in 2100 is then about 1.8 times larger than in the hierarchist utopia (*Figure 14.8*). On the one hand, the relatively large growth in population and the economy results in a large increase of domestic and industrial water demand. On the other, market pricing results in the development and application of high-tech water-saving technology and more efficient water use. Industrial water supply by 2100 greatly exceeds agricultural water supply, even more than in the hierarchist utopia. Industrial

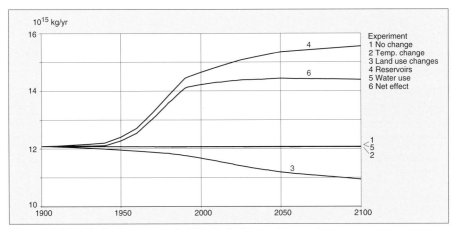

Figure 14.6 Stable fresh-water availability in the hierarchist utopia.

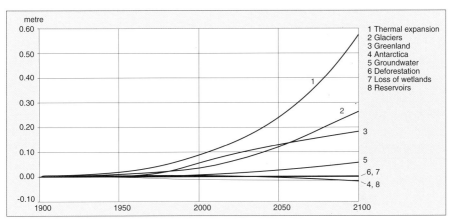

Figure 14.7 Individual contributions to sea-level rise in the hierarchist utopia. The net effect is given in Figure 14.9.

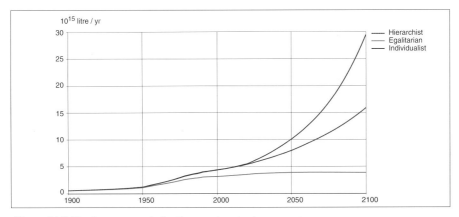

Figure 14.8 Total water supply in three water utopias.

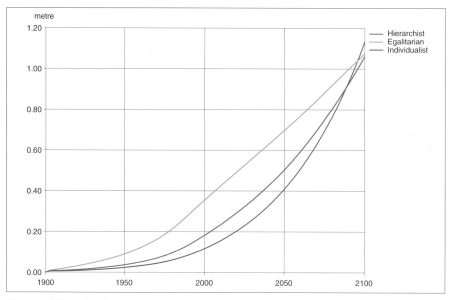

Figure 14.9 Sea-level rise in the three water utopias.

water-use efficiency will become higher than in the hierarchist utopia, but this does not compensate the effect of the rapid industrial growth. Water supply costs increase in the same order of magnitude as in the hierarchist utopia. The fact that water use in the individualist utopia is larger would suggest that water would be scarcer and costs would greatly increase, but this is not the case in the individualist utopia which includes a more optimistic estimate of potential water supply. Actually, the fraction of GWP spent in the water sector persistently decreases from the present up to the year 2100.

As in the egalitarian utopia, the individualist utopia has a sea-level rise in 2100 which is about the same as in the hierarchist utopia (*Figure 14.9*). However, here we see the opposite to the egalitarian utopia: the larger contribution from groundwater, due to the higher water use, is compensated by the smaller effect of climate change. We have used here the same irrigation scenario as in the hierarchist utopia. If we accounted for a stronger increase of irrigation, which is likely in the individualist case (Chapter 15), sea-level rise could even be *larger* than presented in *Figure 14.9*. However, at the same time it is likely that temperature increase in an individualist world will be less than in the hierarchist utopia (Chapter 16), which would imply a *smaller* sea-level rise from climate change. An integrated analysis of these effects is given in Chapter 17.

Sensitivity of global sea level

Sea-level rise in the hierarchist utopia is considerably higher than in the central projection of the IPCC (Warrick et al., 1995) due to our relative high estimate of the groundwater contribution. It is therefore interesting to analyse the sensitivity of our projection to the assumptions used. We distinguish between assumptions on the climate-related contributions to sea-level rise and assumptions as to the groundwater contribution. With respect to the first category, we use similar assumptions as the IPCC. As an example of the sensitivity of our model, we have varied the glacier melting parameters and the climate sensitivities and initial imbalances of the ice sheets (applying uncertainty domains as given by Equation 11.1). We then obtain a range for sea-level rise in 2100 of 88-127 cm (*Figure 14.10a*). If we subtract the groundwater contribution of 57 cm, we arrive at a range of 31-70 cm, which falls within the uncertainty range of 20-86 cm for the IS92a scenario given by the IPCC. The IPCC range is larger because it also includes uncertainties in the climate system (the 'climate sensitivity parameter') and thus also includes uncertainties on global temperature increase; the analysis of *Figure 14.10a* is based on one fixed temperature scenario.

To analyse the assumptions behind the groundwater contribution, we have carried out two types of experiments. First, we have varied the price and growth elasticities of water demand and the diffusion rate of new water-saving technologies (again following the method described in Chapter 11). This results in a range for groundwater withdrawal and consequently for sea-level rise as shown in *Figure 14.10b*. Second, we have varied the groundwater outflow characteristic. As a reference, we use a linear relation between groundwater outflow R_{gw} and groundwater storage S_{gw}:

$$R_{gw}(t) = S_{gw}(t) / k \qquad (14.1)$$

Parameter k is the lag time of the groundwater reservoir. As an alternative we have used the more general equation:

$$R_{gw}(t) = (S_{gw}(t))^p / k \qquad (14.2)$$

For most river basins in the world, parameter p falls somewhere in between 1 and 2 (Shaw, 1994). The resulting sea-level rise for p = 2 is shown in *Figure 14.10c*. We can see that the projection of sea-level rise is not particularly sensitive to the assumption of a linear groundwater reservoir if we compare this to other assumptions made. We have also varied the initial groundwater storage. In the reference case we use a figure of 4×10^6 km^3, which – according to L'vovich (1977) and L'vovich and White (1990) – is the part of the total groundwater that is in active exchange with the surface. *Figure 14.10c* also shows the sea-level rise projections if we assume an initial groundwater storage that is a quarter of L'vovich's estimate.

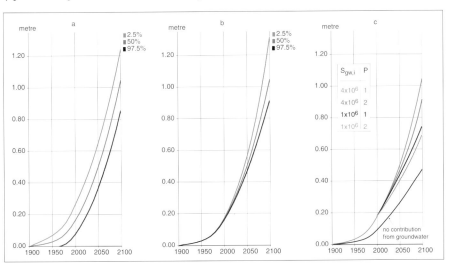

Figure 14.10 Sensitivity of global sea level in the hierarchist utopia to the response of glaciers and ice sheets (a), the extent of global groundwater withdrawal (b) and the groundwater outflow characteristic (c).

14.6 Possible water futures: the broader scope

In the previous section, we discussed only a very limited selection of possible water futures, the so-called water utopias. Here, we present a larger set, also including water dystopias. We focus on water scarcity and supply costs and analyse the results of 27 simulations with the AQUA submodel, each of which is characterised by:

- a certain world view on water,
- a certain style of water management, and
- a certain socio-economic and environmental 'background'.

The latter refers to the input from the other TARGETS submodels. First, from a hierarchist 'background' we vary world view and management style for water, obtaining an area of possible futures as shown in *Figure 14.11*. For ease of survey, we have chosen to show only two indicators (water scarcity and water costs) for only one point in the future (the year 2100). Second, we assume an egalitarian background with respect to the economy and population, like in the egalitarian water utopia in section 14.5, and again vary the world view and management style. Finally, the same is done for an individualist background. We now have three areas of possible futures, each belonging to a certain type of background. From the extent of these areas (*Figure 14.11*), we can conclude that, *given a certain external development, there is still a broad range of possible estimations of future water scarcity*. This is greatest for an individualist background, which is understandable when we realise that population and economic growth are largest in the individualist background, which is dramatic from an egalitarian perception of water scarcity. Taken the other way around, for example, an individualist perception of water scarcity with an egalitarian background, is much less disastrous.

The same results can also be presented in a different way which puts more emphasis on the importance of the 'background' of the analysis (*Figure 14.12*). From this representation, it is seen that *given a fixed and coherent set of assumptions about the world view and management style for water, it is still impossible to make a strong statement about future water scarcity because of the influence of external developments*. This is strongest in the case of an egalitarian world view, which corresponds to the fact that an egalitarian world is the most vulnerable to perturbations.

From the two conclusions above, we can draw a third conclusion: the assessment of future water scarcity depends significantly on both assumptions *within the analysis* and assumptions on *the background of the analysis*. On the one hand, this justifies the use of a dynamic water model and on the other, an integrated approach. The dynamic water model can be used to 'play' with the assumptions on water. The integrated approach enables us to vary the assumptions with respect to socio-economic development and environmental change in a much broader sense.

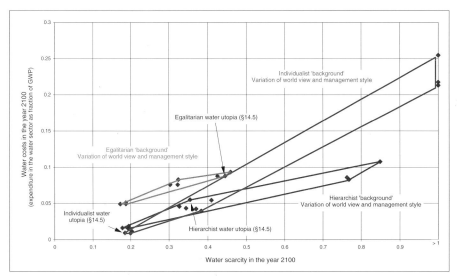

Figure 14.11 Possible water futures (I). The dots indicate simulation results for the year 2100 for a set of 27 experiments. Each area encloses possible futures using a fixed socio-economic and environmental 'background' of analysis and is set up by the nine possible permutations of world view / management style.

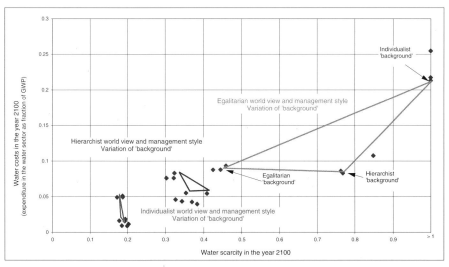

Figure 14.12 Possible water futures (II). The dots refer to the same experiments as represented in the figure above. Each area shows possible futures if the socio-economic and environmental 'background' of analysis are varied, and a fixed and homogeneous set of assumptions within AQUA is used.

14.7 Risk assessment

For possible water futures, we have tentatively identified a low, moderate and high risk area (*Figure 14.13*), in line with earlier classifications of vulnerability for water scarcity (Kulshreshtha, 1993). High risk means that major impacts of water scarcity on population and economic development will probably occur. Moderate risk means that these impacts can be anticipated by breaking with 'business as usual'. Low risk means that the future will not bring water scarcity problems that significantly exceed the problems we encounter today. Although we have drawn the lines between the different risk areas without making a distinction for the perspectives, the fact that water scarcity is defined differently for each perspective implies that the physical meaning of a certain risk line is different for each perspective. Exceeding the water scarcity line of 50% means for the egalitarian that *total water supply* will be more than half the *stable runoff* in inhabited areas, which is unacceptable. Individualists, however, enter the high risk area (according to their own perception) only if *water consumption* becomes more than half the *total runoff*.

Despite the different perceptions, some of the possible futures are undesirable from any perspective (see *Figure 14.13*). One of these undesirable futures (in the upper right corner of the figure) is the case with an individualist background and management style, while water dynamics prove to be as egalitarians expect. An

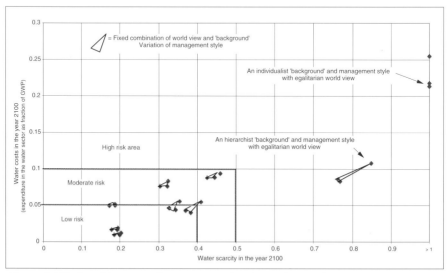

Figure 14.13 Possible water futures (III). The dots refer to the same experiments as represented in Figures 14.11 and 14.12. The nine clusters of experiments show the effect of a variation of management style. The possible water futures are divided into areas of low, moderate and high risk.

egalitarian style of water management in this case means only a marginal improvement of the situation. The problem here is that population and economic activities grow relatively fast, while society fails to face the coming water crisis. In this case, sooner or later, increasing water scarcity and costs will seriously affect population (in terms of health) and the economy (in terms of stagnation). It is unlikely that the situation in the upper right corner of the figure will actually be reached at all, since more than 20% of GWP is spent on water; long before, this development path will probably have resulted in an 'overshoot and collapse' of population and economic growth, triggered, for example, by an unexpected series of dry years. This doomsday scenario warns us to be careful in assessing water resources during rapid growth and adequately respond to early warning signals. It also suggests that *if a serious water crisis occurs*, this will be due to unsustainable socio-economic development rather than to mismanagement of our water resources. Solutions should be sought in demographic and economic measures rather than in constructing more dams or in supplying more water from one source than another. Measures like the latter bring only marginal changes in the situation. The same holds, but to a lesser extent, for the case of a hierarchist world in which water dynamics turn out to be as the egalitarian expects (see also *Figure 14.13*). Here too, an egalitarian management style will provide some improvement but by far not enough.

14.8 Water policy and global change

Let us revert to our central question: *Can the world population be provided in the long run with sufficient clean fresh water without threatening the ecological functions of water?* Although the experiments show that there is not one single answer, it has become clear that under some preconditions, risks are much smaller than under others. Recognising that we can, to some extent, influence the preconditions that will shape the future, we have analysed how different priority settings in policy making can lower the risks of future water scarcity (Hoekstra, 1997). The results of this analysis are presented in *Table 14.5*.

The first result is the conclusion that water policies can influence the extent of future water scarcity only to a limited extent. This has already followed from *Figure 14.13*, where we saw that, under certain circumstances (i.e. a certain background and world view) the style of water management does not strongly influence the extent of future water scarcity. This means that 'water problems' can be solved only partially through 'water policy'. Socio-economic policies that influence population and (the type of) economic growth are much more effective.

Although the room for 'doing well' in water policy is limited, it would make things worse not to use it. It depends on the type of water policies that will be put into practice whether this room for management will be effectively used or not. A shift from water-supply towards a water-demand policy is needed all over the world. In

	Past[a,b]	Future[a,c]
General		
• economic & population policies	+	++++
• water policy	++++	+
Water policy		
• water demand policy	+	++++
• water supply policy	++++	+
Water demand policy		
• water pricing (removing subsidies)	+	+++
• water supply technology	+++	+
• water education	+	+
Water supply policy		
• infrastructure policy	+++	+
• water-quality policy	+	++
• land & soil policy	+	++
• climate policy	~ 0	+

[a] The priorities have a merely comparative value: from relatively low priority (+) to relatively high priority (++++).
[b] The priorities of the past are our own estimates and are added to be contrasted with the future priorities proposed.
[c] The future priorities are proposed on the basis of the results of experiments with the AQUA submodel and should be considered indicative.

Table 14.5 Priority setting to safeguard water supply.

developing countries, where water supply infrastructure is likely to be extended most intensively, any supply policy should at least be accompanied by demand policy. An important policy ingredient would be to stop subsidising water in all water-use sectors. Although for different reasons, this fits within all perceptions on what is proper water management. The experiments discussed in the previous sections all have some water pricing policy to remove subsidies (*Table 14.4*). If water subsidies would not be removed at all, this would lead to a much higher water use. In the hierarchist water utopia discussed in section 14.5, this would result in 40% more water use in the year 2100; in the individualist water utopia, water use would even double. An active water supply policy against the market mechanism is only acceptable in some cases of drinking-water supply. Subsidies in public water supply are only acceptable if they benefit people who would otherwise not be able to afford safe water. Since domestic water use is low compared to other uses, the aim of increasing the public water supply coverage in developing countries conflicts little with the aim of saving water. The aim of 'drinking water for all' therefore justifies high investments in public water supply.

Another ingredient of 'good' water policy is to push water reuse, especially in industries. Industrial water use has had the least attention from policy-makers to date. In many of the possible water futures, however, industry will become, by far, the largest water user. As for irrigation, the application of more efficient techniques can reduce water consumption to some extent, but more important is to critically consider plans for new irrigation schemes. A central theme will be the weighing of irrigation against other possibilities to enlarge food production.

The figures from *Table 14.5* should be taken as indicative and with the necessary precaution. We realize that we strain the actual situation in the world by not distinguishing between different regions and by simply dividing time into 'past' and 'future'. The actual situation is that different regions are in different stages of water policy-making due to differences in, for example, population density, socio-economic development and natural availability of water. However, we feel that the elementary processes that lead to water scarcity are similar all over the world and that the transformation of water policy strategies towards broader policy strategies and from supply towards demand policies are, sooner or later, necessities worldwide.

The previous sections have illustrated that a variation of basic assumptions for water and in the socio-economic and environmental context can affect future forecasts considerably. One can add to this that none of the assumptions used is totally unrealistic. However, the projections have been based on runs with the AQUA submodel as a stand-alone tool. Given the importance of external developments and the possible backlash of a water crisis on the population and economy, we recognise the possible added value of running AQUA as an integral part of TARGETS. Results of the integrated experiments are discussed in Chapters 17 and 18.

Food for the future

*"When we change the way we grow food, we change our food,
we change society, we change our values."*
Masanobu Fukuoka, The One Straw Revolution (1975)

15 FOOD FOR THE FUTURE

Bart J. Strengers, Michel G.J. den Elzen, Heko W. Köster, Henk B.M. Hilderink and Marjolein B.A. van Asselt

In this chapter we use the TERRA submodel to explore whether malnutrition and food insecurity can be eliminated while safeguarding the productive potential and broader environmental functions of agricultural resources for future generations. This is done within the context of the three cultural perspectives. The food problem is explained not so much as a problem of production but as one of availability and distribution. The submodel simulations are, however, largely confined to aggregate food demand and supply. Costs and environmental trade-offs are assessed in both utopian and dystopian worlds to determine under what conditions the planet will be able to feed its future population. We explore perspective-based scenarios for population and GWP, the surface area available for cropping, the use of irrigation, fertilisers and other inputs, wood production, reforestation, and the effects of changes in atmospheric CO_2 and temperature.

15.1 Introduction

Currently, sufficient food is produced to feed the world population, yet at the same time more than 1,000 million people cannot afford or do not have the possibility to buy enoughfood to live healthy and productive lives. More than 500 million are chronically undernourished (FAO, 1993a). Malnutrition and food insecurity are not so much related to food production but rather to the unequal distribution of available food (IFPRI, 1995). This is caused by socio-economic factors such as poverty, the political situation, deficient infrastructure and (food) trade. Enough food can also be produced in the future, at least technically speaking: research has shown that, depending on which agricultural production system is practised, 30 to 72 million ktons grain equivalent could be produced annually[1] (Luyten, 1995). This is 8 to 18 times more than current global food production which is 4 million ktons. Even in the highest population scenarios, this theoretical maximum for global food production is sufficient to cover the highest estimates of world food demand, i.e. 24.3 million ktons of grain equivalent in 2100[2].

---

1 Ranging from High External Input or HEI agriculture, characterised by a high degree of mechanisation and use of significant amounts of fertiliser and biocides, to Low External Input or LEI agriculture, characterised by restricted use of mechanisation, legume crops providing all nitrogen and no use of biocides.

2 If the high UN population growth scenario (UN, 1995) is combined with an affluent diet, global food demand will amount to 17.3 million ktons in 2040 (Luyten, 1995). Extrapolation of both assumptions results in a global food demand of 24.3 million ktons by 17,600 million people in 2100.

In principle, food production depends on renewable resources. Some argue that the productive potential of these resources might be seriously threatened by over-exploitation and mismanagement, resulting in degradation of arable land, (ground) water pollution, water logging, depletion of fishing-grounds, and by climate change (Brown and Kane, 1995; Harrison, 1992). For example, the UNEP has estimated that in 1988 6-7 Mha of cropland per year were being lost world-wide through soil erosion, nearly 0.5% of the world's cropland, and another 1.5 Mha of irrigated cropland to salinisation or waterlogging (UNEP, 1984), i.e. about 0.65% of the irrigated area. These threats may lead to a situation in which, food production falls below the level needed to feed the growing population locally or even globally. Hence, the main question that we will consider here is (FAO, 1993a): *Can malnutrition and food insecurity be eliminated while safeguarding the productive potential and broader environmental functions of agricultural resources for future generations?*

In this chapter we use the TERRA submodel to explore part of the answer to this question in the context of the perspectives outlined in Chapter 10. As will be evident from the model description in Chapter 7, we are not able to make statements about changes in food accessibility and distribution. In line with the approach of the FAO, we apply a fixed food distribution function within the Population and Health submodel to assess the consequences of global food availability (see Chapter 4). In section 15.2 the main issues addressed by the TERRA model are described. In section 15.3 the three perspectives are examined in relation to the controversy formulated above. In 15.4 we present the assumptions and outcomes of model simulations for three utopian land and food scenarios. In section 15.5 a larger set of experiments is presented, giving some insights into the dynamics of the TERRA submodel in a dystopian world. In section 15.6, the model outcomes are related to risks with respect to food production, food security, the production potential of arable land and ecosystem functioning. The final section of this chapter contains our conclusions.

15.2 Main issues and uncertainties

In this section, we limit ourselves to the three most important controversial issues behind the aforementioned main question:

- future animal and vegetable food demand;
- the type of agriculture to produce the required agricultural products (food and feed);
- the impact of climate change on food production.

Important issues such as the impact of trade, overgrazing, the availability of capital and energy and salinisation of arable land due to sea-level rise (Harrison, 1992; IPCC, 1995a; Tolba and El-Kholy, 1992) are not yet included in the current version of our model and are therefore left out of the discussion. The question of land availability for commercial biofuels is touched upon in Chapters 17 and 18.

The future animal and vegetable food demands per capita and their relative contribution to the overall diet are uncertain, but one of the main determinants is income: higher real income levels often imply higher demands and an increasing share of animal products in the diet (FAO, 1993a; Scrimshaw and Taylor, 1980). Opinions differ on the relative importance of income and on the extent to which the total food demand per capita and the fraction of animal products can be steered into a certain 'desired' direction. The second issue is about how agricultural production should or can be organised in order to meet global demand. We reduce the question by only considering the surface area to be used for cropland, the use of irrigation water, and the application of fertilisers and other inputs. As with water-related issues, there are distinctly different ways of interpreting past developments and evaluating future options. Some argue that prices of water, fertilisers and land are, or at least should be, the main determinants of the type of agriculture (Harris, 1990; Stevens and Jabara, 1988). Others emphasise that high-tech intensive agriculture should be promoted so that high yields per hectare are obtained and not all land has to be used for agriculture which makes it possible to preserve nature in certain areas (UNEP/RIVM, 1997; WRR, 1992). In this view 'nature' and 'agriculture' are considered as separate entities. Still others point out that agriculture is dependent on the functioning of natural ecosystems and should therefore be integrated with nature. This integration cannot be combined with, for example, an abundant use of chemical fertilisers and pesticides (Brown et al., 1996; Reijntjes et al., 1992).

With respect to the third issue, the impacts of climate change, there is general agreement that CO_2 fertilisation leads to higher potential yields if other limiting factors such as shortage of nutrients and water, are absent. A global temperature rise is expected to raise potential yields in high-latitude regions which benefit from longer growing periods and increased productivity, while other regions either do not benefit significantly or lose productivity (Cramer and Solomon, 1993; Leemans and Solomon, 1993). Besides, potential food production will be affected because climate change affects most major soil processes such as mineralisation and salinisation. Weeds, insects, and pathogen-mediated plant diseases are affected by climate change and atmospheric constituents. Resultant changes in the geographic distribution of these crop pests and their vigour in current ranges will likely affect crops (IPCC, 1995a). Studies on the impact of climate change on actual food production show very different outcomes, both between different crops and regions: from −96% for soybean on sites in south east USA (Rosenzweig et al., 1994) to +234% for wheat in Canada (Brulacich et al., 1994). To a large extent, these differences are caused by assuming different levels of adaptation: in principle many technological adaptations are possible to benefit from the CO_2 fertilisation effect (using new crops or crop varieties, improving irrigation systems, adding fertilisers, etc.), but the socio-economic capability for adaptation differs for different types of agricultural systems, especially in the developing countries. Therefore, the impacts of climate change highly depend on the extent to which one expects it is possible or desirable to adapt agricultural practices (IPCC, 1995a).

15.3 Perspectives on land and food

This section briefly describes, as far as future food demand, food production, land use and land cover are concerned, the three perspectives from the cultural theory: the hierarchist, the egalitarian and the individualist. The *hierarchists* will estimate future food demand levels on the basis of authoritative institutions, e.g. FAO. Both for ethical and political reasons, they will try to ensure a 'decent standard of living for all', which implies covering the food demand at affordable prices. In most developing countries the problem is not excessive but insufficient fertiliser use, and the major challenge is to promote a balanced and efficient use of fertilisers at farm and community levels to intensify agriculture in a sustainable manner (IFPRI, 1995). Hierarchists will favour a planning approach towards irrigation, de- and reforestation and introduction of intensive agricultural techniques. Reforestation is considered as an option to protect fragile soils against degradation. In a situation where the agricultural production is abundant, as nowadays in the developed countries, farmers should be (financially) stimulated to take part of their cropland out of production. Longer term issues like erosion and water logging are considered to be serious but any policy has to be founded on the dominant views of experts in the field. Conservation of natural forests and grasslands is important but secondary compared to food security. Sustainable forest management is promoted if it is does not interfere with economic soundness and national sovereignty. With respect to climate change, the hierarchist emphasises the overall net effect on global food production as being very uncertain and a focus on global production that does not address the potentially serious consequences of large differences at local and regional scales (IPCC, 1995a). However, because climate change might affect local food security, the need of and options for preventive and adaptive action should be explored.

Egalitarians strongly feel responsibility for the future of the Earth and the generations to come. People will change their behaviour voluntarily into a more sustainable one if institutions can be redirected towards sustainable development goals and the greed of the market can be controlled. This results, among other things, in a trend to a more frugal life-style and a more equal distribution of goods and food. With respect to food, demand shifts in the direction of a relatively low-calorie, more vegetarian diet, for example because the use of food crops as feed is considered undesirable. Highly intensive agriculture inevitably undermines sustainable food production in the long run: far more energy (in terms of fossil fuels for fertilisers, machines and transport) is put in than comes out and the use of only a few crop varieties in large mono-cultures results in nutrient mining, erosion, heavy use of pesticides followed by insect resistance, etc. Hence, agriculture will increasingly be oriented towards low-input organic agriculture, highly harmonious with nature. This may require more land, leaving less space to preserve natural forests. Nevertheless, the forests should not be depleted for ethical, ecological, climatological, and long-term economical reasons. Therefore wood production from (tropical) forests should be based on principles of sustainable forestry (FAO, 1994; Walters, 1988). Climate change poses the risk of serious disruptions of local food supply systems. One cannot rely on positive effects: increases in yield from CO_2 fertilisation will probably be minimal due to other limiting factors such as a lack of nutrients (N/P/K) and water and an increasing frequency of extreme events can cause catastrophies of an unprecedented size. Furthermore, the egalitarian emphasises that vulnerability to climate change does not depend on physical and biological responses only, but also on socio-economic characteristics. Low-income populations depending on isolated agricultural systems are particularly vulnerable to hunger and hardship. The capability to adapt is considered to be quite limited.

For the *individualist* food demand is in the first place the result of the free choice of the individual, within the context of the free market, and therefore governments should not interfere in the form of subsidies, regulations and the like. Food shortages can and do occur but they are never more than temporary, which is proven by the fact that food prices have been continuously going down and food insecurity and malnutrition are decreasing, both in relative and absolute terms (FAO, 1993a; Myers and Simon, 1994). With respect to agriculture it is emphasised that high-tech, highly intensive production modes are possible – and necessary – to meet the high food demands of a large world population in the future. The market mechanism will ensure an economically optimal distribution between the use of fertilisers (and other inputs), the expansion of arable land and irrigation. More efficient use of inputs and better management techniques can reduce the environmental risks and side-effects to acceptable levels (Luyten, 1995). The individualist is optimistic about the impacts of climate change on food production: market-driven technological adaptations will limit the negative effects and advantage will be taken of beneficial changes in climate. Preservation of natural areas and stimulating sustainable forest management has no priority. When forests become scarce, wood prices will rise and alternatives, both in demand (e.g. use of synthetic materials) and production (e.g. wood plantations), will be developed through market forces.

The implementation of these three perspectives in the TERRA submodel cannot be other than rather limited and patchy, especially because the model has no explicit representation of social, economic and ecological dynamics. Food demands are derived from per capita income; agricultural practices are mainly determined by scenarios for clearing, irrigation and inputs on irrigated arable land. Price mechanisms have not been modelled and have, therefore, no impact on the type of agriculture that arise in the model; technology is largely represented by an

Variable description	W/M	Hierarchist	Egalitarian	Individualist
Demand				
Relation between income and animal and veg. food dem.	W	middle	weak	strong
Saturation level (kcal/cap per day)	W	3470	3470	4250
Maximum animal food demand (kcal/cap per day)	W	1260	1260	1760
Amount of food crops used as feed (in 2100)	W	32%	2%	50%
Production				
Fraction of wood production associated with deforestation	M	middle	low	high
Potential arable land (Gha)	W	3.3	2.8	3.8
Cleared area in developing region in Mha (1990-2100)	M	473	485	670
Irrigated area in Mha in 2100	M	350	320	460
Maximum level of N-fertiliser use (kg N/ha per crop cycle)	M	200	125	250
Inputs on irrigated arable land	M	middle	low	high
Environment				
Reforestation in developed region in Mha (1990-2100)	M	430	640	300
Recuperative power of degraded land	W	middle	low	high
Positive CO_2 fertilisation effect	W	middle	none	high
Negative effect of temperature increase	W	middle	high	none

Table 15.1 *The implementation of the perspectives in TERRA. The second column indicates whether a variable is related to the World view or Management style.*

exogenous factor. Food shortages only cause an increase in the use of inputs on rainfed arable land, superimposed on the scenario-dictated land-use changes. To approximate the perspectives as well as possible, we have done the following.

For the hierarchist, the values of the perspective related variables have been set such that a reference scenario, based on the IS92a scenario, is reproduced (see Chapter 11). We use the IPCC-IS92a scenarios on GNP and population, N-fertiliser use and deforestation in the developing countries (Leggett *et al.*, 1992). Leggett does not give projections of future changes in arable land and the area of forests in developed countries. Therefore, these have been derived from the Conventional Wisdom scenario of IMAGE2.0 (Alcamo, 1994), which is consistent with the IS92a scenario. There are different interpretations of historical land use, land cover and food intake from FAO data and the processes that underlie their changes. Hence, an unambiguous model calibration is difficult, or even impossible (see Chapter 7). This also affects the scenario experiments because perspective-based assumptions are, at least partly, derived from historical calibration.

The implementation in the TERRA submodel of the egalitarian and individualist viewpoints is difficult because the key aspects that are important in these views, like markets and income (in)equalities are absent in the current version of the model. Differentiation between the egalitarian and the individualist utopia is realised by assuming divergent trajectories or values for the variables mentioned in Table 15.1.

Perspectives on land and food in literature

The FAO (1993a) states that "...the key concern will be how to ensure transition from a world of rapidly growing population and many people chronically undernourished to one of slow or very low demographic growth free from chronic undernutrition, with the least possible adverse effects on resources and the environment. Provided technological and other policy options are put in place, the prospects are for an easing of pressures []." This balanced way of formulating the issue is rather representative for the hierarchist world view. It is also found in the assessment of Working Group II of the IPCC (1995a), regarding the impacts of climate change: "Adaptation to climate change is likely; the extent depends on the affordability of adaptive measures, access to technology, and biophysical constraints []. Many current agricultural and resource policies are likely to discourage effective adaptation and are a source of current land degradation and resource misuse. [] Additional costs of agricultural production under climate change could create a serious burden for some developing countries."

In the book 'Beyond the limits' (Meadows et al., 1991, pp. 53) an egalitarian concern for the future is expressed: "If the flow of food through the human society were more efficient, less wasteful, and more evenly distributed, it would not be necessary to grow more. More food could be grown, however, and it could be done sustainably. But those are hypothetical statements. The present reality is that in many parts of the world the sources of food – land, soils, water, soil nutrients – are falling and the sinks of pollutants from agriculture are overflowing []. Unless rapid changes are made the Earth's exponentially growing population will have to continue to try to feed itself from a degrading agricultural resource base". In similar vein the 'State of the World 1996' (Brown et al., 1996) report confronts the official expectation of increasing food production with the observed trends: "Unfortunately, most national political leaders do not even seem to be aware of the fundamental shifts occurring in the world food economy, largely because the official projections by the World Bank and the FAO are essentially extrapolations of the past. [] The risk of relying on these extrapolations is that they are essentially 'no problem' projections, departing further and further from reality with each passing year. The World Bank shows the world grain harvest climbing from 1.78 billion tons in 1990 to 1.97 in 2000. But instead, the 1995 harvest, at 1.69 billion tons, is 90 million tons below the 1990 harvest. FAO projections using a similar approach yield similar results. They, too, are fast departing from reality."

An individualist pur-sang is the economist Julian Simon. In the book 'Scarcity or abundance?' (Myers and Simon, 1994) he vigorously defends a positive reading of the past: "The long-run price of food relative to wages is now only perhaps a tenth of what it was two centuries ago, due to increased productivity. Famine deaths have decreased during the past century even in absolute terms, let alone relative to population. Africa's food production is down, but this clearly stems from civil wars and the collectivisation of agriculture, which periodic droughts have made more murderous. [] Pick any year in the future and I'll bet that food prices are lower and that the nutrition that people get – in calories per person – are better throughout the world then they are now." The prospect of biotechnology are also judged differently. Naisbitt and Aburdene (1990) state: "What took agricultural scientists and farmers decades to do, using conventional plant breeding techniques, can now be accomplished in months or years, using genetic engineering techniques. [] With each new discovery it becomes easier to manipulate genes. (pp. 268) In a few decades it may be possible to insert genes that change an animal's fertility, size, and behavior – altering salmon migration patterns for example. It's just a matter of time – and not much of that." (pp. 274). They do note, however, that ethical concerns among the public at large may impede these developments significantly.

Soil conservation measures (as part of the management style) are assumed to have a high priority in all perspectives and therefore, in a utopia, land degradation is controlled and has no negative impact on food production. Obviously, different opinions exist on to establish appropriate conservation programmes and one would expect that problems might arise in a dystopian world: for example, 'egalitarian' conservation measures could turn out to be less effective in an individualist world. This kind of discrepancy has not been modelled in the present version of TERRA, because it turned out that the simulated abandonment rates are very sensitive to small changes in SOM and that the CYCLES submodel does not generate stable SOM outcomes if it is run together with TERRA.

Finally, the Learning Factor (see Equation 7.2) – which is associated with the level of technology, efficiency and other not explicitly simulated factors that affect agricultural yields – is assumed to be equal in all perspectives and is used as a calibration factor for the reference scenario. Since the effect of changes in the area distribution among the LGP-classes on potential yield is found to be small (<1%), we use also for this parameter, the same values for the three perspectives. This small impact is disputable if it is compared with other studies (Kane and Reilly, 1992; Rosenzweig and Parry, 1994), but is a result of the chosen level of aggregation and not taking the geographical shift of LGP classes into account.

15.4 Simulation results for the three utopias

In this section, we present the outcomes of the simulation experiments with the TERRA submodel for three utopian worlds, i.e. worlds in which the world view and the management style are the same (see Chapter 10). The world view is reflected in, for example, the elasticity parameters describing the relation between income and food demand. Clearing, irrigation and reforestation rates are part of the management style (see *Table 15.1*). We focus on the TERRA submodel in isolation and, therefore, no feedbacks through the other submodels are involved. All world view and management style-related input variables from the other submodels into TERRA, the so-called 'background', are chosen according to the utopian perspective in TERRA, except for the input variables from the CYCLES submodel, which are hierarchist. Therefore the utopian experiments presented here are in fact 'semi'-utopias in contrast with 'full' utopias in which all background variables are chosen according to one perspective (see Chapter 11). As described in Chapter 7, the input variables are population and GWP, the surface area needed for biomass plantations, irrigation costs, erosivity, soil fertility (Soil Organic Matter and the Q-factor), temperature increase and atmospheric CO_2 concentration. In our analysis of the utopias we focus on the following aspects: food demand and production, land and inputs for agriculture, de/reforestation and climate change impact.

Reference case: the hierarchist utopia

In the hierarchist scenario, food demand in the *developed* region had almost reached the saturation level of 3470 kcal/cap per day in 1990. Furthermore, because GWP per capita in the developed region steadily grows from $ ~15,000 in 1990 to $ ~68,000 by

15 FOOD FOR THE FUTURE

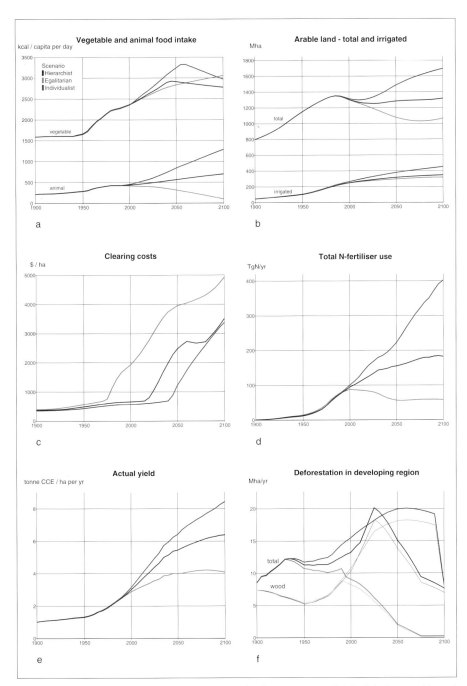

Figure 15.1 The food and land developments for the three utopias with (a) global vegetable and animal food demand, (b) changes in the area of arable land, (c) marginal clearing costs in the developing region, (d) total fertiliser use, (e) yields and (f) deforestation rates in the developing region.

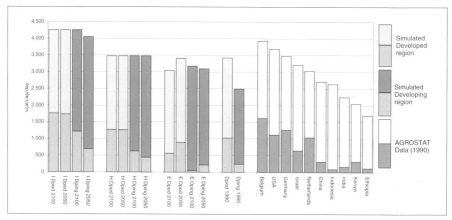

Figure 15.2 Comparison of utopian food demands in 2050 and 2100 with food intake data of several countries in the world in 1990, based on AGROSTAT data. The lower part of each column refers to animal food and the upper part to vegetable food. The first 12 labels on the x-axis are composed of three parts: the first character indicates the utopian perspective (H=Hierarchist, E=Egalitarian and I=Individualist) followed by the region (Dped=Developed region, Dping=Developing region) or country and finally the year (1990, 2050 or 2100).

the end of the next century, the fraction of the animal food demand will continue to rise to the assumed maximum level in the second half of the next century and will remain constant thereafter. The resulting average diet in the developed region in 2100 is comparable to that in West Germany in 1990 (see *Figure 15.2*). The average food intake in the *developing* countries was in 1990 2500 kcal/cap per day of which only 10% consisted of animal products. Food demand will rise to ~2,700 kcal/cap per day in 2010, which is in line with the food intake projections of the FAO. From then on it slowly increases to saturation levels around 2050. Thereafter, per capita food demand does not increase, but rising incomes from $ ~900 per cap per yr in 1990 to $ ~14,500 per cap per yr in 2100, result in an increasing share of animal products. In *Figure 15.1a* the resulting *global* average animal and vegetable food demand per capita are shown. As to the use of agricultural products for feed, we assume that feed fractions and feed conversion factors will remain constant.

The future clearing rate has been chosen such that the expansion of arable land (see *Figure 15.1b*) is in line with the Conventional Wisdom scenario of IMAGE2.0[3] which does, however, not distinguish between rainfed and irrigated arable land. Therefore for the developed countries we adopt the common view that the irrigated area will not or hardly increase after 1990 (Leach, 1995), which results in a simulated stabilisation around 60 Mha. The irrigation rate in the developing region is

---

3 The expansion of arable land in the developing region from 710 Mha in 1990 to more than 900 in 2100 dominates the decrease in the developed region from 630 Mha in 1990 to 405 Mha in 2100.

based on FAO estimates, China excluded. FAO expects a total growth of irrigated arable land of 23 Mha or 19% between 1990 and 2010, where it is assumed that appropriate measures are taken more or less immediately to prevent further losses of existing irrigated land through salinisation (FAO, 1993a). Since population growth is the main driving force in expansion of irrigated arable land (Harrison, 1992; Wolter and Kandiah, 1996), this trend is extrapolated in line with the population growth, which results in an irrigated area in the developing region of 290 Mha in 2100. This simulated total area of irrigated arable land is within acceptable ranges in the sense that it is nearly 90% of the total area which can be irrigated according to estimates of the World Bank (1994). They point out that increasing costs prohibit the full usage of this potential.

As explained in Chapters 7 and 8, both clearing and irrigation costs are expected to rise as land and water become scarce. Hence, a low estimate of the total area of potential arable land will lead to an earlier rise in clearing costs. For the hierarchist utopia, the common estimate of 3.3 Gha potential arable land, of which 2.0 Gha is situated in the developing region, has been adopted. Degraded land and reforested land are *not* considered as potential arable land. Therefore, during the simulation period, it declines to 1.0 and 1.8 Gha in the developed and the developing region respectively. The proportion of the potential arable land that remains uncultivated in 2100 is large enough (more than 50% in the developed and about 40% in the developing region) to keep the marginal clearing costs within 'acceptable' levels (see *Figure 15.1c*). The small differences in the past are caused by different estimates of the amount of potential arable land (see *Table 15.1*) and thus in the remaining uncultivated potential arable land ratio, which determines the marginal clearing costs (see Figure 7.7). Clearing costs in the developed region are not shown: they hardly change after 1990, because no clearing is taking place. The same holds for the irrigation costs which are mainly based on water costs and water-use efficiency, i.e. the amount of water needed to irrigate one hectare of land (see Chapter 14). 'Acceptable' means that the relative agricultural expenditures as a share of GWP decrease from 5.5% in 1990 to less than 1.6% in 2100. However, more empirical research is required to give these and subsequent cost estimates a firmer foundation.

Based on (i) the projected food demand, which is in balance with the food supply in an utopian world, (ii) the area of rainfed and irrigated arable land (*Figure 15.1b*) and (iii) the assumed fallow fractions, the global N-fertiliser use and yields more than double in the period from 1990 to 2100 (see *Figure 15.1d* and *Figure 15.1e*). The total agricultural production in the hierarchist scenario amounts to 7.2 million kton CCE/yr by 2100 where about a third is used to feed livestock[4]. Because fertiliser use has been selected equal to the IS92a scenario values, we have to assume that the

---

4   For an explanation of CCE, see Chapter 7, footnote 3.

efficiency of fertilisers and of other inputs increases by 30% to 40% over the next century in order to satisfy food demand. This is done by adjusting the Learning Factor (see Equation 7.2). As mentioned earlier, the resulting trajectory for this factor is also applied in the egalitarian and the individualist utopia.

Following the Conventional Wisdom scenarios[5] of IMAGE2.0 a large reforestation programme has been set up in the developed countries. The forest area will increase from ~1900 Mha in 1990 to nearly 2400 Mha in 2100. The main part of this reforestation is assumed to take place between 1990 and 2030, with about 50% on former cropland (and the other 50% on grassland). Wood production is simulated only for the developing region (see Chapter 7). The level of sustainability of the associated logging activities can be regulated by a parameter which indicates the fraction of the roundwood production associated with deforestation. In the hierarchist utopia we assigned a value to this parameter such that the deforestation rates in the IS92a scenario are reproduced. As a result, less than 600 Mha or 28% of the 1990 forest area of 2130 Mha is left in the developing region by the end of the next century[6]. This decline is mainly related to wood production (see *Figure 15.1f*). In the individualist utopia almost no natural forests in the developing countries are left by the end of the next century. Therefore, the deforestation rate sharply declines. Small differences between the utopias in the past can be explained from different estimates of potential arable land. This affects the allocation of the clearing policy from natural forest and grassland. Empirical estimates indicate that about 30% of deforestation is caused by the direct effects of logging activities (UNEP, 1991), but we also include in our simulation indirect consequences: logging often opens up the forest for fuelwood gatherers and the landless who perform the actual deforestation (Allen and Barnes, 1985; Harrison, 1992; Tolba and El-Kholy, 1992).

In TERRA, the effect of climate change is addressed on the global scale, where both positive (CO_2-fertilisation) and negative effects (heat stress) might occur. The hierarchist view with respect to climate change is implemented by assuming the 'best guess' values for the parameter which determines the intensity of the CO_2 fertilisation effect (the β-factor in equation 7.3) and for the aggregated multiplier effect of temperature increase on potential yield (see Figure 7.9). In the hierarchist utopia, the atmospheric CO_2 concentration rises to about twice present levels (see Chapter 11), which leads in TERRA to an increase in global average potential yield of 35% in line with the mean value response of +30% for C_3 crops at doubled (700 ppmv) CO_2 concentrations under controlled conditions (Rogers and Dahlmann, 1993). This positive effect will partially be reduced by the negative effects of a

---

5 As stated in section 15.3, the Conventional Wisdom-scenario is used if no projections are available in the IS92a scenario.

6 The IS92a deforestation scenario is adjusted for reforestation efforts. Consequently, reforestation in the developing region is accounted for in TERRA implicitly.

Sensitivity analysis
As for the other submodels, we performed a number of sensitivity experiments for the hierarchist scenario. It turned out that the submodel outcomes (expressed in terms of some key variables such as food demand, agricultural yields, etc.) are the most sensitive for the following parameters. First of all, small changes in the learning factor result in large changes in fertiliser use. This was one of the reasons for using the same time-series for all three perspectives for this parameter. Secondly, the abandonment of rainfed arable land turned out to be very sensitive for small changes in the percentage of SOM in the topsoil.
Therefore, as described earlier, we decided in the present version of TERRA to use fixed abandonment time-series in all model experiments. Another important factor for the model outcomes is the initial amount of potential arable land. A high estimate leads to a higher fertility (or slower marginalisation) of newly cleared land and a slower increase of clearing costs. Finally, a lower ceiling on fertiliser use often results in food shortages if no policy scenarios (irrigation, clearing) are adapted.

temperature increase of ~2.5 °C, resulting in an overall potential yield increase of 32% in the developed and 23% in the developing region[7].

The egalitarian utopia
In the egalitarian scenario, GWP per capita in the developed region peaks in 2050 at $ ~28,000 per cap per yr; almost a doubling of the 1990 level. In the developing region, there is a slow but steady increase of income towards $ ~5,500 per cap in 2100, which is about six times more than in 1990. Based on these income trajectories, food demand in the developed region rises to the assumed saturation level of 3470 kcal/cap per day in the beginning of the next century. Based on our assumptions, the egalitarian has to wait until 2040 before this food demand voluntarily starts to decrease and reaches a level of slightly more than 3000 kcal/cap per day in 2100. In the developing region, total food demand per capita continuously increases to a higher level than in the developed region: ~3170 kcal/cap per day in 2100. The result is a *global* food demand per capita in 2100 of ~3150 kcal/cap per day. Due to income differences the consumption level of animal products is much lower in the developing region: only 30 kcal/cap per day versus 560 kcal/cap per day in the developed region. Because almost 90% of people will live in the developing region in 2100, this results in a *global* demand per capita for animal based food of nearly 100 kcal/cap per day (see *Figure 15.1a*). The per capita animal and vegetable food demands in the developed region in 2100 are comparable to the intake levels of Israel in 1990 (see *Figure 15.2*). The food package as simulated for the developing region in 2100 does not exist in the world today because at this high level of vegetable food intake all countries have a higher animal food intake – a reflection of the egalitarian emphasis on a more vegetarian diet. In the egalitarian utopia the world population is relatively small and almost no

---

7 These estimates are rather high compared to climate impact studies that show small to moderate changes in global agricultural production: at most +11% in the developed countries and –6% in the developing countries in 2060 (Kane and Reilly, 1992; Rosenzweig and Parry, 1994). Our too optimistic estimates result from the crop model in which only the direct effects of temperature increase, based on temperature curves, are considered (see Chapter 7). The effect of LGP changes are hardly taken into account.

crop products are used as feed. This results in a relatively low demand for agricultural products, less than half of that in the hierarchist utopia.

In the egalitarian utopia a considerable fraction of the low productive and usually fragile soils cannot be used as cropland. The estimate of potential arable land is correspondingly low: 2.8 Gha of which 1.7 Gha is situated in the developing countries. In the developed countries there is no land clearing, just as in the hierarchist utopia, because the existing arable land can provide enough food. Technological developments result in higher yields. Combined with a decreasing population size in the developed region and a decreasing food demand per capita, this leads to a situation of surplus production[8] which allows the removal of a considerable amount of arable land from production while achieving low-intensity agriculture. Until 2050 the clearing rate in the developing countries sharply declines to 3 Mha/yr (from 7 Mha/yr in 1990), but then it becomes unavoidable to either clear more land at the sacrifice of forests or use more inputs per ha. In our implementation, the egalitarian considers low-input (organic) agriculture more important than preserving forests (see section 15.3). Therefore, land clearing for agriculture stabilises at 2.5 Mha/yr during the second half of the 21st century at a level, well above the clearing rates in the hierarchist and individualist utopia. Only in this way can total fertiliser use be stabilised. Despite the low estimate of the total potential arable land, the fraction of it remaining is high (80%) which brings the clearing costs to a relatively low level, although higher than in the other utopias (see *Figure 15.1c*). With respect to irrigation, we assume a smaller expansion than in the hierarchist utopia in line with the smaller population growth. The water supply costs per litre stabilise in the second half of the next century and become slightly lower than in the hierarchist utopia, but due to a high water use efficiency, irrigation costs are only half of the hierarchist cost level. Nevertheless, as GWP is 75% lower in the egalitarian utopia by the end of the next century, total expenditures on agriculture decrease to 2.6% of GWP in 2100 compared to 1.6% in the hierarchist utopia.

The resulting global area of rainfed and irrigated arable land in the egalitarian utopia and the intensity level of its use in terms of fertilisers and yields are shown in *Figure 15.1b, 15.1d* and *15.1e*, respectively. These actual yields are about 15% higher than needed to cover the food demand because the atmospheric CO_2 concentration and the temperature increase in the hierarchist input scenario from the CYCLES submodel (730 ppmv and 2.5 °C) result, in the egalitarian utopia of TERRA, in a positive net effect of climate change on potential yields of +21% and +15% in the developed and the developing regions, respectively. This is about 10% lower than in the hierarchist utopia because the egalitarian is less optimistic about the CO_2 fertilisation effect and more pessimistic about the effects of a higher temperature. However, in section 15.5, we will use an egalitarian instead of a

---

[8] In the real world this might result in an increase of trade from the developed to the developing countries. In the current version of TERRA, trade is assumed to be constant after 1990 (5.2% of the agricultural production in the developed countries)

hierarchist input scenario from CYCLES, which results in a so-called 'full' utopia. In that case, an atmospheric CO_2 concentration of only 500 ppmv and a temperature increase of 3.6 °C has, in combination with a lower soil fertility due to decreasing Soil Organic Matter percentages, almost no climate effect on potential yields in the developed region (+1%) and a considerable negative net effect in the developing region (–15%).

Finally, in the egalitarian utopia, the rainfed arable land which can be taken out of production in the developed region, is reforested. The forested area in this region increases from 1920 Mha in 1990 to ~2600 Mha in 2100; about 200 Mha more than in the hierarchist utopia. Because forestry management and wood production are fully based on principles of sustainability by 2075, deforestation rates in the developing region are much lower than in the hierarchist utopia (see *Figure 15.1f*). As a result the amount of forests in the developing region stabilises by 2050 at ~1700 Mha, or 80% of the forested area in 1990.

The individualist utopia

In the individualist utopia, the saturation level of food is much higher than in the hierarchist utopia: 4,250 kcal/cap per day of which a maximum of 1750 kcal/cap per day consists of animal products. These assumptions have been adopted from IMAGE2.0 (Alcamo, 1994). Although they may seem very high at first sight, they are not unrealistic. For example in Belgium and Ireland, 1990 food consumption levels were more than 3900 kcal/cap per day containing more than 1500 kcal from animal products (see *Figure 15.2*). Based on the individualist assumptions about elasticity parameters and an income rising to an average of $ ~135,000 per cap per yr in 2100, the saturation level and the maximum food demand from animal products are reached in 2045 in the developed region. In the developing region, the saturation level is reached in 2060 with an intake of animal products of ~800 kcal/cap per day and an income level of $ ~9,000 per cap per yr. After 2060, as incomes continue to rise to more than $ 28,000 per cap per yr, and diets become more sophisticated, the intake of animal products increases to ~1200 kcal/cap per day. The resulting average global vegetable and animal food demands are shown in *Figure 15.1a*. The global production needed to cover these high food demands for a relatively large population[9] is, in 2100, about two-third more than in the hierarchist utopia. This large increase is for more than one-fourth caused by the assumption that animals are increasingly fed by food crops from agriculture.

Without using explicit cost-minimisation, we assume that the use of fertilisers is often economically more attractive than land clearing. Therefore, fertiliser use rises to a level close to the maximum of 250 kgN/ha per crop cycle in 2100. Despite these intensification efforts, the arable land area has to be expanded to almost 1700 Mha from the 1990 value of 1350 Mha. As in the hierarchist utopia, the irrigation policy

---

9 13,300 million people in 2100; 11,700 million living in the developing countries.

has been related to the population growth which results in an irrigated area of 460 Mha (*Figure 15.1b*). This is in agreement with optimistic estimates as can be found in a study of Harris (1990), but it is almost 20% higher than the most optimistic estimate of the World Bank (1994). Water costs remain slightly lower than in the hierarchist utopia. Combined with a higher water-use efficiency, irrigation costs are 25% lower than in the hierarchist utopia.

The global average yield in the individualist utopia increases to more than 8 tonne CCE/ha/yr (*Figure 15.1e*), which is slightly more than the global theoretical maximum of 7.6 tonne CCE/ha per yr as can be derived from the study of Luyten (1995) and Penning de Vries (1995). This high yield is realised by a combination of irrigation, fertiliser use, positive climate change effects and technological development. In the first part of the next century, clearing costs remain low in the individualist utopia because of an optimistic assessment of the potential arable land: 3.8 Gha (Luyten, 1995). Nevertheless, the relatively high expansion of arable land (see *Figure 15.1b*) results in clearing costs which rise to hierarchist levels in the last part of the next century (see *Figure 15.1c*). Due to the high economic growth in the individualist utopia, total agricultural expenditures decrease to only 0.9% of GWP.

Because reforestation is given a low priority, the expansion of the reforested area in the developed countries is almost stopped in 2025 and remains at zero afterwards. As in the other utopian worlds, no explicit reforestation efforts have been modelled in the developing region. Wood demand and consequently wood production in the developing countries is assumed to be relatively high in the individualist utopia due to a large population. Because wood production is assumed to become hardly more sustainable after 1990, almost no forest is left in the developing countries by the end of the next century. The deforestation rate sharply declines (see *Figure 15.1f*) and a wood supply problem might occur, which, however, is not explicitly modelled in TERRA: it is assumed that wood demand can always be met by alternatives, i.e. plantations and sustainable wood production from temperate forests.

In the individualist utopia, the impact of climate change is evaluated optimistically: the effect of CO_2 fertilisation is barely positive and temperature increase is assumed to have no negative impact at all as adaptation can be made if necessary. In 2100, this results in an increase of the potential yields of 42% and 47% in the developing and the developed regions, respectively. Here, again, using an individualist instead of a hierarchist input scenario from the CYCLES submodel gives a different outcome: the atmospheric CO_2 concentration in the individualist input scenario is 630 ppmv instead of 730 ppmv in 2100, which results in a slightly lower potential yield increase in both regions. In the individualist input scenario from CYCLES, the temperature increase is 1.1 °C instead of 2.6 °C which, however, has no effect on potential yields because temperature increase is assumed to have no negative impact.

15.5 Uncertainties and dystopias: some more model experiments

As in the previous chapters, we have also performed simulations with the TERRA submodel for different combinations of world view and management style with regard to land and food developments. These experiments show how TERRA responds to a discordance between world view and management style, given a certain background perspective. Three world views combined with three management styles and combined with three possible backgrounds gives a total of 3×3×3=27 simulation experiments. Contrary to the experiments in section 15.4, the input scenario from the CYCLES submodel is now always related to the same perspective as the input scenarios from the other submodels. In other words, perspective-related scenarios are used for the inherent soil fertility (Q-factor), atmospheric CO_2 concentration and temperature increase. Basically, there are three possible outcomes for the TERRA submodel in these model experiments:

- A new balance between food and feed demand is found by adjusting the input level on rainfed arable land.
- A situation of surplus production occurs when inputs on rainfed arable land cannot be lowered any further, i.e. when they are equal to zero. In reality, a situation with surplus production is unlikely to persist over prolonged periods

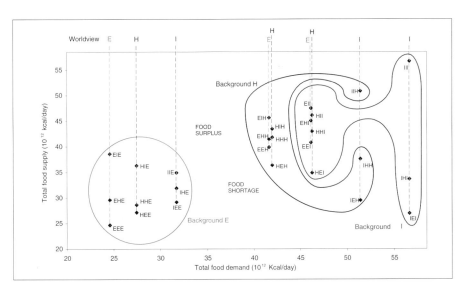

Figure 15.3 Total food demand vs. supply in 2100 for 27 experiments, grouped as to background as indicated by the third letter. Each group of three experiments with the same background and world view, indicated by the first letter, are connected by a vertical dotted line. The second letter corresponds with the management style in the TERRA submodel.

337

and would be countered by response actions by farmers and governments through prices, subsidy etc. However, in TERRA it will persist because the clearing policy, the irrigation policy and inputs on irrigated land are dictated by scenario assumptions.
- A situation of food shortage occurs when inputs on rainfed arable land run up against the assumed (perspective-related) maximum. Again, in reality, such a situation would be responded to in various ways. For example, food prices would rise, which results in lower demands and/or a higher efficiency leading to higher yields and/or more widespread use of irrigation and the like.

Because we run the TERRA submodel with a fixed background, i.e. with fixed inputs from the other submodels without interactive feedbacks, the simulated food shortage or surplus has no influence on e.g. population. In *Figure 15.3*, the total food demand, i.e. animal plus vegetable food demand for all people in both economic regions, in 2100 is plotted against the total food supply for all 27 experiments. Each point represents the outcome of one experiment. The first character of each label indicates the world view in TERRA (Individualist, Hierarchist or Egalitarian), the second corresponds to the management style and the third with the background. If a point lies on the diagonal, then the food demand is in balance with the food supply. As one would expect, this is the case in the so-called 'full' utopias (HHH, EEE and III). Points above the diagonal correspond to a surplus production, below it to a food shortage.

Total food demand depends on two factors: the *background*, which determines for example the total number of people and income, and the *world view* in TERRA, which affects the relation between income and vegetable and animal food demand per capita (see Equation 7.1a). Whether this food demand is met by total food supply depends on the *management style* in TERRA, which is rather inflexible and therefore both overproduction and shortages occur. First, we focus on total food demand in 2100, which is indicated along the *x*-axis of *Figure 15.3*. All nine experiments that correspond with the egalitarian background result in a food demand which is lower than the food demand in any one of the other eighteen experiments because the egalitarian background corresponds to the smallest population and the lowest incomes. Within this set of nine experiments, the egalitarian world view (EEE, EHE and EIE) results in the lowest and the individualist world view (IEE, IHE, and IIE) in the highest food demand. The hierarchist world view is somewhere in the middle (HEE, HHE and HIE). These differences are caused by the world-view-related assumptions with respect to the income-food demand elasticity parameters.

With the hierarchist background, we see almost the same food demand in the egalitarian world view (EEH, EHH and EIH) and the hierarchist world view (HEH, HHH and HIH) in 2100. The income scenario corresponding to this background causes the food intake in the egalitarian world view to nearly reach the saturation level of 3470 kcal/cap per day in 2100; for the hierarchist world view this level is reached 50 years earlier due to the above mentioned assumptions on income-elasticity. The corresponding agricultural production in terms of tonne CCE/yr, however, is 23% lower for the egalitarian world view because it assumes an almost vegetarian diet and almost no food crops being used as feed. In the individualist world view the food demand per capita continues to grow with income for a longer time because the saturation level is assumed to be 20% higher. These demands can be met if an agricultural production of 5.7, 7.2, and 11.2 million kton CCE/yr is supplied in the egalitarian (IEH), hierarchist (IHH) and individualist (IIH) world view, respectively. With the individualist background, we see the same pattern as described above: in both the egalitarian and the hierarchist world view the saturation level is (almost) reached by 2100 and total food demands (in kcal) hardly differ. It is, however, higher than with a hierarchist background because of a larger population

Uncertainty analysis

To asses the uncertainty in the outcomes of the TERRA submodel we performed an UNCSAM analysis. Following the procedure described in Chapter 11, all input variables and parameters used for the differentiation of the three perspectives are varied uniformly throughout the domains. Considering the submodel sensitivity for the learning factor, we have also included this parameter in the UNCSAM analysis. In line with section 15.4, the background variables, which are not varied in this analysis, are chosen according to the utopian perspective in TERRA, except for the input-variables from the CYCLES submodel which are hierarchist. *Table 15.2* shows the uncertainty ranges for a few key output variables.

With respect to the first three output variables, which are related to the intensity and scale of the agricultural sector, the utopian values of the hierarchist are, more or less, in the middle of the uncertainty range. The explanation is that the hierarchist values of most parameters are in the middle of the range of values used for the implementation of the three perspectives and that the TERRA submodel output depends almost lineairly on most input parameters. This also explains why the egalitarian utopia is found close to and below the 2.5 percentile values and the individualist close to and above the 97.5 percentile values.

As described in section 15.4, the egalitarian and the individualist utopias represent rather extreme and opposite positions in forest clearing and the degree of sustainable forest management. Therefore, in the egalitarian utopia, the area of natural forest remaining in the developing region is far above the 97.5 percentile value and the individualist far below the 2.5 percentile.

Variable value in 2100	Hierarchist		Egalitarian		Individualist	
	<2.5%	>97.5%	<2.5%	>97.5%	<2.5%	>97.5%
Actual yield (tonne CCE/ha/yr)	4.6	7.6	3.5	5.3	5.3	8.4
	(6.4)		(4.1)		(8.4)	
Fertiliser use (TgN/yr)	101	273	52	150	107	286
	(183)		(58)		(402)	
Vegetable production (10^6 ktonne CCE/yr)	6.3	8.7	4.3	5.7	7.0	9.7
	(7.2)		(3.5)		(12)	
Natural forest in developing region (Mha)	325	1250	681	1410	255	1230
	(595)		(1700)		(50)	

Table 15.2 Uncertainty ranges for the three utopias (utopia values in parentheses).

(13,300 vs. 11,800 million). If the individualist world view also holds for the land and food situation, food demand per capita grows to 4250 kcal/cap per day which results in a food demand of almost 57×10^{12} kcal/day or 12 million ktonne CCE/yr of which 50% is used as feed[10].

Whether food demand can be met by food supply depends for each combination of world view and background on the management style in TERRA. As *Figure 15.3* shows, experiments with the individualist management style lead to the highest food supply and the egalitarian management style to the lowest for each group of three experiments with the same background and world view. The experiment EIE in the three experiments in the left-hand side of *Figure 15.3* shows the largest surplus production. Experiment IEI in the lower right corner shows the largest food shortage. This shortage is much larger than the surplus production in the first case because the individualist has more adaptability in terms of management: the application of N-fertilisers and other inputs on rainfed arable land can be lowered much further than it can be raised in a egalitarian management style. However, it appears that in a individualist management style, reducing the N-fertiliser level to zero on rainfed arable land is not enough to prevent a surplus production: the area of arable land and the input level on irrigated arable land is still too high. In TERRA, the individualist manager anticipates a high food demand. If it fails to occur, this nearly always leads to a food surplus. This is confirmed by *Figure 15.3*: nearly all experiments with an individualist management style, indicated by the second character in each label, result in a surplus production. In the real world more response actions would be taken, for example, reducing the clearing and/or irrigation activities.

In our experiments, the egalitarian manager considers the desire to establish a low-intensive agriculture to be more important than feeding the world population, which results in (large) food shortages. In the egalitarian utopia (EEE) the input level on arable land is close to what is considered as an acceptable maximum. An increase of total food demand beyond the utopian level soon leads to a food shortage and becomes more profound in the case of a larger and/or richer world population. In *Figure 15.3*, this increasing food shortage is shown by the group of nine experiments with an egalitarian management style, going from a 'full' egalitarian utopia (EEE) to an egalitarian management style confronted with an individualist world view and background (IEI). In this worst-case scenario, the egalitarian manager is confronted with an individualist food demand of 13,300 million 'rich' people – a task which cannot be performed within the egalitarian constraints. Finally, as expected, the hierarchist management style can be found in the middle: food shortages and food surpluses are less profound because the hierarchist has more flexibility in terms of raising or lowering the agricultural production by changing the input level on rainfed

---

10 In terms of Grain Equivalents (GEs), used in the global food study of Luyten (1995), 12.0 million kton CCE is about 18.9 million kton GE. In Luyten's study, this is close to the food requirement in the case of an 'affluent diet' which is 20.4 million kton GE based on a food requirement of 4.2 kg GE/cap/yr.

arable land. Nevertheless, if the hierarchist management style is confronted with an individualist world view (i.e. a high animal food demand per capita and a high fraction of the agricultural production being used for feed), then it results in a rather large food shortage if the world population grows to the hierarchist or individualist level of 11,800 and 13,300 million, respectively (IHH and IHI).

15.6 Risk assessment

The model outcomes, as presented in the previous sections, can be related to risks. In order to do so, three aspects will be distinguished: (i) actual food production and food security; (ii) production potential in terms of the amount of arable land and its inherent fertility; and (iii) ecosystems. With respect to the first, we are able to make risk-related statements by comparing global food production to global food demand. The second aspect is related to longer term risks: a continuous decrease of inherent soil fertility will undermine food production and thus food security in the long run[11]. Little can be said with respect to ecosystem-related risks: climate change has, in our model, a direct impact on simulated food production only. Indirect impacts, such as a change in the stability of ecosystems due to (the rate of) climate change, are absent. Only the decrease of forested area can be used as a proxy for bio-diversity and ecosystem functioning. As far as possible, all three aspects will be discussed now in the context of the submodel outcomes as presented in sections 15.4 and 15.5.

In the egalitarian (semi-)utopia there is a relatively low food demand due to a small population with low incomes and an emphasis on a vegetarian diet. As mentioned before, this low demand can hardly be met within the context of the egalitarian constraints with respect to fertiliser use and forest preservation. The egalitarian risk aversion with respect to ecosystems turns out to be risk seeking in terms of accepting increasing pressures on food security and therefore on social stability. In our simulation of the egalitarian semi-utopia (see section 15.4), this dilemma is 'solved' by assuming a relatively high rate of land clearing after 2050 (see *Table 15.1*) combined with an input level close to the acceptable maximum. However, in the experiments of section 15.5, a simulated increase of total food demand beyond the utopian level (that is, with a hierarchist or individualist background) combined with a high temperature increase, low atmospheric CO_2 concentration and declining soil fertility (that is, with an egalitarian world view), soon leads to reaching the ceiling on fertiliser use and thus to a food shortage because clearing and irrigation policies are governed by exogenous time-series. This shortage becomes a serious problem in the case of a larger and/or richer world population (IEH and IEI in *Figure*

---

11 In all experiments with TERRA and TARGETS, it is assumed that abandonment of arable land due to land degradation decreases to zero by the end of the next century. We made this choice because our modelling approach, as described in section 7.2.5., was not sufficient to model abandonment adequately.

Who is to blame for starvation?

Some of the points in *Figure 15.3* can be enlivened by a narrative which captures some of its features. A quite dystopian world is associated with the point IEI: an individualist view of how the world functions is confronted with egalitarian policy measures. The story behind it could be that a high but unbalanced growth in economic welfare leads to fertility and mortality patterns which cause a rather high population growth. The market ideology is dominant and the Western life-style, with its high meat diet, is spreading among the more affluent citizens of the developing world. Savings from the elimination of harmful subsidies on water, pesticides and fertilisers are used to promote egalitarian policy measures to protect farmland, conserve soils, save water, and manage pests. However, these measures are not sufficient to produce the food required to feed a large number of people, many of whom have adopted Western-style consumption patterns.

This results in an increasing pressure on the food system, in the form of rising prices, local shortages and the first signs of soil deterioration, which is not countered by an expansion of rainfed and irrigated arable land. One of the reasons for this is that egalitarian groups, backed up by the media, successfully point out that clearing of (rain)forest for agriculture is impossible as it will destroy the basis of life on Earth. Hence, large areas are protected as nature reserves and for tourism. Also, the construction of large water reservoirs to expand the irrigated area meets with huge resistance. In the end, the spread of a Western-style diet in combination with policies oriented towards sustainable agriculture and preservation of nature, causes severe food production crises.

Who is to blame? There are at least two stories, each with its own rationality. On the one hand, the individualist will argue that egalitarian NGOs, backed up by the media and popular discontent, have thwarted the necessary and promising steps. As a result, increases in agricultural productivity have been delayed. Forests cannot be used productively so they do not generate the revenues the governments in poor countries need to support innovations in agriculture. Egalitarian groups, on the other hand, will blame the unrestrained egoism of the market and the complacency and inertia of vested interest groups. They will realise that they partly have been successful, but at the same time point out that sound environmental agricultural policies have to coincide with stringent measures to curb population and economic growth and with measures that discourage Western consumption patterns. As they point out: "Perhaps the greatest way to increase food use efficiency is to reduce the world's consumption of meat" (Brown *et al.*, 1996). In terms of *Figure 15.3*, this is the difference between point IEI (with a food supply-demand ratio of less than 0.5) and point EEI (with a ratio of almost 0.9).

15.3). It is possible to adapt the clearing and irrigation policies but then the food shortage is 'solved' by accepting higher risks with respect to food production through a further deterioration of the environment which show up in the model as a further decline of forested area in the developing countries.

In the individualist utopia, it is no problem to meet high food demands. It is possible to produce enough food by establishing an intensive, high-tech agricultural sector. Environmental trade-offs will not occur because it is possible to adapt to, for example, climate change. Soil fertility is maintained by well-considered soil management practices. Tropical forests may disappear but this not considered as a problem. A simulated decrease of total food demand below the utopian level can lead to a food surplus which becomes more profound in the case of a smaller and/or poorer world population (that is, with an egalitarian or hierarchist background, see *Figure 15.3*). However, this surplus production can be limited as the individualist has more adaptability in terms of management. For the individualist management style, low future food risks exist; even situations of food surpluses can occur, especially with an egalitarian background. On the other hand, the individualist management style can be associated with high risks with respect to ecosystems which, however, never come true in the TERRA submodel.

In the hierarchist utopia, it is possible to meet the global food demand. With respect to soil fertility, little or no decrease occurs in the next century and the impact of climate change is somewhere in between the individualist and the egalitarian utopia. However, the hierarchist manager does not go as far as the individualist with respect to deforestation and intensification of agriculture. This results in food shortages if food demand develops as expected in the individualist world view and becomes more serious in the case of a richer and larger world population, that is, with an individualist background. In terms of risk, the hierarchist manager is concluded to be situated between the egalitarian and the individualist for all three aspects.

15.7 Conclusions

In this chapter we present a variety of future projections with respect to land-use changes and food production. As a basis for our hierarchist utopia or 'reference scenario', we have used the IS92a scenario of the IPCC combined with the Conventional Wisdom scenario of IMAGE2.0 to obtain a consistent set of time-series for key variables. GWP and population are the input variables, which determine food demand; N fertiliser use, forests and arable land in the developed and the developing world are reproduced by the TERRA submodel whenever a specific technology scenario is assumed.

Experiments have shown that in the egalitarian and individualist semi-utopias global food demand can be met. Although little confidence can be assigned to simulated costs, it can be concluded that expenditures on agriculture tend to go down in all utopias if expressed as a fraction of total GWP. Furthermore, egalitarian managers experience considerable pressure to give up their reservations with respect to land clearing and intensive agriculture. Dystopian futures, in which an egalitarian management style in TERRA is combined with a world that develops according to the hierarchist or individualist world view, soon lead to (major) food shortages and thus imply high risks with respect to food security and social stability.

The individualist manager experiences no problems at all in the next century (unless a food surplus is considered a problem). However, this is a somewhat distorted picture as the individualist accepts high risks with respect to the environment (ecosystem stability, bio-diversity and land fertility). As environmental feedbacks are so far only implemented to a limited extent in the TARGETS model, these risks turn out to be of minor importance in our simulation experiments. In Chapter 18, we discuss these risk-related topics in an integrated context.

Human disturbance of the global biogeochemical cycles
16

*"The same climatic record that renders the fatalist fatalistic,
and the egalitarians fearful, renders the individualist cheerful."*
J. Adams, Risk (1995)

16 HUMAN DISTURBANCE OF THE GLOBAL BIOGEOCHEMICAL CYCLES

Michel G.J. den Elzen, Arthur H.W. Beusen, Jan Rotmans and Marjolein B.A. van Asselt

The key question underlying the controversies addressed in this chapter is to what extent the global biogeochemical cycles and the climate system are being disturbed by anthropogenic processes. The CYCLES submodel is used to explore the influence of various model routes on future projections of global environmental change. These routes are characterised by specific assumptions about the key processes within the global carbon (C) and nitrogen (N) cycle and the climate system. This creates the possibility to assess various emission scenarios, paving the way for the more integrated experiments described in Chapters 17 and 18. It is reiterated that the current state of scientific knowledge with respect to global C and N cycles and climate change is still beset with major uncertainties.

16.1 Introduction

Although scientific knowledge of global biogeochemical cycles is increasing rapidly, there are still major gaps. Subjective interpretation of these gaps results in different assessments of the rate, magnitude and impacts of human-induced changes in global cycles. The climate debate exemplifies the kind of intellectual battle that can take place, given uncertainties about the global biogeochemical cycles of the basic elements carbon (C), nitrogen (N), phosphorus (P) and sulphur (S), and the climate system. In general, the controversies within the scientific community on global biogeochemical cycles focus on the relationships among the physical, biological and chemical processes comprising the complex dynamics of the global biogeochemical cycling as well as the role of the various feedbacks. This chapter focuses primarily on the main uncertainties within the controversies and various perspectives on the carbon and nitrogen cycle, i.e. the fate of the anthropogenic emissions of C and N compounds, and on the climate system. To investigate these issues, the CYCLES submodel (den Elzen *et al.*, 1995; 1997) is used. Following the approach of perspective-based model routes as outlined in Chapter 10, we make specific clusters of assumptions about the biospheric, oceanic and atmospheric processes and the feedback processes within the global carbon-nitrogen cycle and climate model. Based on these model routes, future projections of the fate of anthropogenic C and N compounds in the global environment, and of global mean surface temperature changes, are made and then analysed.

This chapter is structured as follows. Section 16.2 describes briefly the controversies on and uncertainties within the carbon and nitrogen cycle, and the climate system. Different interpretations of these uncertainties underlie the main controversies, which in section 16.3 is translated into perspective-based model routes. This enables us to perform experiments addressing these controversies. Three utopian images of the future are presented in sections 16.4, and also compared with the IPCC projections. A risk analysis is described in section 16.5 that forms the basis for an evaluation of the controversies postulated at the beginning of the chapter. Section 16.6 gives the conclusions of this chapter.

16.2 Controversies and uncertainties

16.2.1 The controversies

Global carbon and nitrogen cycles
For the global carbon and nitrogen cycles, the uncertainties in our knowledge of the present sources of, and sinks for, the anthropogenically produced C and N compounds are considerable (see Table 8.1 and 8.2). In fact, there is an imbalance between the sources and sinks, which leads to the so-called missing carbon and nitrogen sink (Galloway *et al.*, 1995; Schimel *et al.*, 1995). The ratio between anthropogenic and natural fluxes differs for the two biogeochemical cycles. For the carbon cycle, the anthropogenic perturbation (about 7 GtC/yr throughout the 1980s) is relatively small compared to the natural fluxes (about 60 GtC/yr atmosphere-terrestrial biosphere, and about 90 GtC/yr oceans-atmosphere) (Schimel *et al.*, 1995). For the global nitrogen cycle, the human-induced fixation of nitrogen (about 150 TgN/yr) is of the same order as the natural fixation (Galloway *et al.*, 1995). However, there is strong evidence that the anthropogenic increments lead to a disturbance in the balance of both cycles, causing environmental problems like climate change, soil acidification and water pollution (see *Figure 8.1*). Current views differ as to the significance, irreversibility and impact of the perturbing signal. The major controversies all hover around the question: *Are the global biogeochemical cycles being seriously and irreversibly disturbed by human perturbations and what are the future consequences for humans and the environment?*

Climate system
The Earth's surface air temperature is modified by a natural greenhouse effect. Without this natural process the global mean surface air temperature of the Earth would be some 33 °C lower. Since the beginning of the industrial age, anthropogenic emissions of most greenhouse gases have increased substantially. The likely result of this human-induced change in the atmospheric energy balance is an alteration of the

Earth's climate, commonly referred to as the 'enhanced greenhouse effect'. The magnitude of the warming process, the exact rate of change and the regional distribution, however, are largely unknown (IPCC, 1995a). Furthermore, various feedbacks within the climate system may either stimulate or impede an initial warming process. The climate change issue attracts enormous interest in a wide range of scientific fields (von Storch and Hasselman, 1994) as well as in decision-making communities, where it fuels furious political debate. The present discussion on anthropogenic climate change and its consequences are not entirely new (Stehr *et al.*, 1995). At the end of the 19th and the beginning of the 20th century discussions arose among geographers, meteorologists and climatologists '*avant-la-lettre*' about climate variability and human-induced climate change. At the time the majority of 'climatologists' and meteorologists were convinced that global climate was more or less constant; some of them argued that climate was not a stable phenomenon. Notwithstanding an impressive accumulation of knowledge, uncertainties have increased during the 1980s and 1990s (Arrhenius, 1896; Brückner, 1889; Hann, 1903), leaving the debate unresolved. The fundamental controversy pertaining to the climate debate can therefore be summarised in the question: *Is the global climate being disturbed in a serious and irreversible way, and if so to what extent, at what rate of change and with what regional pattern, and what are the human and environmental consequences?*

16.2.2 The uncertainties

The selection of the main uncertainties as described in this section is based on the methodology described in van Asselt and Rotmans (1995). Three measures to characterise the uncertainties are used: magnitude, degree, and time variability and importance. Magnitude refers to the relative contribution of the uncertainty of a specific process to the overall uncertainty of a key indicator, such as the atmospheric CO_2 concentration. The criterion of degree denotes the absolute uncertainty, whereas time variability refers to the change in uncertainty in historical perspective. The uncertainties are finally ranked by determining the importance of the uncertainty, using the indices magnitude, degree and time variability in that order of priority.

The global carbon cycle
Within the global carbon cycle, the main uncertainties about the components of the carbon budget are the oceanic and terrestrial sinks, with its main primary and secondary processes given in *Table 16.1*. The primary oceanic process refers to the atmospheric CO_2 exchange with the oceans, which depends on oceanic dynamics, and ocean/atmosphere mixing processes. The primary terrestrial biospheric process is its atmospheric CO_2 exchange, i.e. the emissions associated with land-use changes. The secondary processes are the biogeochemical feedbacks. Here, the N

Processes	Uncertainty	Magnitude	Degree	Time variability	Importance
Ocean					
• Primary processes	oceanic CO_2 uptake[a]	middle	middle	low	middle
• Secondary processes	temperature feedbacks	small	high	low	middle
Terrestrial biosphere					
• Primary processes	terrestrial CO_2 uptake[b]	large	middle	high	high
• Secondary processes	CO_2 fertilisation	large	high	high	high
	temperature feedbacks	large	middle	middle	high
	soil moisture changes	large	middle	high	high
	migration of ecosystem	middle	middle	middle	middle
N cycle					
• Secondary processes					
	N fertilisation	large	middle	middle	high
	temperature feedbacks	large	middle	middle	high

a The carbon dioxide exchange with the oceans depends on the processes of oceanic dynamics (convection, diffusion and precipitation of dead organic matter), and oceanic/atmospheric mixing processes.

b The carbon dioxide exchange with the terrestrial biosphere depends mainly on the net terrestrial emissions associated with land-use changes, and to a lesser extent on terrestrial soil processes (decay, soil respiration, humification).

Table 16.1 Importance of uncertainties in the global carbon cycle.

fertilisation and temperature feedback effect on terrestrial nitrogen cycling are also considered, since these constitute a major part of the missing carbon sink (Schimel *et al.*, 1995). The values of the indicators magnitude, degree and time variability for the uncertainties are based on the scientific knowledge, as summarised in IPCC reports. It shows that the main uncertainties in the carbon cycle are related to the terrestrial feedback processes that are used to balance the past carbon budget, i.e. the CO_2 and N fertilisation and climate feedbacks.

The global nitrogen cycle
The fate of anthropogenic nitrogen inputs is surrounded with many uncertainties. We only know where some of these inputs remain: in the oceans and atmosphere (N_2O) (Galloway *et al.*, 1995) (see Table 8.2). The remainder is either stored on continents in groundwater, soils or vegetation, or denitrified to N_2 (atmosphere). The main primary processes describing the oceanic, terrestrial and atmospheric sources and sinks, are given in *Table 16.2*. Secondary processes, i.e. feedbacks, are not considered here, since they do not affect the exchange fluxes between the compartments. The uncertainties in the nitrogen accumulation in the ocean mainly depend on the anthropogenic river inputs (Galloway *et al.*, 1995; Meybeck, 1982). The uncertainties in the terrestrial nitrogen mass balance often arise from our incomplete knowledge of the processes of denitrification and leaching (Galloway *et*

Uncertainty		Magnitude	Degree	Time variability	Importance
Ocean					
• Sources	N fixation	small	high	middle	middle
	deposition (NH_4^+, NO_x)	middle	middle	middle	middle
	anthropogenic river inputs	middle	high	high	high
• Sinks	denitrification	large	high	high	high
Land					
• Sources	N fixation	small	high	middle	middle
	deposition (NH_4^+, NO_x)	large	middle	high	high
• Sinks	denitrification	large	middle	high	high
	runoff to surf. water	middle	middle	middle	middle
	leaching to groundwater	large	high	high	high
	harvesting pathway	middle	middle	high	high
Atmospheric chemistry					
	N_2O build-up (lifetime)	large	middle	high	high
	NO related to trop. O_3 production	large	middle	high	high

Table 16.2 *Importance of uncertainties in terms of primary processes in the global nitrogen cycle.*

al., 1995). Also uncertain is the pathway of anthropogenic nitrogen inputs from fertilisers and legume production into the environment. In the atmosphere, the main uncertainties arise from the stratospheric break-up of N_2O into N_2, which is associated with the atmospheric lifetime of N_2O. *Table 16.2* gives the major uncertainty factors, which are more or less coupled to the fate of anthropogenic nitrogen inputs. The major uncertainties in the natural processes of denitrification and N fixation are not considered in this study.

The climate system

The geophysical feedbacks, i.e. the water vapour feedback and cloud distribution and optical properties, form the major source of uncertainty, next to the direct and indirect radiative effects of aerosols. The uncertainties in the direct radiative effects of aerosols are associated with the scattering efficiency (fraction of sunlight being scattered), and the back scattering (fraction of scattered sunlight, being reflected back to space) (Charlson *et al.*, 1992). The radiative feedback related to the stratospheric ozone depletion is seen to be a less important source of uncertainty due to its minor and declining contribution to the total radiative forcing.

Table 16.3 gives the main selected uncertainties in the climate system, which are based on IPCC reports. Time variability can be illustrated by examining estimations of the climate sensitivity, i.e. the equilibrium global surface temperature increase due to a doubling of CO_2 levels during the past decades. It is related to the geophysical feedbacks and its estimation is based on calculations with General Circulation

Processes	Uncertainty	Magnitude	Degree	Time variability	Importance
Radiation					
• Primary processes	greenhouse gases	large	low	low	low
	tropospheric ozone	middle	middle	middle	middle
	aerosols; direct effects	large	middle	middle	high
	aerosols; indirect effects	large	middle	middle	high
• Secondary processes	radiative effects of stratospheric ozone depletion	small	middle	middle	low
Climatology					
• Geophysical feedbacks	water vapour	large	middle	middle	high
	snow-ice albedo	middle	middle	middle	middle
	clouds	large	high	high	high

Table 16.3 *Importance of uncertainties regarding the climate system.*

Models (GCMs) between 1975 and 1992 and on estimates by climate experts (Morgan and Keith, 1995). It appears that over the last few decades the central estimate of climate sensitivity has fallen from 3.5 °C in the mid-eighties to 2.5 °C (-30%) (IPCC, 1992, 1994) at present, which indicates a high time variability (den Elzen, 1993).

A summary of the crucial uncertainties for the global carbon and nitrogen cycles and the climate system is given in the *Table 16.4*. This list is used in the implementation of the perspective-based model routes.

The global carbon cycle
- terrestrial CO_2 exchange (terrestrial emissions from land-use changes)
- CO_2 fertilisation
- temperature feedbacks on soil respiration and mineralisation
- N fertilisation
- temperature feedbacks on N uptake and net primary production

The global nitrogen cycle
- denitrification (and related leaching)
- pathway of harvested products into the environment, i.e. fresh surface water (and related to the oceanic river inputs).
- deposition of the oxidation products of NO_x and NH_3
- lifetime of N_2O

The climate system
- geophysical feedbacks (water vapour, snow-ice albedo, clouds)
- aerosols; direct effects

Table 16.4 *Crucial uncertainties in the global carbon and nitrogen cycles, and the climate system (in arbritary order) as included in the model.*

Comparison of our study with that of Morgan and Keith (1995)

Does our list of crucial uncertainties coincide with the uncertainties as listed by other researchers? To answer this question, we compare our classification with the work of Morgan and Keith (1995). Their aim was to articulate subjective expert judgement by interviewing 16 climate experts of different schools of thought. Part of this effort was to rank the factors contributing to the experts' uncertainty. Uncertainties in the nitrogen cycle were not considered in their survey. Three different classification procedures (frequency of items, weighed frequency and weighed sum of frequency) all yield the same top five factors across the whole expert set.

Comparison of the two lists requires some aggregation. Uncertainties related to clouds and indirect and direct effects of aerosols in our list, are subsumed under their heading 'cloud optical properties'. Soil moisture changes are to be considered as crucial components of what Morgan and Keith call 'CO_2 exchange with terrestrial biota'. Morgan and Keith's list now compares well with our ranking: 'cloud optical properties', CO_2 fertilisation, temperature feedbacks on soil respiration and net primary production, are in the top five of their and our list. Different from our approach is that Morgan and Keith also consider the specific uncertainty factors in the boundary conditions, modelling details, and detailed physical processes (only considered in GCMs). These uncertainty factors cannot be assessed with our simple box model, and are therefore not taken into account in the coming uncertainty analysis.

16.3 Perspectives on the global carbon and nitrogen cycle and climate system

Following the approach outlined in Chapter 10, the uncertainties for the global biogeochemical cycles of C and N, and the climate system, are interpreted from three perspectives: the hierarchist, the egalitarian and the individualist. The CYCLES submodel does not contain any policy variables, so only the world view is of interest. The key items for the world view with regard to the biogeochemical cycles, and climate system and their assessment, are summarised in *Table 16.5*.

The hierarchist world view

Hierarchists believe that the global biogeochemical cycles might be seriously disturbed by irreversible human activities. Consistent with their inherent tendency to control, they tend to ignore speculations about possible processes and feedbacks aside and confine themselves to the ones of which the probability of occurrence and magnitude are to a certain extent known. They interpret uncertainties in line with the estimates of prominent scientific experts and institutions, for instance the IPCC for climate change. Taking into account the current IPCC estimates, hierarchists expect that the CO_2 and N fertilisation are the dominant factors for balancing the past carbon budget, and thereby form important dampening mechanisms of the future atmospheric CO_2 concentration. The overall future effect of the temperature feedbacks is a slightly positive one (IPCC, 1995a). Based on current insights, it is believed that a major part of the 'missing' anthropogenic nitrogen (25-75%) accumulates in the atmosphere (Galloway *et al.*, 1995). The hierarchist suports the

Crucial uncertainties	Hierarchist	Egalitarian	Individualist
Perception of global biogeochemical cycles	biogeochemical cycles might get seriously disturbed	biogeochemical cycles are already irreversibly disturbed	biogeochemical cycles can adapt to the disturbances
C cycle overall effect of terrestrial feedbacks	moderate CO_2 and N fertilisation	amplifying temperature feedbacks	strong CO_2 and N fertilisation
N cycle accumulation of 'missing nitrogen'	moderate N storage on land, i.e. biomass and soils	high N storage on land, i.e. groundwater soils (acidification)	high N storage in the atmosphere, and low N storage on land
perception of climate change	serious threat	catastrophic threat	environmental system can adapt to climate change
Climate system geophysical feedbacks radiative effects	amplifying effect moderate cooling of aerosols	strong amplifying effect low cooling effect	minor amplifying effect strong cooling effect

Table 16.5 *Perspective-based interpretation of the global biogeochemical controversy.*

scientific view that a significant part of the remainder accumulates in the soils and biomass of the temperate forests in the Northern Hemisphere because of the carbon sequestration from the added anthropogenic nitrogen deposition (N fertilisation) (Schimel *et al.*, 1995). Geophysical feedbacks, especially water vapour, are assumed to have an amplifying effect on the initial warming process induced by anthropogenic emissions of greenhouse gases. The radiative forcing related to sulphate aerosols is believed to explain most of the difference between the observed temperature increase, and the simulated temperature increase due to the increases in the concentrations of greenhouse gases.

The prevailing view of governments on climate change is that the Business-As-Usual scenario of the IPCC (IS92a, see Chapter 13) gives reason for concern. The majority of scientists seem convinced that climate change is likely to occur in the long term if anthropogenic emissions of greenhouse gases continue to increase (Broekner, 1987; IPCC, 1994; 1995b; Schneider, 1989). However, many government institutions may feel that scientists exaggerate the consequences and the potential to adapt, and underestimate the consequences of strict climate policy measures.

The egalitarian world view

According to the egalitarian myth, minor human-induced changes already have a major influence on the behaviour of the environmental system. Egalitarians fear that nature is increasingly disrupted by the growing pressures on the environment and they think that adaptation of natural systems to new situations is not or only partly

Perspectives on climate

Simon, an economist and outspoken sceptic on environmentalism, stated this about the climate change issue: "My guess is that global warming will simply be another transient concern, barely worthy of consideration ten years from now. [] Indeed, many of the same persons who were then warning about global cooling are the same climatologists who are now warning of global warming ..", and about the acid rain issue: "In Europe, the supposed effects of acid rain in reducing forests and trees growth have turned out to be without foundation; forests are larger, and trees are growing more rapidly, than in the first half of this century", and "The acid rain scare re-teaches an important lesson: it is quick and easy to raise a false alarm, but to quell the alarm is difficult and slow" (Myers and Simon, 1994).

A similar voice is heard from the Marshall Institute: "This most recent Marshall Institute review of scientific evidence on climate change confirms the earlier conclusion that predictions of an anthropogenic global warming have been greatly exaggerated, and that the human contribution to global warming over the course of the 21st century will be less than one degree Celsius and probably only a few tenths of a degree. Spread over a century, a temperature rise of this magnitude will be lost in the noise of natural climate fluctuations."

Greenpeace stated in the Climate Time Bomb Catalogue that: "Whilst no single extreme event, nor sequence of events, can be attributed to the greenhouse effect, Greenpeace believes that the totality of the planet's climatic experience since 1990 indicates that the first footprint of climate change as a result of the human-enhanced greenhouse effect is now becoming clear. We are not alone in this view. The respected Enquete Commission (the German Bundestag's advisory body on climate change) stated in its 1992 report that 'Our planet is already warming at an increasing rate. The first signs of climate change are already measurable and noticeable. Hence there is no reason any more to delay urgently required actions.' "

Also Barney (1993), leader of the global 2000 project in 1980, sees reasons for concern: "As a result of the increasing concentrations of the greenhouse gases, the temperature of the entire planet is expected to begin increasing soon. The best estimate currently available of global temperature change comes from the Intergovernmental Panel on Climate Change (IPCC)... On a planetary scale, an increase of 2.5 °C has enormous significance... Furthermore, the pollutants causing this global disaster are expected to continue accumulating in the atmosphere for at least several decades... The impacts on human settlements and the whole community of life are large."

---

possible. Egalitarians expect environmental problems to be aggravated by amplifying feedbacks, although dampening feedbacks may delay serious disasters. Speculative positive feedbacks with possibly catastrophic impacts are considered, even if they are at present strongly disputed within the scientific community. On the other hand, potential negative feedbacks tend to be ignored in this risk-averse world view.

Egalitarians argue that the global carbon cycle is already seriously and irreversibly disturbed, and believe that the net effect of the terrestrial feedbacks is dominated by the positive temperature feedback on soil respiration. The temperature feedbacks affecting net primary production will be positive within a small range: slight temperature changes will result in an increase in the net primary production; only radical changes will strongly reduce it. This results in a situation in which substantial climate change forces a decrease in the terrestrial uptake of CO_2, thereby amplifying the initial warming process. Egalitarians reject the idea that carbon dioxide and nitrogen fertilisation effects will act as balancing processes. In a similar way, egalitarians will expect serious environmental consequences from the human-induced disturbances of the nitrogen cycle.

Finally, egalitarians believe that present and anticipated future pressures are likely to result in major, probably catastrophic, climate change. They support scientists who claim that climate change is already well under way and that it is one of the major threats to humans and the environment (Flavin, 1991; Hansen *et al.*, 1988; Houghton and Woodwell, 1989; Leggett, 1990). Positive geophysical feedbacks due to water vapour and clouds are expected to dominate the response of the climate system to initial disturbances. Cooling effects from aerosols are considered to be of minor importance; an increased concentration of aerosols in the atmosphere is undesirable because its sources, the SO_2 emissions from fossil fuel combustion, are associated with acidification.

The individualist world view
Consistent with the individualist's perception of nature, the environmental system is robust in reacting to disturbances and has the ability to adapt and evolve. Individualists do not believe that human activities result in irreversible environmental catastrophes. The environmental system itself is able to cope with the fluctuations or humans are ingenious enough to find solutions. This attitude makes that individualists emphasise dampening feedbacks, even if they are speculative. However, amplifying effects, if they occur, are considered negligible.

Disturbance of the global carbon cycle due to human activities is thought to be minor; negative feedbacks such as CO_2 and N fertilisation will dominate the overall effect of terrestrial feedbacks. Consequently, the net primary production of terrestrial ecosystems, and also the food supply, may significantly increase. No undesirable impacts are expected from changes in the N fluxes.

The climate system is considered to be resilient and self-regulating, providing fairly stable mean climatic conditions. Scientists who adhere to this point of view (Balling, 1995; Lindzen, 1990; Ray and Guzzo, 1990) deny the possibility of persistent transformations of the climate. It is argued, for example, that the radiative effects of aerosols might offset an eventual global warming and that geophysical feedbacks will only slightly amplify any initial climate change. Climate change is not associated with catastrophical impacts on human society, also because humans have shown an amazing capacity to adapt.

The qualitative descriptions presented above are used to quantify multiple perspective-based model routes in the CYCLES submodel. The assumptions concerning the choice of the model parameters for the three perspectives are based on sensitivity analyses as described in den Elzen (1993) and den Elzen *et al.* (1997).

16.4 Three images of the future: utopias

In this section we present the results of model experiments in which all input scenarios of the CYCLES submodel from the other TARGETS submodels are in line with the utopian world view implemented in CYCLES. We refer to these scenarios as non-integrated utopias, i.e. futures in which there is a match between the world view of biogeochemical cycles and the climate system, and the world view and management style in the background (see Chapter 11). Based on these model routes, future projections of the fate of anthropogenic C and N compounds in the global environment, and of global mean surface temperature changes, are made and analysed. For the global environmental impacts we focus on the environmental issues: climate change (atmospheric CO_2 concentration, and absolute and relative temperature increase), soil acidification (yearly N accumulation on land, i.e. in the soils of natural ecosystems), water pollution (N storage in groundwater, and total N transported by the world rivers to the oceans).

The reference case: the hierarchist utopia
The central estimates for the feedback parameters are used for the hierarchist utopia, leading to a well-balanced past carbon budget (see Chapter 8). This utopia presents a future in which the atmospheric CO_2 concentration increases to about 735 ppmv by 2100, almost three times the pre-industrial concentration of 280 ppmv *(Figure 16.1)*. This is mainly due to the high CO_2 emissions from fossil fuel combustion. The final concentration level is about 5% higher than the central IPCC-1995 projection for the IS92a emission scenario. The CO_2 emissions associated with land-use changes are given in *Figure 16.2*. Overall, the terrestrial biosphere acts as a sink for carbon due to the dominance of the terrestrial feedbacks, i.e. the CO_2 and N fertilisation, over the emissions from land-use changes *(Figure 16.3)*.

For this utopia, the human-induced disturbances of the global nitrogen cycle correspond with increasing fertiliser use and nitrogen emissions from fossil fuel combustion for the future period (1990-2100). This leads to a total anthropogenic N fixation of about 330 TgN/yr by 2100, more than a doubling of the present levels *(Figure 16.4a)*. The resulting increase in the accumulation of nitrogen in the atmosphere, the oceans, and on land causes environmental impacts far beyond the present levels in terms of water pollution and acidification *(Figure 16.4d-f)*. The accumulation of nitrogen in the soils of natural ecosystems is influenced by the deposition of oxidation products of NO_x and NH_3. It is a measure of soil acidification and is projected to increase in the future, leading to a large area suffering from caustic acidification effects. For the industrialised region, the present levels remain high due to stabilising future anthropogenic NO_x and NH_3 emissions. For the developing region, these high levels are reached before the second half of the next century because here these emissions keep growing. Groundwater pollution in terms of N accumulation reaches levels twice the present ones by 2100. Water pollution, in

16 HUMAN DISTURBANCE OF THE GLOBAL BIOGEOCHEMICAL CYCLES

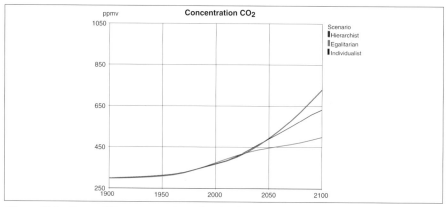

Figure 16.1 Future CO_2 concentrations for the three utopias over the period 1900-2100.

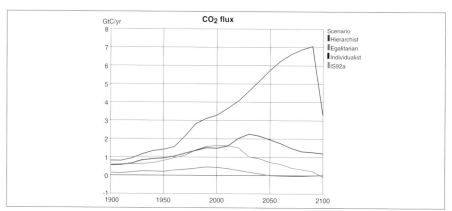

Figure 16.2 The direct emissions associated with land-use changes for the three utopias over the period 1900-2100, and the IS92a emission scenario. The sudden drop in the individualist utopia stems from deforestation in the developing region (Figure 15.1f).

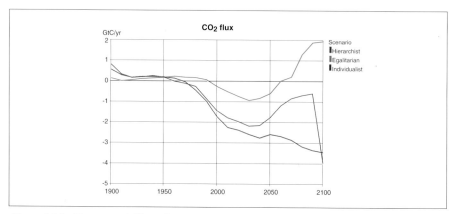

Figure 16.3 Net terrestrial biospheric emissions (direct emissions associated with land-use changes minus net ecosystem production) for the three utopias over the period 1900-2100.

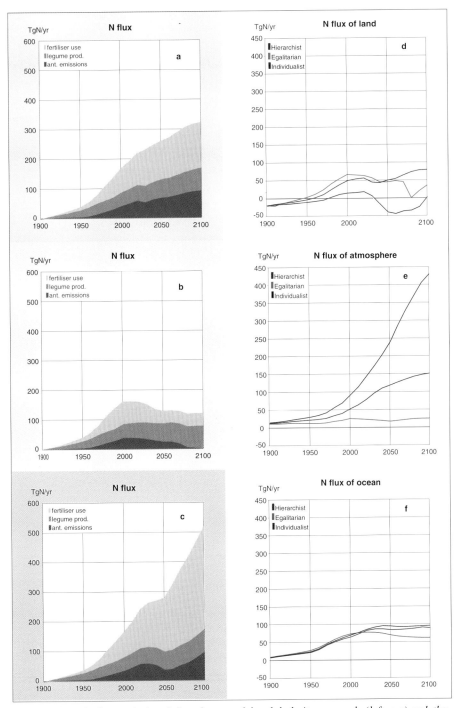

Figure 16.4 The human-induced disturbances of the global nitrogen cycle (left, a-c) and the components of the global nitrogen budget: atmosphere, land and ocean (right, d-f) for the three utopias.

tonnes of nitrate, transported by rivers to the oceans on a global scale also reaches twice the present levels. Ongoing denitrification from cattle and agricultural soils (direct sources), as well as sewage and leaching (indirect sources), result in the situation that most of the anthropogenic N inputs end up in the atmosphere, where it has no environmental impacts except for N_2O.

In this hierarchist utopia, the radiative forcing associated with the greenhouse gases and sulphate aerosols shows an increasing trend over the whole future period, comparable with the central IPCC-1995 projection. The temperature increase over this period falls within the range of historical temperature observations of between 0.3 and 0.6 °C over the period 1900-1990, and is about 2.2 °C for the period 1990-2100 (IPCC, 1994; 1995a) *(Figure 16.5a)*. These values are almost equal to the central IPCC-1995 projection. The relative temperature increase, expressed in temperature increase per decade, remains above 0.1 °C per decade for the whole period 1990-2100 and rises to more than 0.25 °C per decade after 2050 *(Figure 16.5b)*.

With reference to the global biogeochemical cycles controversy, the hierarchist utopia is a future in which the human-induced disturbance of the global carbon cycle reaches levels 2 to 3 times the pre-industrial levels. The disturbance of the global nitrogen cycle may exceed the present high levels. The absolute and relative temperature increases are associated with climate impacts (IPCC, 1995a). According to most scientists, this future may therefore be confronted with severe environmental impacts in terms of climate change, water pollution and soil acidification during the next century, which does not sound utopian. However, many governmental and business interests may choose to consider these impacts as 'acceptable' in their attempts to remain in power and ensure profits. One characteristic response could be to assess the cost of adaptation and conclude that they are small enough to be of no economic significance. Typically, such a hierarchist position would be under siege from the egalitarian side. Governments will almost certainly be a battleground for the forces of the individualist and the egalitarian world view for the decades to come, on such aspects as the interpretation of climatological observations.

The egalitarian utopia
From the egalitarian perspective, the defined model routes for the terrestrial feedback processes do not necessarily lead to a balanced past carbon budget. Therefore the CO_2 emissions associated with land-use changes are used as a balancing mechanism, since they are poorly known (1.6 ± 1.0 GtC/yr for the 1980s), and form a major component of the carbon budget (~25%). The main uncertainties in this flux arise from our incomplete knowledge of the biomass content of the tropical forests, which for this utopia is reduced by 40% in order to balance the low terrestrial CO_2 uptake due to the small CO_2 and N fertilisation effect. The CO_2 emissions from land-use changes over the 1980s correspond with the lower IPCC estimates *(Figure 16.2)*. These emissions are, for the whole period 1990-2100, much lower than those

for the hierarchist utopia because of the low biomass content of the tropical forests and the future halt of deforestation. Overall, the terrestrial biosphere acts as a small net sink or source, except for the second half of the 21st century, when the temperature feedback on soil respiration leads to additional terrestrial CO_2 releases *(Figure 16.3)*. The CO_2 emissions from fossil fuel combustion for this utopia decline to about 55% of the 1990 levels by 2100 (about 3.3 GtC/yr by 2100; see Chapter 13). Despite these low CO_2 emissions, the rate of concentration increase is higher than the present rate for the period 1990-2100, and the concentration does not stabilise; this is a direct result of the magnitude of the terrestrial biospheric flux.

In our implementation of this utopia, the projected disturbance of the global nitrogen cycle for the 1900-1990 period does not significantly differ from the scarce observed data. The atmospheric and oceanic accumulation is within the scientifically accepted range, and leads to a simulated missing nitrogen sink hardly varying from its central estimate *(Figure 16.4d-f)*. The oceanic N sources themselves, i.e. deposition and river inputs, differ from central estimates, but arewithin the scientifically accepted ranges. For example, the riverine discharge of anthropogenic dissolved N is about 40 TgN/yr in 1990 (central estimate 25 TgN/yr), which is an acceptable upper limit based on the available data (Duce *et al.*, 1991; Meybeck, 1982). The low future anthropogenic nitrogen inputs from fertiliser use and emissions from fossil fuel combustion are still responsible for high nitrogen accumulation on land in comparison with the hierarchist utopia. This is due to the low atmospheric losses by denitrification and high leaching to the groundwater. Soil acidification (N accumulation in the soils of natural ecosystems) for the industrialised world decreases below the present levels for the future period due to decreasing anthropogenic NO_x emissions from fossil fuel combustion. Groundwater pollution in terms of nitrogen storage increases to three times the present levels by 2100, whereas the water pollution from nitrogen stabilises to levels just above the present levels due to a stabilisation in food consumption.

The temperature increase over the period 1900-1990 for this utopia falls above the IPCC range. Because the contribution of anthropogenic factors to the present temperature increase is controversial, we accept the simulated temperature rise of 0.8 °C over the 1900-1990 period as being characteristic for the egalitarian world view *(Figure 16.5a)*. The small radiative cooling by sulphate aerosols, combined with the high climate sensitivity, leads to a high temperature increase for the 21st century, up to 3.5 °C by 2100, despite the much lower radiative forcing by greenhouse gases, including CO_2. The rate of temperature increase even exceeds 0.2 °C per decade for the whole future period. These projections even exceed those of the hierarchist utopia, despite the lower atmospheric CO_2 concentrations. This somewhat unexpected outcome is the consequence of our assumption that egalitarians view the climate system as extremely sensitive; this result will actually reinforce their fear that the hierarchist assessment of the impacts from a Business-as-Usual emission scenario are irresponsibly optimistic.

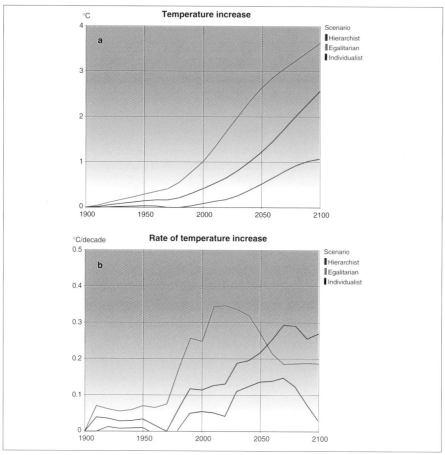

Figure 16.5 Absolute (a) and relative (b) global mean surface temperature increase for the three utopias. Because the contribution of the anthropogenic factors to temperature increase is controversial, we accept that our 1900-1990 simulated temperature increase, based on anthropogenic factors only, falls outside the range observed by IPCC (1994; 1995a).

What is the egalitarian position with respect to the global biogeochemical cycles controversy? Our model experiments suggest that, despite the high CO_2 emission reductions for the global carbon cycle, the atmospheric CO_2 concentration does not stabilise within the time horizon of our analysis. For the global nitrogen cycle, even in this utopia there is ongoing water pollution but soil acidification decreases. Because the climate system in the egalitarian view is quite sensitive, the expected temperature increase and its rate of increase are high. These model results will affirm the egalitarian conviction that only rigorous policy intervention can protect large parts of humankind from catastrophical environmental disruptions.

Reproduction of the IPCC-IS92a scenario
Here we will focus on the comparison between the IPCC-1995 central projection (IS92a emission scenario), and our projection for the hierarchist utopia. Since our hierarchist scenarios for the anthropogenic CO_2 emission and land-use changes are comparable with the IPCC IS92a scenario and the main model parameters within the CYCLES submodel are based on the central IPCC estimates. This comparison gives an indication of whether our model is able to reproduce the IPCC projection.

The simulated atmospheric CO_2 concentration is about 5% higher than the central IPCC-1995 projection due to small differences in: (i) the anthropogenic CO_2 emissions, and (ii) the calculated carbon budgets for the 1980s. Our simulated CO_2 emissions associated with land-use changes show a trend comparable to the trend over the historical period presented by Schimel et al. (1995). As to the future period (1990-2100) the simulated trend is, especially for the second half of the 21st century, much higher than in the IS92a emission scenario (Figure 16.2). One reason for the latter is that our emission flux does not account for the CO_2 uptake by the biomass plantations, whereas the flux presented by the IS92a emission scenario does. The CO_2 emissions from fossil fuel combustion are also somewhat higher (about 5%) than in the IS92a scenario. When we calculate the atmospheric CO_2 concentration under the conditions of the IS92a scenario, as described in den Elzen et al. (1997), our central CO_2 concentration projection is about equal to the IPCC-1995 estimate ($\pm$ 705 ppmv). However, this match is somewhat misleading since our model uses the old IPCC-1994 carbon budget. Hence it should be compared with the IPCC-1994 projection, which is about 25 ppmv lower by 2100 (680 ppmv). The $\pm$ 5% difference found can only be explained by the differences in the carbon balancing mechanisms. The IPCC-1995 estimate is based on calculations with global carbon cycle models, which all have a balanced carbon budget using only the CO_2 fertilisation feedback. This condition was prescribed in the modelling exercise for the calculations of the IPCC-1992 emission scenarios (Enting et al., 1994). However, in our model the balance is obtained through a mixture of the CO_2 and N fertilisation, and temperature-related feedbacks.

There are now two possible explanations for our higher projections, as already recognised in the IPCC-1994 Scientific Assessment report. First, the N fertilisation effect does not increase in the future, whereas the CO_2 fertilisation feedback does. Secondly, the temperature-related feedbacks can lead to additional atmospheric CO_2 releases due to increased soil respiration.

The individualist utopia

From the individualist perspective, the biomass content of the tropical forests is about 40% above the reference value in order to balance the high terrestrial CO_2 uptake by the dominating CO_2 and N fertilisation processes in natural ecosystems. The CO_2 emissions from land-use changes over the 1980s now correspond with the high IPCC estimates *(Figure 16.2)*. In the future these emissions will show continuous increase up to three times the present emissions due to ongoing deforestation of the tropical forests (Chapter 15). Combining these high emissions with the high CO_2 uptake due to the CO_2 and N fertilisation effects finally leads to overall negative terrestrial biospheric emissions somewhat higher than those for the hierarchist utopia, which leads to a balanced past carbon budget *(Figure 16.3)*. The atmospheric CO_2 concentration greatly increases to about 640 ppmv by 2100 *(Figure 16.1)*, despite a stabilisation of the CO_2 emissions from fossil fuel combustion at a level of 12 GtC/yr (twice present levels) from 2025. However, this is still below the hierarchist level since the CO_2 emissions from fossil fuel combustion for the hierarchist utopia are much higher (see Chapter 13).

For the global nitrogen cycle, this utopia also gives a well-balanced nitrogen budget within the scientific uncertainties. The anthropogenic nitrogen inputs in absolute terms are equal to those of the hierarchist utopia, although for this utopia they are the net result of greater fertiliser use and lower emissions of N compounds from fossil fuel combustion *(Figure 16.4c)*. The nitrogen accumulation in the atmosphere is the highest for this dystopia over the whole period 1900-2100. However, it is assumed to have no environmental impacts besides the increasing N_2O concentration. N_2O reaches levels which are about 5% higher than those for the hierarchist utopia. Due to high atmospheric losses, the land compartment even becomes a nitrogen sink. Soil acidification decreases below present levels over the next century. Water pollution in terms of nitrogen load in the world rivers increases the present levels two to threefold, which forms the only important environmental impact in this utopia.

The temperature increase over the 1900-1990 period is extremely low (about 0.1 °C) and falls below the IPCC range. The contribution of the present temperature increase by anthropogenic factors is controversial; according to this perspective it is believed that the present increase is mainly induced by natural factors, such as changes in the solar constant. The final temperature increase is about 1°C by 2100, mainly because of the combined effect of more cooling by sulphate aerosols, lower radiative forcing by CO_2 and a low climate sensitivity *(Figure 16.5a-b)*.

The high-growth world of the individualist is characterised by high CO_2 emissions from fossil fuel combustion and land-use changes but also by a high terrestrial biospheric uptake by the dominating CO_2 and N fertilisation processes, which together leads to a significant rise in the atmospheric CO_2 concentration. Because successful innovations in efficiency and renewables, it is still below the hierarchist path. For the global nitrogen cycle, the anthropogenic nitrogen inputs are mainly stored in the atmosphere by denitrification, with no environmental impacts. Only water pollution shows an increasing pattern. With respect to climate change, the temperature increase is low as the climate sensitivity is assumed to be small. The rate of temperature increase is also slow. This individualist utopia sees a world in which large pressures from the human system cause only minor negative impacts which, moreover, can be accommodated as their pace is slow.

16.5 Risk assessment: dystopias

As in the previous chapters, we have also performed simulations with the CYCLES submodel for three world views combined with three possible backgrounds. To characterise the experiments, we refer to the world view on biogeochemical cycling and the climate system only.

The atmospheric CO_2 concentrations for the different experiments are presented in *Figure 16.6a-c*, in which the solid lines refer to the utopias, and the dashed lines to the dystopias. The individualist and hierarchist experiments give almost similar CO_2

concentrations for the same background scenario because there is only a small difference in the net terrestrial biospheric emissions (*Figure 16.3*). Most projections exceed or approach their perspective-related target level (high risks), including the hierarchist utopia if one accepts the 650 ppmv target. For the egalitarian experiments, the low terrestrial carbon uptake in combination with the high CO_2 emission pathways of the hierarchist or individualist leads to high CO_2 concentration levels, far exceeding the egalitarian target level. Only with the low CO_2 emissions of

Risk-related targets

This section uses a methodology for assessing the risks involved in emission scenarios as described in Rotmans and den Elzen (1993a), den Elzen (1993) and Alcamo *et al.* (1996). The methodology makes use of long-term targets which represent certain limits on indicators for the functioning of the environmental system such as temperature rise for climate change. They should be aimed at by policy-makers in order to limit the associated impacts and risks to acceptable levels. If these targets are not met, the probability of negative consequences increases rapidly. A future in which these targets are met reflects a world in which the anticipated risks of environmental change tend to be within acceptable levels. Since the targets themselves are a reflection of one's attitude to risk and world view, we define a risk-avoiding (low risks) and risk-taking (high risks) set of targets to function as indicators representative for the egalitarian and the hierarchist/individualist attitude on risks.

The four indicators described in *Table 16.6* are used. The first, the atmospheric CO_2 concentration represents the state of the global carbon cycle. For the atmospheric CO_2 concentration, the IPCC (1995b) used target levels between 450 and 750 ppmv to calculate pathways leading to a stabilisation of the CO_2 concentration at these levels. Here, we follow Wigley *et al.* (1995) in adopting the target levels, 450 and 650 ppmv, as the lower and upper limits for the risk evaluation of the utopias and dystopias. For evaluating the state of the global nitrogen cycle, the indicator of the yearly N accumulation on land or N flux to land (subsequently to be referred to as N accumulation on land) has been chosen as the second indicator since this is associated with a variety of environmental impacts such as acidification and water pollution. However, this indicator should be interpreted cautiously, since it is the net result of N inputs from fertiliser use and anthropogenic emission accumulation on land, the atmospheric nitrogen releases associated with land-use changes. The yearly N accumulation on the continents (land) is compared with its present (1990) accumulation of 42 TgN/yr (high risk), and its accumulation of about 20 TgN/yr (low risk) in 1970. The last two, the absolute and relative global average surface temperature increases, represent climate change impacts. For these, we use high and low-risk targets for the lower and upper limits proposed in Alcamo *et al.* (1996), namely, an absolute temperature increase of 1 °C and 2 °C above the pre-industrial global mean temperature, and a relative temperature increase of 0.1 °C and 0.2 °C per decade (*Table 16.6*).

Beyond these targets, there is an increasing risk that ecosystems will be unable to adapt in time to changes in temperature and precipitation patterns, and that sensitive coastal areas are damaged, with sudden changes in the climate system expected.

Indicator	Target value	
	Risk-avoiding	Risk-taking
Atmospheric CO_2 concentration	450 ppmv	650 ppmv
N accumulation on land	20 TgN/yr	42 TgN/yr
Absolute global average temperature increase	1°C	2 °C
Rate of temperature increase per decade	0.1 °C per decade	0.2 °C per decade

Table 16.6 A risk-avoiding and risk-taking set of target levels or goals on indicators of the global carbon and nitrogen cycle and the climate system as representatives of the egalitarian and the hierarchist/individualist attitude against risks.

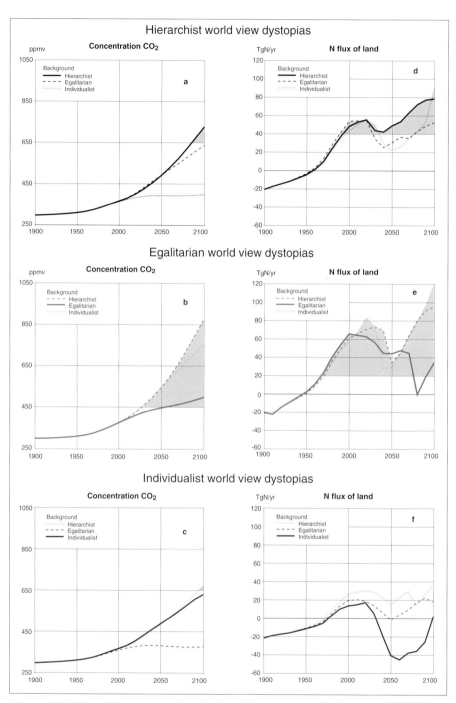

Figure 16.6 Future CO_2 concentrations (a-c) and nitrogen accumulation on land (d-f) for the three dystopias (solid lines refer to the utopias; shaded area indicates high risks where the related target is exceeded).

the egalitarian background, does the CO_2 concentration remain below the hierarchist risk level, and even below the egalitarian risk level for a hierarchist and individualist world on carbon cycling.

The consequences of the future anthropogenic nitrogen inputs in terms of yearly N accumulation on land for the nine experiments are given in *Figure 16.6d-f*. Since nitrogen accumulation on land is the net effect of fertiliser use, N deposition coming from anthropogenic emissions and land-use changes, the differences in the projections are difficult to explain. For the hierarchist projections, the nitrogen accumulation on land is mainly caused by the N deposition of the anthropogenic emissions from fossil fuel combustion. The target level is exceeded for the most of the 1990-2100 period, which implies environmental impacts i.e. water pollution and acidification beyond the present levels. The egalitarian projections are even somewhat higher due to the storage in the groundwater. For the individualist dystopias the nitrogen accumulation on land remains below the target level, and are associated with low risks. Only the experiments with low NO_x emissions of the egalitarian background are associated with low risks.

The projections on the rate of global temperature change and the absolute temperature rise show clear differences for the nine experiments, mostly because of the assumptions about the climate system (*Figure 16.7a-c*). Because of the assumed extremely sensitive climate system, the egalitarian projections are all associated with high climate risks. For the individualist dystopias, the climate system adapts to its anthropogenic disturbances, resulting in low climate risks for the next century. The hierarchist dystopias are also associated with high climate risks, except the one with

Statistical uncertainty analysis

A statistical uncertainty analysis using UNCSAM is also performed on the CYCLES submodel. Following the procedure described in Chapter 11, all input variables and parameters used are varied uniformly throughout the domains. The 'unbalanced' uncertainty range[1] varies from 660 to 795 ppmv, whereas the 'balanced' range is 690-730 ppmv (*Figure 16.8a*). This should be compared with the projection range of experiments associated with a hierarchist background scenario. The latter spectrum even comprises the unbalanced range, which teaches us that a systematic variation in the model parameters as in the perspective approach can apparently lead to a broader uncertainty range than the one obtained with a statistical uncertainty method.

The uncertainty range for the N accumulation on land varies between −10 and 105 TgN/yr by 2100 (*Figure 16.8b*). This is mainly due to the high uncertainties in denitrification of agricultural soils. The uncertainty range of the N accumulation in the soils of natural ecosystems (coupled to the N fertilisation feedback) is much less due to the small uncertainties in the retention of nitrogen inputs by natural ecosystems. The final uncertainty range for the absolute and relative temperature changes is caused mainly by the uncertainty in the climate sensitivity ([1.5,4.5] °C) which results from our lack of knowledge on the geophysical feedbacks (*Figure 16.8c*).

---

1 As a selection criteria to obtain 'balanced' experiments, samples were not accepted in which the simulated atmospheric CO_2 concentration in 1990 differed by more than 1% (a chosen accepted small difference between observed and simulated values) from the observed value. All samples were accepted for the 'unbalanced' experiments.

16 HUMAN DISTURBANCE OF THE GLOBAL BIOGEOCHEMICAL CYCLES

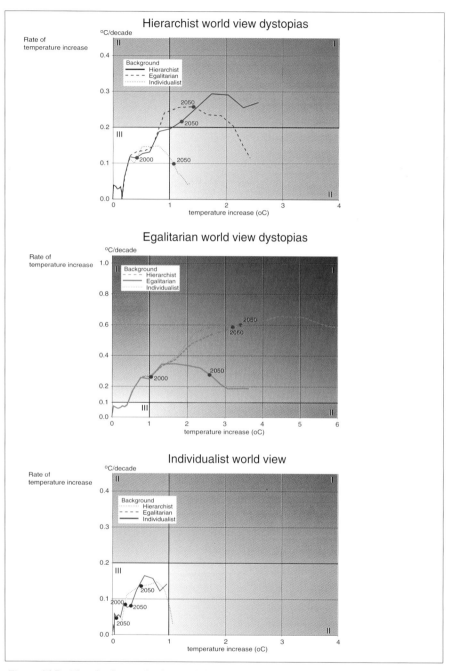

Figure 16.7 The absolute and relative global mean surface temperature increase for the three dystopias related to absolute and relative temperature target level (solid lines refer to the utopias):
I   maximum climate risk area: both climate targets are exceeded;
II high climate risk area: one climate target is exceeded;
III low climate risk area: no climate target is exceeded.

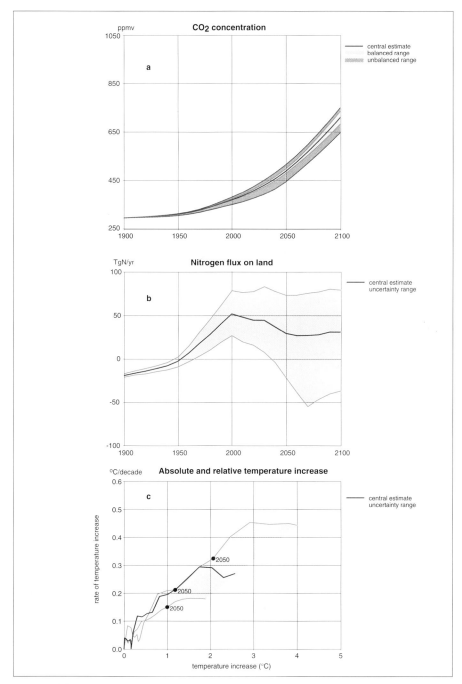

Figure 16.8 Future CO_2 concentrations (a), nitrogen accumulation on land (b) and absolute and relative temperature increases (c) over the period 1900-2100 for the reference scenario. Uncertainty range represents the '95%-confidence level' range between the 2.5 and 97.5 percentiles.

the low emissions for the egalitarian background. In general emissions reductions like in the egalitarian background scenario, lead after 2010-2020 to a change from an increasing to decreasing rate of temperature increase.

16.6 Conclusions

Our analysis shows a wide variety of future patterns of the fate of anthropogenic C and N compounds in the global environment, and of global temperature changes. The IPCC atmospheric CO_2 concentration projections are reproduced with only marginal differences due to a difference in the carbon-balancing mechanism. The CO_2 concentration projections of the utopias and dystopias exceed, in most cases, the target CO_2 concentration levels which we have tentatively associated with each of the three world views. To stabilise CO_2 concentrations before 2100 at a level below the risk-target levels, future reductions in CO_2 emissions as proposed in the egalitarian background scenario are necessary. However, given the egalitarian world view concerning biogeochemical cycling, even such a stabilisation cannot be achieved with these emissions reductions.

A major source of uncertainty within the global nitrogen cycle is rooted in our lack of knowledge about denitrification rates. Different estimates of future denitrification rates related to the different world views result in a wide variety of projections of future nitrogen accumulation on land. The egalitarian and hierarchist dystopias all result in severe environmental impacts, whereas for the individualist perspective no impacts were found. Only the experiments with low NO_x emissions using the egalitarian background are associated with low risks.

The temperature projections depend on assumptions about climate sensitivity and the radiative effects of sulphate aerosols, both of which are related to world view. The egalitarian projections are all associated with high climate risks for the next century while the individualist dystopias result in low climate risks. The hierarchist dystopias are also associated with high climate risks, except the one with the low emissions for the egalitarian background scenario. In general, emission reductions like those in the egalitarian background scenario lead to a change from an increasing to a decreasing rate of temperature increase after 2010-2020.

17 The larger picture: utopian worlds

"The optimist thinks we live in the best of all possible worlds. The pessimist fears [s]he's right."

17 THE LARGER PICTURE: UTOPIAN FUTURES

Bert J.M. de Vries, Jan Rotmans, Arthur H.W. Beusen, Henk B.M. Hilderink, Michel G.J. den Elzen, Arjen Y. Hoekstra and Bart J. Strengers

This chapter synthesises the insights gained from the model experiments made in the previous five chapters. The hierarchist utopia examined in Chapter 11 is only one possible future. We now explore the consequences of two other utopian futures: the egalitarian and the individualist. A selection of conditional forecasts from integrated simulation experiments with population, food, water and energy supplies, land use, global temperature and sea-level rise are presented. One way of looking at the model outcomes is by focusing on the various transitions which characterise the development of the human-environment system. Extending the time horizon of the model simulations into the 22nd century yields additional insights into the relation between the human and the environmental system.

17.1 Introduction

The main goal of the TARGETS1.0 model is to place possible developments within the subsystems of the world in an integrated perspective. In Chapters 12 to 16, simulation results of experiments with the TARGETS1.0 submodels are discussed in isolation, while in Chapter 11 the results of an integrated simulation experiment for the hierarchist utopia are presented. In this chapter, we pursue this analysis further and include the other two perspectives. In this way we elaborate on the various controversies which have been raised in the preceding chapters: can a large population be maintained at an adequate health level and will there be enough energy, water and food without overburdening the natural environment? We start with the integrated utopias which are based on assumptions about world view and management style taken from a single perspective for all submodels (see Table 11.1). After a comparative overview of the three utopias and a comparison with the non-integrated experiments, we present them in the context of transitions. Although the experiments are set up with the year 2100 as the time horizon, we also present some results for an analysis until the year 2200 because some crucial aspects of the long-term dynamics only become manifest after 2100. A few key aspects of the utopian futures are illustrated with the indicator framework presented in Chapter 9.

17.2 Inclusion of feedbacks in integrated experiments

As explained in Chapter 11, we have constructed the hierarchist utopia along the lines of so-called reference scenarios. The IS92a-scenario of the IPCC (Leggett *et al.*, 1992) and the associated population and economic growth scenarios of the UN and the World Bank are in particular used to represent a hierarchist outlook on the future of the world. This allows a comparison of the model experiments with rather widely used quantitative expectations about future global trends. We will first focus on the question of whether the feedback relationships in the TARGETS1.0 model as a whole significantly affect the simulation results. The three diverging scenarios for Gross World Product (GWP) are presented in Chapter 11 (Figure 11.3). They are exogenous and hence the same for integrated and non-integrated experiments[1]. All submodel results presented thus far, however, are without feedback interactions operating between the submodels. The integrated experiments presented in this chapter, however, do take the feedback relationships into account. For this reason, we summarise in *Table 17.1* some key results of the integrated versus the non-integrated experiments.

As part of a trial-and-error procedure, the submodel experiments have provided the results from which the integrated hierarchist utopia has been constructed. The Population and Health submodel was run with exogenous time-series from hierarchist projections for food intake per capita, the fraction of people with access to safe drinking water and for the average temperature rise. The population size arising from experiments with three perspectives is used in the subsequent runs with the Energy submodel TIME. This has been repeated for the AQUA, the TERRA and the CYCLES submodels in the way indicated in Table 11.1. These results are referred to as 'chain approach' and they are given in the first three rows of *Table 17.1*. These results have provided the inputs for the construction of the integrated hierarchist utopia, the results of which are given in the fourth row of *Table 17.1*.

This experimental set-up makes it possible to see how the egalitarian and the individualist utopias emerge in the process of sequentially introducing the corresponding utopian world view and management style (the last column of *Table 17.1*). It is useful to compare the results of the chain approach in *Table 17.1* with the integrated results, the difference being the inclusion of interactions and feedbacks. Such a comparison leads to the following conclusions:

- The population sizes simulated in Chapter 12 differ only slightly from those in the integrated utopias; the differences in GWP per capita are therefore small. Apparently, the feedbacks from food intake, safe water access and temperature rise are quite limited in the utopian worlds.

---

1 This is not completely correct because GWP includes food expenditures which makes it also a function of food production.

17 THE LARGER PICTURE: UTOPIAN FUTURES

Value in 2100	Population (10⁹)	GWP/cap ($/cap)	CO$_2$ emission (GtC/yr)	Food intake (kcal/cap/day)	CO$_2$ concentration (ppmv)	Temperature change (°C)
Chain approach						
Hierarchist	11.75	21,145	21.0	3470	728	2.6
Egalitarian	7.95	7,585	3.4	3150	498	3.6
Individualist	13.25	41,225	12.2	4250	630	1.1
Integrated experiments with TARGETS1.0						
Hierarchist	11.65	21,305	21.0	3460	730	2.6
Egalitarian	7.45	8,460	3.5	3065	495	3.7
Individualist	13.30	41,335	12.1	3905	635	1.1

Table 17.1 Summary of results for experiments in Chapters 12-16 and integrated experiments.

- Including the feedbacks hardly influences the CO$_2$ emissions because population and welfare, the main determinants of energy use, do not change significantly. Moreover, the somewhat higher GWP per capita level compensates for the slightly lower population.
- In the integrated utopia for the egalitarian future food intake per capita is 3% lower and for the individualist future 8% higher than without the feedbacks. This reflects the egalitarian view that climate change has a negative effect on yields and the individualist view that the effect of climate change turns out to be positive.
- Because CO$_2$ emissions do not change much when the feedbacks are included, the CO$_2$ concentration in the integrated utopias are almost the same as those presented in Chapter 16. The same holds for the temperature change.

Because the submodels have in the first place been implemented on the basis of hierarchist 'middle-of-the-road' projections, one would expect the integrated hierarchist utopia to give similar results to experiments in which all submodels are fed with the relevant time-series taken from the integrated hierarchist utopia. Comparison of the results has shown that the difference for the important model variables is less than 5%. This suggests a certain consistency in the projections which have been used. However, the inclusion of feedbacks for the egalitarian and the individualist utopias as well, leads only to minor changes in the 2100 values of key variables such as population and life expectancy. One explanation is that in all three utopias the levels of food intake and water use are so high that a drop of 5-15% due to feedbacks from the environmental system have only a fairly limited impact. Of course, such a conclusion can only be drawn with the utmost caution because we do not explicitly deal with such factors as the distribution of economic welfare. Our hypothesis is that feedbacks only become really important in the dystopias discussed in the next chapter.

17.3 Wishful thinking: three utopian futures

In Chapter 11 the hierarchist utopia gives a flavour of the kind of insights that can be derived from integrated simulation experiments with the TARGETS1.0 model. It is a sketch of a world in which population growth gradually slows down, driven by the processes of modernisation. Material welfare accelerates the forces which have reduced fertility and increased health expectancy in the past in the industrialised regions. The process of economic growth in this utopian world is supported by adequate investments in well-managed food, water and energy-supply systems. The transitions discussed in Chapter 2 are successfully followed. Environmental pressures increase in some respects but can be kept within limits, which at least at the global level may be considered acceptable from a somewhat optimistic hierarchist perspective. A 'decent standard of living' is possible for all, although it takes time to achieve it. Let us now look at all three integrated utopias.

Population in the egalitarian utopia stabilises at about 7.5 billion people; in the individualist scenario it keeps growing and reaches 13 billion by 2100 as against 11.6 billion in the hierarchist utopia (*Figure 17.1*). Consequently, welfare in 2100 measured as GWP/capita ranges from $ 7000 per capita in the egalitarian to $ 44,000 per capita in the individualist utopia. Life expectancy at birth ranges from 75 in the egalitarian to 85 years in the individualist utopia. Food consumption develops in line with the population trends: it rises in both the hierarchist and the individualist utopia and stabilises in the egalitarian world at slightly above the present level (*Figure 17.2a*). The stabilisation in the egalitarian world reflects a low food intake and a decline in the fraction of animal-based food. This is partly the result of necessity: the quite low ceiling on fertiliser input per hectare in the egalitarian utopia is already reached by 2050; food consumption, both in absolute and per capita terms, declines. However, in this world the consequences of an impending food shortage would be ameliorated by more even distribution. After 2080 the individualist utopia also faces a limit in fertiliser effectiveness and per capita food in the developing world starts declining. In such a world a possible response would be a reduction in processing losses and a stimulus for biotechnological innovations through market forces. Food demand is neigther fully met in the hierarchist nor the egalitarian utopia because the ceilings on fertiliser input have been reached.

Total water supply is 28% (individualist) and 72% (egalitarian) lower than in the hierarchist utopia (*Figure 17.2b*) due to different levels of economic activity and population in combination with a less conservative appraisal of efficiency improvement and the removal of subsidies or even the introduction of taxes. The large difference in water supply between the individualist and egalitarian scenarios is due to much lower irrigation requirements (less food supply) and much less growth in industrial water demand (increased efficiency) in the latter. Nevertheless, in the egalitarian world the same fraction of the potential supply is used because the

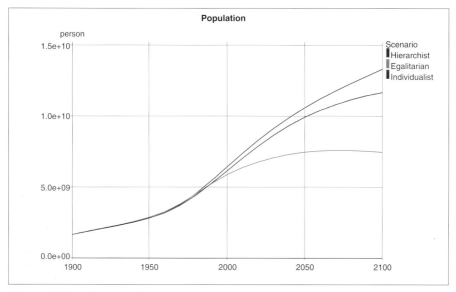

Figure 17.1 Simulated population size for the three integrated utopias (see Figure 11.4a for comparison).

potential supply is thought to be low. For similar reasons the flux of untreated waste water in both utopias is lower than in the hierarchist one. The different appreciation of water quality suggests the change in water quality over the next century in the individualist world is of the same order of magnitude as the change in the past century. In the egalitarian utopia water quality is perceived to be poor and will deteriorate for another 80 years after which it starts to improve.

In both the hierarchist and the individualist utopias, primary energy supply continues to grow during the next century, the difference being relatively small (*Figure 17.2c*). High levels of economic and population growth cause much higher end-use energy demand in the individualist world, but the development of energy efficiency innovations restrains the growth in primary energy use – it is even below the hierarchist level in the second half of the next century. The low growth in population and economy in the egalitarian utopia lead, in combination with vigorous energy conservation measures and efficiency innovations, to a stabilisation in primary energy at 1980 levels by the year 2100. There are no energy shortages because alternative options like electricity from nuclear/solar and biomass-derived fuels take over from increasingly expensive oil and gas. Land for biofuels is not a constraint and other constraints – such as public resistance to nuclear power, for example – do notoperate in any of these utopias.

The resulting pressures on the environmental system disturb the biogeochemical cycles of the basic elements C, N, P and S, leading to various environmental impacts. Due to the enhanced greenhouse effect the global average surface temperature is

17 THE LARGER PICTURE: UTOPIAN FUTURES

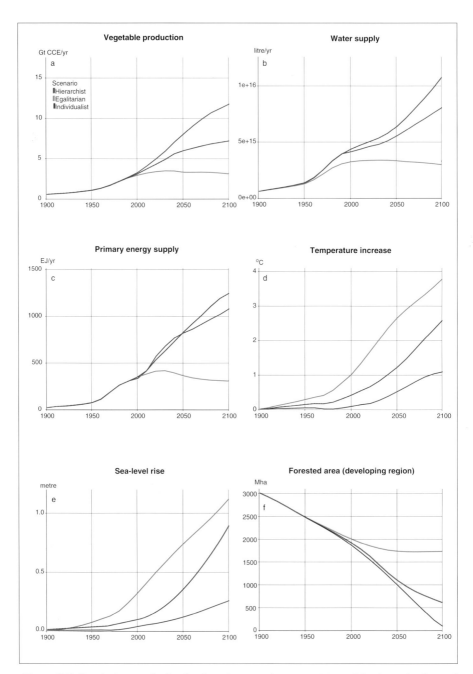

Figure 17.2 Simulation results for the three integrated utopias: (a) total food supply, (b) total water supply, (c) primary energy supply, (d) rise in global average surface temperature, (e) rise of global average sea level and (f) forested area (developing region) (see Figure 11.5a, 11.6a, 11.7a and 11.8 for comparison)

expected to rise by 1.1 °C to 3.7 °C (*Figure 17.2d*). Sea levels are projected to rise 25 cm in the individualist and up to about 80 cm in the hierarchist, and over 120 cm in the egalitarian utopia (*Figure 17.2e*). As explained in Chapter 16, the divergence in the assessment of the climate system's vulnerability leads to much smaller temperature and sea-level changes in the individualist utopia than in the egalitarian utopia, despite the much higher population and economic activity levels in the former. Because of the pressures on land resources due to demand for food, feed and wood and non-sustainable forestry practices, the forested area in the developing world will decrease at a much faster rate in the individualist utopia than in the egalitarian world, so that at the end of next century, almost no primary forest is left in developing regions (*Figure 17.2f*). In the egalitarian utopia, the forested area in developing regions stabilises at a little above half the 1900 level.

17.4 Economic growth in the three utopias

In each of the three utopias there is a great deal of response behaviour in the form of price-induced changes in investments and exogenous scenarios. One important response is the allocation of the available industrial output to the various sectors. For the allocation of required investment goods among the various economic sectors, including food production, water and energy, we use the simple mechanism explained in Chapters 3 and 11. We first look at the production side. Based on a presumed relationship between the desired per capita service output and the per capita industrial output, the investments in the service sector capital stock are calculated. The service output is calculated, using a constant capital-output ratio for the service sector capital stocks. We then choose the gross investment ratio, i.e. the proportion of industrial output that is re-invested in industrial capital in such a way that the sum of service output, agricultural output and consumption follows the exogenous GWP trajectory (see Figure 11.3)[2]. It should be emphasised that the time-paths calculated for these variables merely serve as activity levels for the food, water and energy submodels in a way which allows us to relate them to economic growth scenarios by other institutions. We now assume that all investments required for the food, water and energy sector in the utopian scenarios are also delivered. *Figure 17.3a* shows how industrial output is allocated in the hierarchist utopia among investments in industry, services and food, water and energy-related capital stocks. The remainder is industrial output associated with consumption goods.

Because we are not using an input-output framework to connect the production side and the consumption sides of the economy, we interpret the service output as

---

2 The values in our simulation experiments cannot be compared with historical estimates or economic scenario studies. Hence, to calibrate these time-series to historical data on consumption expenditures and value added in industry and services, we had to use conversion factors. In fact, a large proportion of what is denoted here as service output is labelled consumption in economic statistics.

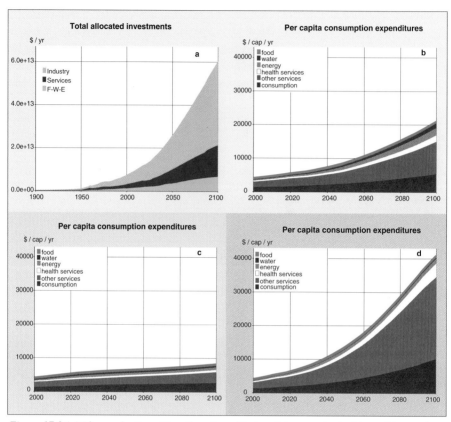

Figure 17.3 (a) The production side: allocation of industrial output in the hierarchist utopia, (b-d) the consumption side: per capita consumer expenditures in the three utopias.

including health service expenditures and expenditures on water and energy[3]. The proportion of the service output which remains after subtraction of expenditures for health, water and energy is referred to as 'other services'. *Figures 17.3b-d* show the thus defined consumption side of the system in the three utopias: per capita levels of expenditures for food, water and energy, for health services, for other services and for consumption. The latter three, along with the GWP per capita paths (Figure 12.1b), are used as driving forces in the various submodels. The difference between GWP and the sum of expenditures on food and total service output is available for what in the present context is defined as consumption.

---

3 Because the investments in the water and energy sectors are not included in the investments in the service sector but the expenditures on water and energy are, there is a discrepancy between the fixed capital-output ratio for the service sector on the production side and the capital-output ratio as derived from the output on the consumption side.

Endogenous and exogenous responses

As explained in Chapter 3, each submodel includes response behaviour. Some important response mechanisms are part of the endogenous model dynamics, for example the response to depletion of water and fossil fuel resources in the form of higher costs/prices and a consequent reduction of demand. Other responses are relationships derived from correlations between historical data, e.g. the decline in the growth of health service expenditures or increased water supply coverage as a function of GWP per capita. Another element of the response dynamics consists of exogenous time-paths which are explicitly related to a particular perspective/management style. Examples are the increased availability of contraceptive methods, forest clearance and irrigation policies, the removal of water subsidies and implementation of a carbon tax. Also the allocation of the available investments to the three subsystems of food, water and energy is a kind of exogenously fixed response.

Many of these response relationships are discussed in the previous chapters. Because such response actions change the inputs and outputs of the submodels, they also affect the integrated simulation results. For example, a rise in energy prices influences CO_2 emissions which, through a change in CO_2 concentration and temperature, impact upon food production and hence on the level of fertiliser use. This in turn influences mortality and thus, through population, energy demand. There are several more such causal loop diagrams (see Chapters 2 and 18). Finally, there are important response mechanisms which will undoubtedly take place in the real world in some form or another but which are not taken into account in our simulations. An important example here is the damage costs which are incurred for adaptation to rising sea level. Other examples are the complex responses to food, water and energy shortages in the form of trade, innovation, migration or possibly war.

Let us now have a closer look at the investments in the food, water and energy sectors (indicated with FWE), as these are the sectors which are modelled in detail. They form the crucial links between the human and the environmental system. For the hierarchist utopia the absolute levels are shown in Chapter 11 (Figure 11.10a). For the three utopias the investment levels to maintain the FWE sectors differ by a factor of 3 by 2100. In a situation of abundance – the individualist world – resources can be exploited with an almost negligible fraction of total economic output: the proportion declines from 3.6% in 1990 to 0.8% in 2100. In the egalitarian world of scarcity, this proportion initially increases from the estimate of 7% in 1990, to fall later on to about 4% in 2100 . For the hierarchist, it falls after an initial rise from 4.6% in 1990 to 2.8% in 2100. How are these investments allocated among the nine categories within the FWE sectors? As represented in *Figure 17.4a-c* the proportion of investments for agriculture declines in all three utopias, mainly because food demand is saturated and inputs are limited. Also, the proportion of investments in water supply continue to increase in all three utopias. This is caused especially by the continuing exponential growth in industrial water demand. The proportion of investments in the energy system increases in the first half of the next century because of the exploitation of more expensive oil and gas occurrences, the penetration of initially expensive renewables and the relatively rapid growth in of electricity use. Later on, the hierarchist prefers to use rather cheap coal, whereas for the egalitarian and individualist there is a rather massive switch to renewable resources because coal is no longer subsidised and renewable sources become

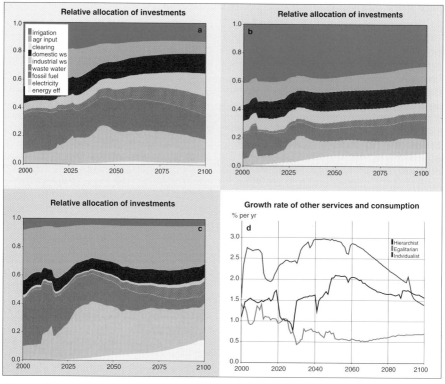

Figure 17.4 (a-c) Distribution of investment allocation to the food-water-energy sectors in the three integrated utopias for the period 2000-2100, (d) maximum allowable annual growth rate of per capita expenditures on other services and consumption. The differences in the year 2000 are the consequence of using different model routes to explain the past.

available at rapidly declining costs. The proportion of energy efficiency investments rises steeply after 2060 in the individualist world; it increases much earlier in the egalitarian world.

Figure 17.4d shows the annual growth rate of 'other' services and consumption as shown in *Figure 17.3b-d*. The trend in these growth rates indicates how much of economic growth is available as 'free' consumption per capita if the health and food, water and energy sectors are maintained as prescribed in the corresponding submodels. We use the growth rate in other services and consumption as an indicator of the system's potential to sustain growth in welfare while maintaining the life support systems. As stated in Chapter 11, this is only meant as a crude indication of how the maintenance of the three important support systems interferes with welfare-related economic production. As one would expect, the permitted growth rate for consumption is highest in the individualist world; in the order of 2-3% per yr. However, it starts declining after 2050 to values below 2% per yr as the

Cultural dynamics and the plausibility of utopias and dystopias

There is another interesting aspect to the (im)plausibility of the utopias. It is associated with the dynamic interactions among the three perspectives and an integral part of Cultural Theory (Thompson et al., 1990; Chapter 10). Utopias may be rather implausible because the dominant style of management is confronted with increasing opposition as it is experienced as ineffective or unfair or both. This may happen not because it is incorrect with regard to how the world functions, but because it cannot provide convincing evidence or cannot avoid the pitfalls inherent in its style of management.

As indicated in *Figure 17.5*, each management – or governance – style has its threats and opportunities. The hierarchist world may be characterised as 'muddling through': a form of incrementalism which is sometimes the only way to persuade large, hierarchically structured organisations to move in the direction of ambiguous long-term goals (Lindblom, 1959). It is what usually is thought of as 'conventional development'. In its best form it leads to a wisely governed world community with an important role for, among others, organisations which are affiliated to the United Nations. If it fails, and becomes an inefficient bureaucracy with protectionism and excessive centralised control, one of the other types of governance may become dominant. Similarly, the market-oriented individualist and the community-oriented egalitarian style of management face both threats and opportunities.

Society's response will depend on how the failure is perceived. If bureaucratic inertia and loss of individual responsibility is experienced by the majority of people as failed expectations of economic prosperity and freedom, the entrepreneurial dynamism of the individualist may become dominant. Price reforms, privatisation and restructuring bring innovations which revitalise the economy. Multinational firms become dominant actors. If, on the other hand, the dissatisfaction with the hierarchist institutions is with its ineffectiveness in dealing with issues of environment and poverty, egalitarian forces may become stronger. Non-Governmental Organisations (NGOs) become major forces of change and the emphasis shifts from large-scale, global approaches to a community orientation. Life-style changes may drive out technical fixes as the preferred response. Once the dominant management style is individualist or egalitarian, it will also be faced with inherent threats and opportunities. In the egalitarian utopia, for instance, measures to protect the global commons will almost certainly require hierarchist institutions. The individualist utopia is almost certainly characterised by gross inequality in material welfare and access to resources.

Also the dystopias discussed in the next chapter are implausible in as far as there will be societal responses to signs of dystopian tendencies. This would be the case, if, for example, those who support policies to accelerate economic growth and resist energy taxes are confronted with irrefutable evidence of global warming. These kinds of societal response cannot be dealt with in the TARGETS1.0 model. The adaptation experiments discussed in section 18.3 give an indication, however, of the kind of response dynamics which may start to operate if dystopias evolve. An example of the narratives which can be supported by such analyses is given in a recent study of the Scenario Working Group of the Stockholm Environment Institute (Gallopin et al., 1996).

dominant management style		threats		opportunities
	market	survival of the fittest	economic correction	global market
		barbarism		california dreaming
	government	protectionism	'muddling through'	wise guidance
		central control	'Conventional Development'	world federalism
	citizens	isolationism	lifestyle changes	ecocommunity
		ethnocentrism		grassroots solidarity

Figure 17.5 Threats and opportunities within utopias.

burden to maintain society's support systems is increasingly felt. The individualist will not interpret this as dystopian but instead rely on new and as yet unknown options to mitigate these rising costs. In the egalitarian world the maximum permitted growth rate slowly falls to a level of 0.5-0.8% per yr which is sustained for the rest of the century. This reflects the egalitarian perspective which supports more frugal consumption patterns in the face of major environmental and human pressures. The hierarchist path is in-between and shows the same trend as in the individualist scenario, which can largely be attributed to the energy transition. If more industrial output is allocated to consumption goods or to the provision of services, there would be a shortage. One such dystopia is discussed in Chapter 18.

17.5 Transitions in utopian futures

In Chapter 2 we have defined transitions. Here we describe the dynamic paths for the three utopias in terms of transitions. It is best to speak of a family of interrelated transitions which cover a variety of time and spatial scales (e.g. Drake, 1993; Malone, 1993). We will only quantify some aspects of the important transitions at the global level: the health transition, the agricultural transition and the water and energy transition. Some of the other transitions which have been proposed will be touched upon, including the ecological transition, the forestry transition and aspects of underlying technology transitions. Yet other proposed transitions, for example those related to urbanisation and education, are not dealt with at all. *Table 17.2* summarises the indicators used to quantify the transitions dealt with.

Because the economy is merely represented by a scenario generator within TARGETS1.0, little insight will be gained from our model with respect to the economic transition from a largely agriculturally based society to one in which manufacturing and later on the formal service sector play an increasingly dominant role. We will therefore omit the economy from our discussion. Another important transition is the health transition which includes the epidemiological transition (Chapter 4 and e.g. Ness *et al.*, 1993). Population growth slows down with a decline in the crude death rate and, somewhat later, in the crude birth rate, and the proportion of infectious diseases falls as welfare-related diseases take their place. Fertility drops

Model variable	Unit
Crude birth rate, crude death rate	per 1000 people
Disability Adjusted Life Expectancy (DALE)	years
Fertiliser use per ha and per unit of food production	kg N/ha and kg N/kcal per yr
Proportion of animal-based food in total food intake	kcal/kcal
Total irrigated cropland area	Mha
Yield	tonne CCE/ha per yr
Proportion of forested area remaining	ha/ha
Water consumption per capita	litre/cap per yr
Proportion of fresh water in quality class A (high)	-
Secondary energy use per capita	MJ/cap per yr
Primary energy use and emissions per unit of economic output	MJ/$ per yr and kg X/$ per yr (X = CO_2, SO_2, NO_x)

Table 17.2 Indicators of transitions.

because of the more widespread use of contraceptives in combination with rising income (hence lower desired family size) and rising female literacy. Mortality declines as the exposure to health risks and the resulting disease incidence are reduced thanks to improved health services and increasing access to food and safe drinking water. Important elements of the health transition can be seen from the simulation results for the period 1900-1990 (Chapter 4). As to the future, in both the hierarchist and the individualist utopia the crude birth rate falls gradually to 10-11 births per thousand (*Figure 17.6a*). The crude death rate is about constant after 2020 in the hierarchist utopia, but continues to fall in the individualist one. It reflects a continuous rise in the level and effectiveness of medical services and causes population growth to continue beyond the year 2100 in the individualist world. In the egalitarian future welfare-related health risks increase more slowly while poverty and malaria-related risks are significantly higher, mainly because per capita health service expenditures are only about half the hierarchist level (see *Figure 17.3b-c*). The healthy life expectancy rises in all three utopias; in the egalitarian utopia poverty-related morbidity and mortality remain the dominant cause of morbidity and mortality whereas life-style related factors are dominant in the hierarchist utopia (*Figure 17.6b-c* and Figure 12.2[4]). The level of birth and death rates leads to considerable ageing: over 25% of the world population are older than 65 years in 2100 compared to 15% in 1990.

Within the land and food system it is possible to identify several transitions which are interrelated and which are interwoven with the economic and health transition. Basically, they all have to do with the move towards more intensive forms of agriculture: the rate of forest clearing slows down; more fertiliser, pesticides and other inputs are used per hectare; yields rise. Related to this, but with quite different factors involved, is the decline in forested area, which at a later stage may be reversed once regeneration and reforestation exceed deforestation. Most of these transitions are a continuation of past trends, as is evident from the simulation results for the period 1900-1990 (Chapter 7). The trends in the hierarchist utopia are largely along the lines of the transition described above, as can be seen from *Figure 17.7*. Per capita food intake diverges after 2010: by 2050 the level is 3100 kcal/cap per day in the egalitarian utopia and 4100 kcal/cap per day for the individualist utopia (*Figure 17.7a*). In the second half of the next century, the egalitarian consumption of meat is back at the 1950 world average while it has doubled in the individualist utopia. The forested area continues to decline, although to a smaller degree because the rate of clearing for agriculture continues to fall. Fertiliser use per hectare keeps on rising, and so do yields to accommodate the per capita food intake (*Figure 17.7b*). In the individualist utopia, the transition is even more marked. In the egalitarian utopia,

---

4 The difference between the integrated and the non-integrated utopias is quite small, hence we refer to the results presented in Chapter 12.

17 THE LARGER PICTURE: UTOPIAN FUTURES

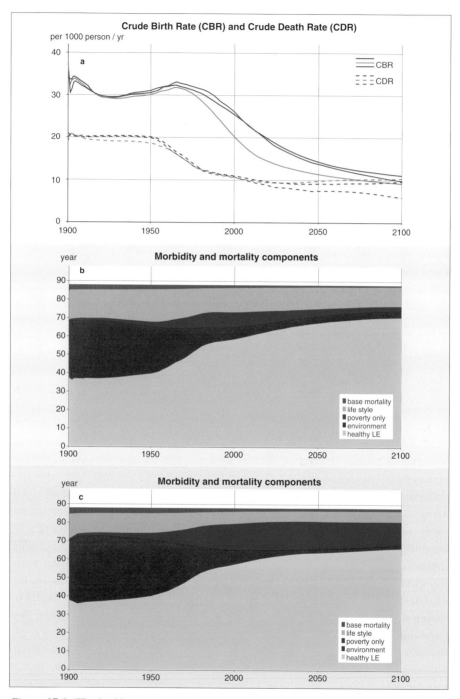

Figure 17.6 The health transition: (a) crude birth rate and crude death rate in the three utopias, (b) factors contributing to morbidity and mortality in the hierarchist and (c) the egalitarian utopia (see Figure 12.1 for comparison).

there are a few trendbreaks to relieve the pressure on the environment by actually redirecting the transition:

- a decline in the proportion of animal-based food in total diets;
- a decline in the amount of fertiliser used per ha and per kcal of food produced, reflecting the advance of less intensive, more organic forms of agriculture;
- a decline in the growth and eventual stabilisation in the irrigated cropland area; and
- a slowdown and eventual stop in net deforestation so that the world's forested area remains constant or even increases.

These trendbreaks are often mentioned as part of a sustainable development strategy. In the individualist utopia and, to a lesser extent in the hierarchist utopia, such a reversal of ongoing trends does not take place because it is considered unnecessary, detrimental to the pursuit of material welfare or unfeasible. As one would expect from utopias, this judgement is correct in as far as the simulation experiments indicate that food demand can largely be met without significant repercussions from the natural system. Only in the egalitarian utopia does the transition to a largely vegetarian diet and away from intensive agriculture occur out of necessity: without it, there would be major food and water shortages. The choice for biodiversity and organic farming – in the form of restrictions on clearing, irrigation and deforestation, and a ceiling on fertiliser use (Chapter 15) – is at the cost of lower per capita food intake. However, the available food is better distributed; in the developing region food intake is 10% below 1990-levels, whereas there is an almost 20% increase by 2100.

For the water and energy systems, one transition element is the trend towards increasing efficiency and a decline in intensity of use, both per person and per unit of economic output. Another element that affects both is part of the ecological transition: for water the reversal in the trend of a deterioration in water quality, for energy a fall in the emissions of carbon, sulphur and nitrogen oxides per unit of energy use. The water transition started this century. Water consumption has grown continuously at increasing rates of growth (Chapter 6). Water quality has steadily deteriorated. However, in a few parts of the world, the second phase of the water transition has already started: stabilisation of water consumption and improvement of water quality, as e.g. in the Rhine basin. One might expect this to occur worldwide during the next century. Due to improvements in efficiency, water consumption per capita and per unit of economic output starts to fall in all three utopias within a few decades from now while access to safe water improves (*Figure 17.8a*). Water consumption continue to grow in the hierarchist and individualist utopias, but the growth rates stabilise. In the egalitarian utopia, consumptive water use reaches a maximum in the first half of the next century (*Figure 17.8b*). The water price increases significantly in all three scenarios but for different reasons. In the individualist world, costs hardly rise but the removal of subsidies causes an increase

17 THE LARGER PICTURE: UTOPIAN FUTURES

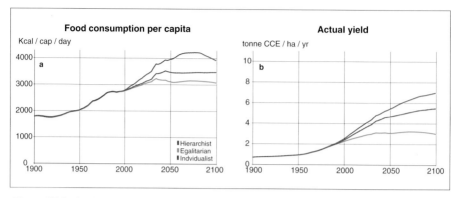

Figure 17.7 The food transition in the three integrated utopias: (a) food intake per capita, (b) actual yield per ha.

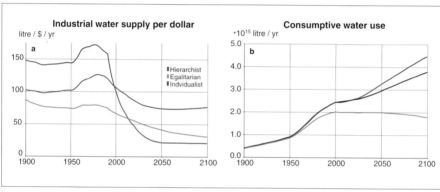

Figure 17.8 The water transition in the three integrated utopias: (a) industrial water use per unit of economic output, (b) consumptive water use. The differences in the period 1900-1990 are the consequences of using different modelroutes to explain the past.

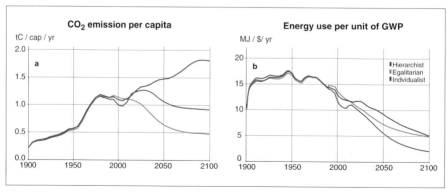

Figure 17.9 The energy transition in the three integrated utopias: (a) CO_2 emission per capita; (b) primary energy use per unit of economic output.

in price by a factor five to ten. On the other hand, water cost in the egalitarian future actually declines from 2040 onwards but removing subsidies or incorporating all costs leads to a similar price increase. Only in the hierarchist utopia do water cost and price rise smoothly and modestly. The trend in all utopias is that the ratio of water consumption to potential water supply is moving towards stabilisation. Climate change has, in our model, only a small effect on potential supply.

The transition in energy use per unit of economic output started in the second half of this century, at least for the world at large (Chapter 5). Per capita energy use has been steadily rising. Will transitions continue or start in the simulated utopias? Despite large autonomous and price-induced improvements in efficiency – the average energy price rises by a factor of 2 to 3 between 1990 and 2100 – secondary fuel and electricity use per capita keeps growing except in the egalitarian future. In combination with the penetration of non-fossil energy-supply options - some 40-50% market penetration in the egalitarian and individualist utopias – the CO_2 emission per capita declines in the egalitarian utopia and later on, also in the individualist utopia (*Figure 17.9a*). Such an onset of the ecological transition is absent in the hierarchist utopia. Another indicator is the energy intensity expressed as primary energy production per unit of economic output. Thanks to a combination of autonomously

Indicators and indices -hierarchist results for population and health, and biogeochemical cycles

As described in Chapter 9, we have attempted to represent the situation depicted in the various submodels by introducing a hierarchy of indicators and indices. *Figure 17.10* depicts the health transition for the hierarchist utopia in the form of the indicator tree discussed in Chapter 9. The lower rows of graphs indicates how environment-related risks (food, water) decline from 1990 onwards because of continuing development, which is here indicated by the percentage of literate women living above the poverty level (socio-economic index). Meanwhile, life-style-related risks continue to increase.

We have also constructed a hierarchy of indicators and indices for the element cycles. *Figure 17.11* presents the hierarchical framework of indicators and indices for the hierarchist scenario. The environmental pressure indices in the bottom most graphs show increasing normalised equivalent emissions. The decreasing trend in CFC-11 and acidification equivalent emissions reflect exogenous scenarios which are partly based on international agreements. The eutrophication equivalents are dominated by domestic sewage and hence follow population growth. The overall pressure index in the middle graph shows that the overall pressures on the biogeochemical cycles will be almost double the present values by the end of the next century.

Except for CO_2 and temperature change, all state indicators already exceed the 'no-effect' levels before 1990 and remain above these levels. These indicators are normalised on the basis of a desired reference situation and therefore express a measure of risk. The chlorine level will remain above the 1985 level of 3 ppbv level, as shown in the normalised chlorine concentration index ('no-effect' level is 2 ppbv). The water pollution index follows the eutrophication equivalents; unlike the situation with regard to climate change and acidification, there are hardly any delays in the system. The soil organic matter in the developing countries will decrease slightly over time due to the combined effect of nutrient mining and erosion of the cultivated land, and the conversions of terrestrial ecosystems into rainfed arable land. The overall CYCLES state index in the upper most graph is a measure of global environmental risk. It exceeds the 'no-effect' level by a factor of 2-3.

Finally, the impact index is described by the inverse of the globally weighed average biological productivity on land and in sea. The biological productivity on land shows a slightly negative trend as a result of the combined effect of an increasing trend in the industrialised countries (N fertilisation effect, and intensification of

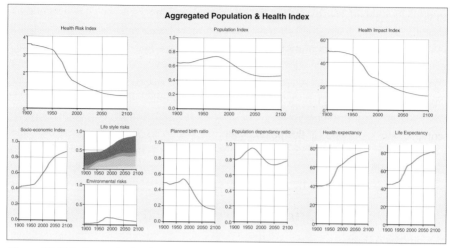

Figure 17.10 Indicator and indice tree for the Population and Health submodel.

cultivated land), and a decreasing trend in the developing countries (deforestation of high productivity terrestrial ecosystems). Marine plankton production increases slightly due to the combined effects of P fertilisation in the oceans and temperature feedbacks. The combined effects of these reverse trends in biological production leads to a stabilisation of the final impact index.

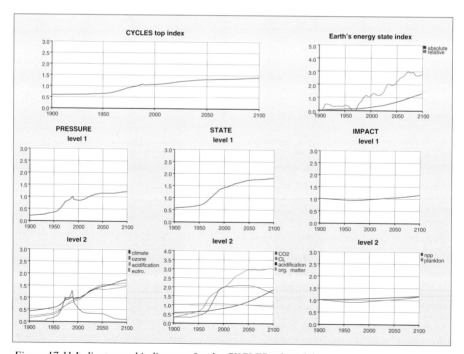

Figure 17.11 Indicator and indice tree for the CYCLES submodel.

and price-induced energy-efficiency improvements, this intensity falls in all three utopias (*Figure 17.9b*).

Other environmental variables of interest are atmospheric CO_2 concentrations and the resulting climate change impacts in the form of (global average) surface temperature increase and sea-level rise. Other indicators are nitrogen fluxes onto land – which are directly related to water quality – and UV radiation. However, because the dynamics of the element cycles and the climate system are slow in comparison with the anthropogenic patterns of change, these environmental variables are not part of a clear transition for the time horizon considered here, in any of the utopias.

17.6 Into the 22nd century: towards a sustainable state?

In the context of transitions towards a sustainable state, the notion of sustainability is – often either explicitly or implicitly – associated with a dynamic equilibrium in which major state variables do not change over time. We have extended the time horizon in the TARGETS1.0 model to the year 2200 to explore the behaviour of the model in the very long term. One would expect to gain insights into the interaction between dynamic processes which operate according to different time-scales. The results are shown in *Figure 17.12a-f*. The blue curves show the evolution of the hierarchist utopia. We assume continued growth in GWP at 1.3 % per yr which, in combination with a population stabilising at about the level in 2100, gives an increase in per capita income to some $ 85,000 per capita – twice the 1990 level in Switzerland. This leads to an increase in pressure on the environmental system through rising water and energy requirements, especially for the continuously growing industrial sector. In energy supply, the potential for reducing the energy-intensity gets exhausted and the transition to non-carbon fuels stops at around 30%. The share of coal remains at about 50% and as a result CO_2 emissions rise to 36 GtC/yr by 2200. Because of an exponential rise in SO_2 and NO_x emissions, temperature change only slowly increases to over 4 °C. This causes a continuing rise in agricultural yields up to 2200. Nevertheless, to satisfy demand for food and wood, forested areas in the developing world vanish before the end of the 22nd century. Because of the inertia in the environmental system, more serious impacts will only show up in the 23rd century. This experiment indicates the relevance of the different time-scales on which the human system and the environmental systems function.

If we apply an egalitarian world view and management style, the population starts declining after 2080 down to about 6 billion people by 2200 despite the much slower increase in economic welfare ($ 21,000 per capita by 2200). Other features of the egalitarian utopia also indicate a transition towards a steady-state situation: energy supply is stable at about the 1990 level, water supply declines to even below

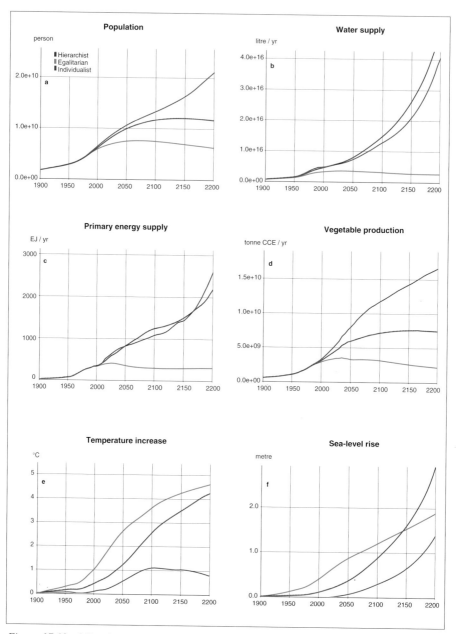

Figure 17.12a-f Simulation results for the three integrated utopias up to the year 2200 (see figure 17.2)

the 1990 level and food demand continues to fall. Nevertheless, the presumably much greater sensitivity of the climate system and the lower SO_2 and NO_x emission paths lead to an almost 2 °C higher temperature rise throughout the 22nd century than in the hierarchist utopia. In this long-term egalitarian perspective, the consequences of centuries of greenhouse gas emissions are alarming indeed.

Taking the assumptions of the individualist utopia, exponential GWP growth pushes economic welfare to a staggering \$ 140,000 per capita by 2200 with a still rising population of 21×10^9. This experiment is at best an illustration of the kind of constraints which become operational at extremely high pressure. Food demand and supply continue to grow to five times the present level to feed more people with diets increasingly animal-based. In this world high-tech options like genetic engineering would have to be developed out of necessity, in a spiral of solving old problems by creating new ones. All arable land will be exploited; a large part of it has to be irrigated. Water demand increases threefold between 2100 and 2200. Energy supply exceeds the hierarchist value after 2150; by the end of the 22nd century so much natural gas has been used that the system hits the limit and is forced to a further intensification of using renewables and coal. Nevertheless, the assumptions on the robustness of nature are such that temperature increase and sea-level rise are still below those in the hierarchist and egalitarian utopias. This is what one would expect for the individualist utopia – but it is also the statement which most evidently stretches the use of our model beyond what it can be decently used for.

17.7 Conclusions

This chapter describes three utopias – worlds in which a certain view on how social and environmental systems function accompanied by consistent and therefore 'adequate' action. If the world is a place of abundance, it can stand increasing pressure. If emerging problems are tackled with pioneering ingenuity, humankind will follow a path that is sometimes dangerous but never catastrophic. If, on the other hand, life is an intricate and vulnerable web of connections, easily destroyed or impoverished, it would be foolish and irresponsible to continue unconstrained growth. In between is a world which can be characterised as one in which cautious institutional control based on expert judgement dominates decision-making.

The three scenarios give an indication of the trade-offs involved. Whereas the individualist scenario scores high in population health and economic welfare, and low in environmental impact, the egalitarian picture is the opposite. The growth in per capita 'net income' that can be sustained over the next century is almost twice as high in the individualist future as in the egalitarian one. The concept of interrelated transitions, as it turns out, can be nicely illustrated with the help of the utopian model experiments. All three utopias experience a transition to lower birth and death rates, higher life expectancy and lower resource intensities. Only in the food transition there is a marked difference between the egalitarian and the other two perspectives. Finally, the model is rather robust in the sense that an extension of the time horizon to the year 2200 yields meaningful results and highlights the difference in time-scale between the human and the environmental systems.

Integrated experiments with the TARGETS1.0 model indicate for the three utopias that the feedbacks, implemented in the present model version, have rather little influence on the behaviour of the overall system. One explanation is that the situation is so utopian in terms of food, water and energy supply that feedbacks from the environment are too small to have a noticeable impact on population-related variables like healthy life expectancy. This suggests that integration is more important in the dystopian scenarios – which are explored in the next chapter.

Our utopian experiments clearly illustrate the divergence in global futures which results from the use of perspective-based model routes. This confirms our hypothesis that the uncertainties related to world view and management style, each with their own rationality, dominate these which are narrowly scientific. The simulation experiments with the three utopias also suggest that there are at least three 'best of all possible worlds', which can be consistently and coherently quantified within the framework of the TARGETS1.0 model. The underlying assumptions – or world views – can be used as a rationalisation for [not] taking particular policy measures. Of course, this conclusion can only be drawn with the utmost caution because of the deficiencies of the TARGETS1.0 model. For example, the rise in sea level in the hierarchist utopia has no explicit negative consequences whatsoever for agriculture, population or the economy. Does this conclusion necessarily lead to pure relativism – the postmodern idea that 'anything goes'? In the next chapter we will extend our exploration by looking at futures that tend towards failure.

Uncertainty and risk: dystopian futures
18

"The new capability of anticipation makes the future also effective in the present."
E. Jantsch, The self-organizing universe (1980)

18 UNCERTAINTY AND RISK: DYSTOPIAN FUTURES

Bert J.M. de Vries, Jan Rotmans, Henk B.M. Hilderink, Michel G.J. den Elzen, Arthur H.W. Beusen, Arjen Y. Hoekstra and Bart J. Strengers

This chapter explores dystopian futures. After a summary of the uncertainties and risks discussed for each of the subsystems, integrated experiments are presented in which world view and management style throughout the world system are at odds. We also investigate the effectiveness of various response options and of the timing of certain policy measures.

18.1 Introduction

In the previous chapter we outlined possible futures which are based on coherent sets of assumptions about how the world system functions and how it is managed. These are called utopias and constitute the diagonal elements in the matrix presented in Figure 10.7. In a way, they are idealised and therefore implausible images of the future. In this chapter we first present some simulation experiments in which dystopian trends are explored with the integrated TARGETS1.0 model. This is a prelude to the next section in which we analyse in more detail images of the future where world view and management style are at odds. These are referred to as integrated dystopias (see Chapter 11) and they are actually more plausible because they contain real-world tensions between diverging world views and management styles. Two major chains which cause feedback loops are presented as a framework discussing some interesting dystopian futures and to give an assessment of associated risks. Finally, we explore the adequacy of response actions in terms of intensity and timing, and the consequences of allocating insufficient investments to the food, water and energy sectors.

One of our intentions is to use the TARGETS modelling framework to explore strategies for a more sustainable future. Using the causal chain approach in the form of the Pressure-State-Impact-Response (PSIR) concept, we focus in the first instance on the question to whether changes in the human system are likely to interfere in undesirable ways with the functioning – now and in the future – of environmental systems. The answer deals with the extent of disturbances, the speed at which they occur, and the degree of irreversibility. If undesirable consequences are anticipated, this places limits on the pressures generated by the number of people and their needs and activities as has been worked out in, for example, Chapter 16. Possible strategies have to assess potential damage and mitigation costs, both of which can only partly be expressed in model terms. Constraints on the pressures involve what economists call 'opportunity costs' or 'forgone benefits', i.e. welfare which has to be sacrificed. Evidently, any answer to the above question is full of uncertainties and not just technical uncertainties which will disappear with progress in scientific research. Many of them will remain with us for a long time, and their appraisal will continue to be controversial. It is along these lines that we now present some dystopias.

18.2 Dystopian tendencies

In Chapters 12 to 16, various combinations of world views and management styles have been investigated with the TARGETS1.0 model. The experiments indicate that a future with a small population with a high life expectancy is desirable on *population and health* developments but not to be expected. On the other hand, a doomsday scenario with an extremely large population, in miserable health conditions is not very plausible either. However, stagnating socio-economic developments and a further deterioration of the environment could lead to increasing morbidity and mortality levels and curb population growth. Much will depend on the nature and effectiveness of governance. Policies related to family planning could turn out to be robust; measures directed towards human development could accelerate the levelling off of population growth. The model experiments on *energy* indicate that the aspiration of a high-growth world has to rely on a combination of technological optimism and energy taxing if emissions are not to exceed 10 GtC/yr. A corollary to this finding is that if such emission levels are considered acceptable, the imposition of a high carbon tax would be a heavy and unnecessary burden in a low-growth world. In general, if carbon emissions are to be reduced, energy policies should focus on more efficiency in meeting end-use demand, the abolition of coal subsidies and support of non-carbon based alternatives.

One of the undesirable *water* futures emerges from a combination of a medium to high population and economic growth scenario, and a water sector which functions and is managed in a prudent, egalitarian style. Water scarcity problems would then by far exceed the problems encountered today, probably in the form of 'overshoot and collapse' situations. The model experiments also suggest that water-related problems can be solved only partially through water policy; socio-economic policies that influence population, and that scale and nature of agriculture and economic activities are much more effective. As with energy, policies should shift from water supply towards water demand and water subsidies should be removed. The only exception would be drinking water: if combined with end-use efficiency, the objective of 'drinking water for all' justifies high investments in public water supply. Re-use of water should be promoted and new irrigation schemes should be carefully planned as part of food policy objectives. With regard to *food*, a management style which emphasises low-input agriculture and nature conservation is most prone to food shortages. The reason is that such an egalitarian management style is aversive to risks in terms of ecosystems, with the argument that their degradation or disappearance will undermine food production in the long run. This has a price in the sense that a low-inputs agriculture may be unable to adequately feed the world population. Hence, opposition against or failures of new Green Revolutions in a medium to high population and economic growth scenario would probably result in serious food shortages which could jeopardise social stability. The vulnerability of the climate system does not alter this conclusion. On the other hand, in our model experiments the individualist world view and management style pose the greatest risks to tropical forests. Finally, the experiments with the CYCLES submodel make clear that the risks associated with increasing greenhouse gas emissions and agricultural inputs is strongly dependent on the world view held about the vulnerability of the *environmental system*. The egalitarian world view which emphasises nature's vulnerability, leads to dystopias with extremely high climate risks. As expected, the climate risks are small in individualist dystopias. It turns out that a stabilisation in the next century of the CO_2 concentration at a level below 450 ppmv is not feasible, not even with an egalitarian, low-growth scenario.

In section 18.3 we will proceed with a systematic, integrated exploration of dystopias, but first we want to investigate a few 'dystopian tendencies' to get a feel of the factors leading to dystopian futures. One possibility is that some elements of the management style cannot be implemented according to the corresponding world view. Another possibility is that a particular management style can be enforced, but it turns out that the underlying world view is incorrect. We construct model-supported narratives around such combinations of views and actions. From a utopian perspective, each of these narratives – or scenarios – represent an unexpected and undesirable development, hence the reference to dystopian tendencies. The three experiments presented here are: (i) a one-third higher economic growth rate in the hierarchist utopia (Hierarchist→Individualist), with a sensitive climate system (Individualist→Egalitarian), (ii) slow technological progress in the individualist utopia (Individualist→Hierarchist), and (iii) resistance to behavioural change and policy measures in the egalitarian utopia (Egalitarian→Individualist).

A hierarchist world with high economic growth and a sensitive climate
Suppose the hierarchist utopia is confronted with a much higher economic growth rate than is expected: 3% per yr instead of 2.3% per yr. This will please those who cherish individualist values, but it worries the hierarchists because the pressure on the environment will increase and the CO_2 concentration will exceed the level which is thought to be acceptable (*Table 18.1*). Indeed, the higher energy demand which is largely met by coal leads to higher CO_2 emissions and hence a higher CO_2 concentration in 2100.

To reduce the risk of climate change posed by such a rise in CO_2 levels, we did a second experiment in which governments respond by subsidising energy conservation and levying a carbon tax. The carbon tax is according to the 'constant emissions' profile calculated by Nordhaus (1994); the subsidy is introduced by assuming that consumers accept a 6-8 year payback time on energy efficiency investments. They are quite effective: by the year 2100 the lower energy use with a lower proportion of coal and faster penetration of biofuels and solar electricity lead to a CO_2 concentration in 2100 slightly lower than in the hierarchist utopia. The rise in temperature is not smaller because SO_2 emissions have dropped by just as much as the CO_2 emissions. As *Table 18.1* shows, this temperature rise would be much higher if the climate system is as sensitive as the egalitarians fear and would cause serious food shortages after 2060. Such an expectation would interfere with the search for more sustainable development, and hence the hierarchists would be forced to introduce more drastic policy measures to curb emissions. These experiments confirm the insights on climate risks gained in Chapter 16.

Disappointing technology in a high-growth individualist world

The individualist utopia is full of technical progress and efficient, market-oriented resource management. What would happen if land and water-related technologies do not meet up to the expectations? Suppose that agricultural yields do not increase in the absence of a second Green Revolution and biofuel yields and water use efficiency hardly improve. The maximum attainable efficiencies in using water for domestic and industrial purposes do not increase either. Such a situation may occur because of large income disparities and no interest on the part of the rich to develop or introduce these technologies. As a result, food supply cannot meet demand – which causes increased mortality and reduced population growth – and water use rises 40-50% above the utopian level. However, the resource is assumed to be so abundant and the response, in the form of waste-water treatment, so effective that this higher water use hardly has an impact (*Table 18.2*).

Values in 2100	Secondary fuel demand (EJ)	Proportion of coal in primary energy use (%)	CO_2 concentration (ppm)
1990	177	31	353
Hierarchist utopia	800	59	720
Idem, 3% per yr GWP-growth	900	59	880
Plus introduction energy conservation subsidies	610	56	800
Plus introduction carbon tax (towards $ 400 per tC in 2100)	540	47	715
Idem with sensitive climate	540	47	711

Table 18.1 Indicators for the experiment with high economic growth, climate policy and a sensitive climate in a hierarchist utopia.

Values in 2100	Population (10^9)	Food per capita (kcal/cap/day)	Penetration of biofuels (%)	Fraction of water in Class A (%)	CO_2 concentration (ppm)
1990	5.3	2700	0	81	353
Individualist utopia	13.3	4250	24	67	630
Halving the rate of agric yield increase	12.8	2600	23	72	624
Plus less learning biofuels	12.7	2490	15	72	648

Table 18.2 Indicators for the experiment with disappointing technology in the individualist utopia.

In a second experiment we assume that innovation with regard to biofuels does not meet the goal of delivering commercially produced biomass below $ 25-30 per GJ. Biofuel penetration will be slower; yet it requires more land due to lower biomass yields which competes with land for food. In combination with the stagnating improvements in agricultural yields, per capita food intake falls from 2040 onwards to levels below the 1980 world average. The resulting malnutrition leads to a 4% lower population in 2100 and thus to a flawed health transition. The CO_2 concentration rises only slightly. Although more fertilisers are needed, after 2020 they no longer increase yields. Water quality actually improves because the decrease in nitrogen in sewage effluents as a result of lower food production exceeds the additional nitrogen flux from more fertiliser use.

This experiment points out a more general finding: the natural system is so robust in the individualist world view that dystopian trends like the ones introduced here hardly have any impact. One reason for this is that some important quality aspects of human life and nature, and some potentially strong feedbacks are not incorporated in the simulations. A narrative for this future could be that the few live in pockets of wealth, while the majority suffer from starvation and lack of safe water and medical services, thus resembling elements of some recent scenarios[1]. This situation could lead to a real catastrophe if the climate turns out to be as sensitive as the egalitarians fear. In that case, a further decline in yields might cause widespread food shortages among the poorest segments of the world population.

Resistance against life-style changes in an egalitarian future

The egalitarian world view implies that the natural system can easily be disturbed by humans in a way which negatively affects human populations and their quality of

---

1 For example, the barbarisation scenario (SEI, 1996) and the possibility of a cultural clash (Huntingdon 1995). See also the article 'The coming anarchy' by Kaplan (1992).

Values in 2100	Population (10^9)	Life expectancy (yr)	Global average food intake (kcal/cap/day)	Animal-based part (%)
1990	5.3	65	2700	15
Egalitarian utopia	7.4	76	3100	5
With non-vegetarian trend	7.0	72	2100	25-29
With ineffective education	7.6	67	1900	30-33

Table 18.3 Indicators for the experiment where there is a failure to realise transitions to vegetarian food and smaller family size.

life or even survival. The paradox in the egalitarian management style is that it prefers individual commitment and responsibility over huge bureaucracies and multinational firms on the one hand, while on the other, strong institutions are needed for the implementation of all the necessary and sometimes drastic changes[2]. Assuming that global governance is adequate to take the policy measures in the desired direction, there is the additional problem that some facets of the egalitarian utopia are very much life-style related. In this experiment we explore the vulnerability of the egalitarian world with respect to two of these: the switch to a largely vegetarian diet and enhanced family planning efforts. What would happen if the world population can neither be persuaded nor forced to make such changes in behaviour?

A failure to confine the animal-based part of the average diet to present levels or below would cause an increase in the number of ruminants and in the associated methane emissions. When the animal-based part rises to 25-30% of total food intake, serious food shortages develop because of the assumed ceilings on fertiliser use, and the assumed clearing and irrigation paths (*Table 18.3*). As a result, life expectancy drops by four years and the population is 0.4 billion lower by 2100. The resulting impact on the natural system is small: the simulated temperature increase is 0.1 °C less. If education efforts fail to reduce the number of children per woman in the fertile age range, average life expectancy drops by another six years and the population rises to 7.6×10^9 by 2100. This aggravates the food crises. Again, however, it is the people that suffer, not the natural system. This experiment indicators how important the life-style component can be in pursuit of certain transitions.

---

[2] This is an oversimplification of the situation because many exponents of an egalitarian world aim at changes which are based on grass roots or spiritually oriented movements; a model like ours cannot explicitly include such processes of change.

18.3 Systematic exploration of dystopias

The previous section provides some indications about the elements in the cause-effect chains which may cause dystopian developments. In this section we explore the dystopias in a more systematic way. We assume that the discordance between the world view and management style will only cause response actions which are endogenous in the model, such as an increase in the costs to provide water and energy. This is an important constraint because it does not allow for response actions such as building dikes or migration as a consequence of sea-level rise, or more or less clearing or irrigating land. In a few cases we explore the consequences of exogenous response: changes in certain policy variables such as land clearing and irrigation are introduced exogenously to represent adaptive behaviour.

In the analysis of utopias in Chapter 17 the feedbacks between the five submodels are concluded to have a rather modest influence. Our hypothesis is that in the dystopian futures these interactions will be much more important, i.e. that the feedbacks are the key elements in the emergence of dystopias. On the basis of a large set of experiments, we have identified two major interacting chains between the human and natural systems which turn out to be the most interesting ones on dystopias: the population-food chain and the energy-climate chain. A third chain is associated with the interaction between water and climate. There are a few more such chains within the TARGETS1.0 model (Figure 2.3). For example, water demand depends on the population level and economic development and has to be met by a supply, which depends on water availability and quality. These give no interesting dystopias with the TARGETS1.0 model. One of the reasons is that effects like rising water costs and sea-level rise do not influence the functioning of the other systems in the model – only more investments are needed.

In the following experiments we distinguish between the human and natural systems. This distinction is not sharp within the AQUA and TERRA submodels; the behavioural variables are taken as elements of the human system. The distinction between world view and management style variables, to which one cannot also apply strict criteria, is made clear in Chapters 12-16. We have explored several dystopias. For instance, what happens when resources turn out to be abundant, technological innovations effective and fast and the climate system quite insensitive to human-induced disturbances, but the world is managed by egalitarians? It turns out that the world as perceived by the individualist has such a robust environmental system and a human system which is so insensitive to policy measures that an egalitarian management style does not make much difference. There is one exception: if the egalitarian wish to implement a low ceiling on fertiliser use is applied, a serious food shortage develops.

The systematic explorations with the TARGETS1.0 model suggest that one situation in particular will lead to a really disastrous future, which will come about if

Interacting chains between the human and natural system

A concise representation of these chains in the form of the eight curve of Chapter 2 is given in Figure 18.1.

The population-food chain
Population size and structure depend on fertility and mortality rates, which are governed by economic growth, food intake and access to safe water and sanitation. The population size and the welfare level (GWP/cap) determine food demand, which should be met by food production. The latter, in terms of vegetable production, depends on the amount of cultivated land and the agricultural yield per hectare. The actual food consumption per capita is one of the main health determinants which affects mortality levels and hence life expectancy. This in turn influences the population size and structure, closing the population-food eight curve.

The energy-climate chain
The population and the sectoral activity levels lead to an energy demand, which is satisfied by exploiting a mixture of energy-supply sources. Part of these sources are fossil fuels which on combustion generate CO_2, NO_x and SO_2. The increasing atmospheric concentration of these gases causes a change in radiative forcing which in turn results in a change in the Earth global-mean surface temperature and precipitation. This climate change influences food production and water availability, which in turn may influence mortality and morbidity patterns of the human population. Then, the eight curve is closed in the sense that the changed size and health status of the human population will again influence energy-demand levels, etc.

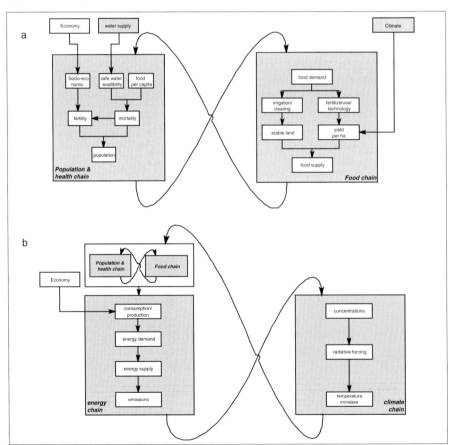

Figure 18.1 (a) The population-food chain and (b) the energy-climate chain.

there is too much pressure from the human system (mainly in the form of greenhouse gas emissions) in combination with a vulnerable environmental system, especially the climate system. After discussing this in some more detail, we pose the question: What risks for human health emerge when the environmental system turns out to function as the egalitarians fear and not as the hierarchists expect? Such a scenario may be called either the *egalitarian nightmare* or the *hierarchist complacency*. From the hierarchist utopia as the reference, we focus on the following three processes:

- processes which are part of the biogeochemical cycles of C and N and the climate feedbacks within the CYCLES submodel (Chapter 16);
- the CO_2 fertilisation feedback and the direct climatic effects on agricultural yields within the TERRA submodel (Chapter 15);
- the climate factors influencing sea-level rise, water-quality distribution and potential water supply (Chapter 14).

Figure 18.2a shows the results for some important model variables of the utopian (hierarchist) and the dystopian runs (see also Chapters 11 and 17) where the sequence of events is: significant temperature rises, after some decades exerting a negative influence on yields, causing food shortages which, in turn, increase mortality and shorten life expectancy. The collapse in the availability of food is evident: by 2100 the global average food intake has dropped to 1500 kcal/cap per yr. This is the result of the strongly negative climate effects of an extremely large temperature increase of about 6 °C on agricultural yields in combination with the modelled rigidities of a fixed ceiling on inputs and no adjustments in clearing and irrigation policies.

Average life expectancy falls from 76 to about 60 years. Because the population is lower than in the hierarchist utopia and the GWP trajectory is fixed, the welfare level expressed as GWP per capita in the second half the next century is higher than in the dystopia. This corresponds with somewhat higher levels of health services, but the resulting fall in mortality is only partly offset by a parallel rise due to falling food availability. The effect of the temperature increase on the other chains is reflected in *Figure 18.2b*. By 2100 sea level has risen 1.6 metres. Other water-related effects are an exponential increase in water supply costs per litre, a dramatic shift in the distribution between water-quality classes and a much lower potential for fresh-water availability. This dystopia indicates how a vulnerable environmental system can interfere with the expectations of smooth transitions with regard to health, food, water and energy.

In discussions about life-style and North-South aspects of the global future, some people say the greatest threat to a more sustainable path is the pursuit of ever more material possessions. Such a scenario focuses on the controversies between the individualist and the egalitarian – a controversy appearing almost everyday in the newspapers. How would the individualist utopia look with an egalitarian view on climate? Interestingly, it turns out that the feedback from climate change on

18 UNCERTAINTY AND RISK: DYSTOPIAN FUTURES

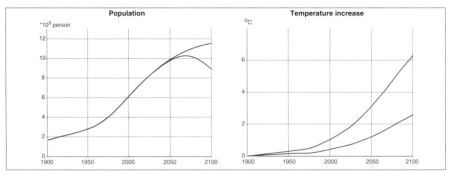

Figure 18.2a The egalitarian nightmare: a vulnerable climate system is confronted with a 'Business-as-Usual' growth of population and economy.

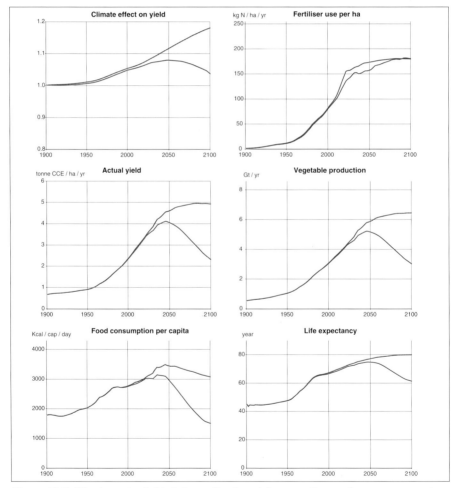

Figure 18.2b The dynamics behind the egalitarian nightmare: climate change causes serious reductions in crop yields which can only partly be offset by increasing fertiliser use. A food shortage develops which causes a significant decline in average life expectancy and population.

Adaptation through exogenous policies

The extreme situation described above is not very plausible because one may expect that [parts of] the human population respond to it with behavioural changes and policy measures. We have performed simulation experiments in which undesirable consequences such as severe food shortages are counteracted. Three response – or adaptation – strategies are explored.

Policy 1: agricultural policy

The land policy is simulated by introducing the individualist management style for food supply: more clearing to increase the total arable land to 1800 Mha in 2100, an increase in irrigation resulting in 600 Mha irrigated area by 2100, and a higher ceiling on the use of fertilisers. It should be noted here that 450 Mha of irrigated area is seen by most experts as a maximum (Chapter 7). As shown in *Figure 18.3*, a major food crisis can be prevented although food availability still decreases rapidly to unacceptably low levels in the last half of the next century. The dystopian food situation which results from the negative climate effects on agricultural production can be alleviated in this way, but after 2050 tensions increase again. The limits to available arable land and marginal effectiveness of nitrogen fertiliser are reached.

Policy 2: population policy

In addition to a land policy, one for population and health is introduced, resulting in a stabilisation of the population at 8.8×10^9. Life expectancy reaches a level of 80 years in 2075 but still decreases to a level of 76 at the end of the next century due to the emerging food shortages. Although significantly lower than in the no-response dystopia, there is still a fall in daily food consumption to about 2400 kcal/cap per yr.

Policy 3: energy policy

To obtain lower CO_2 emissions, we assume the introduction of a high carbon tax combined with an investment programme for commercial biofuel plantations. On introducing such an egalitarian management style for the energy system, the CO_2 emissions are reduced from 21 GtC/yr to 9 GtC/yr in 2100. The resulting CO_2 concentration rises to about 670 ppmv instead of 850 ppmv. The temperature increase is only marginally reduced over this period because the SO_2 emissions are also lower, which dampens the cooling effect of lower CO_2 concentrations (Chapter 16). Yet food availability remains above 1990 levels during the whole century and results in a stabilisation of the population at 9×10^9 people and a life expectancy of about 80, which is comparable to the hierarchist utopia.

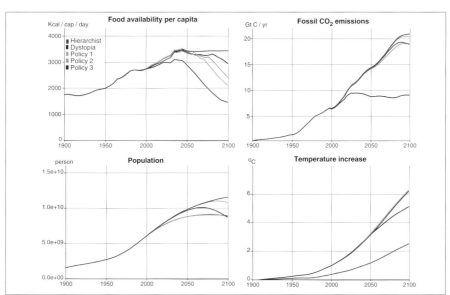

Figure 18.3 The effect of adaptation in the form of response policies: the diversion of a catastrophy by a combination of land, population and energy policies. The measures to reduce the CO_2 emissions are particularly effective.

population, mainly through the food system, is less disastrous than in the hierarchist utopia. This is in agreement with the analyses in Chapter 16. The main reason for this is that the individualist world view as parameterised in our simulation experiments assumes fast technological innovation in the realm of energy efficiency and non-fossil energy sources. This leads to large-scale penetration of cheap non-fossil energy sources[3]. Hence, the carbon-dioxide emissions are much lower than in the hierarchist scenario despite higher population and GWP growth.

18.4 Additional explorations: policy timing and overconsumption

The effectiveness of a carbon tax
One of the core issues to be considered in global climate change is whether the world community should put effort into reducing greenhouse gas emissions, and if so, how much, given the many uncertainties in long-term economic, technological and climate change developments. A furious debate is going on between opponents and proponents of immediate abatement. Again, different perspectives offer different insights. Opponents argue that early efforts at significant mitigation may prove needlessly expensive (Wigley *et al.*, 1996). Reasons given are that investments in economic growth lead to more resources available for mitigation in the future; that a transition to lower carbon intensity is cheaper if it is in tune with replacement of existing capital; that technical progress will lower mitigation costs through time; and that a higher near-term emission path may have greater cumulative emissions but the same final level of atmospheric CO_2 concentration. On the other hand, it is argued that the longer we delay the abatement activity, the more wrenching the abatement is likely to become. This position is based on decision-makers not having foresight and expectations about implementing more stringent mitigation efforts. The relevant question is not so much which perspective is correct but how to send a strong signal to decision-makers that in the near future a more stringent abatement strategy can be implemented.

To support the international negotiations about emission reduction, emission targets have been proposed in relation to future atmospheric CO_2 concentration levels. The IPCC (1995c) calculates CO_2 emission pathways leading to a stabilisation of the atmospheric CO_2 concentration at 450, 550, 650, 750 and 1000 ppmv; Wigley *et al.* (1996) consider the range of 450 ppmv to 650 ppmv (Chapter 16). The IMAGE2 model has been used to estimate emission paths which satisfy certain constraints on climate parameters, the so-called 'safe emission corridors' (Alcamo *et al.*, 1996). We use the TARGETS 1.0 model to put the question of timing in an integrated perspective.

---

[3] This is certainly not true for all individualist-oriented scenarios; some of them, however, are quite explicit about this (e.g. Kassler 1994).

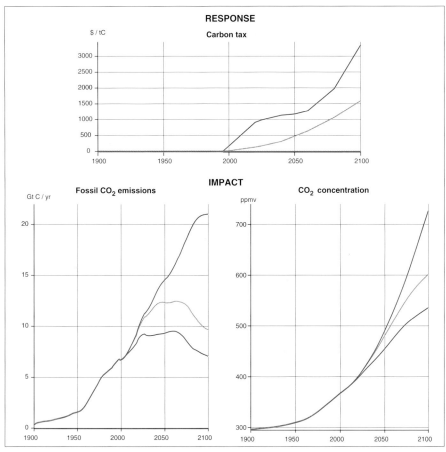

Figure 18.4 The carbon tax required to achieve a CO_2 concentration of 550 ppmv by the year 2100, and the corresponding CO_2 emissions and CO_2 concentration.

First, we look for carbon tax pathways in the hierarchist utopia, which will lead to a levelling off of the CO_2 concentration at about 550 ppmv by 2100[4]. Two such pathways and the resulting CO_2 emission profiles are shown in *Figure 18.4*. To reduce the concentration from 730 ppmv to about 600 ppmv by the year 2100, an exponentially increasing tax towards $ 1500 per tC in 2100 is to be applied. This is three times higher than the carbon tax profiles discussed in the literature (Hourcade et al.,1995)[5]. To bring the concentration below 550 ppmv, a more stringent carbon

---

4 We assume that the CO_2 emissions associated with land-use changes are the same as in the – hierarchist – reference scenario. We also assume a market-independent, RD&D oriented construction programme for NTE-capacity (nuclear, solar) at a rate of 600 MWe/yr.
5 In the egalitarian utopia, we apply a carbon tax of $ 1500 per tC (Chapter 13).

tax is required. In this case, emissions reach a platform of 9-10 GtC/yr in the period 2020-2060, after which they drop to about 7 GtC/yr in 2100. The carbon tax required to achieve these emission reductions first rises in a linear fashion to about $ 1000 per tC around the year 2040. In this phase the re-introduction of coal is discouraged to the extent that the proportion of coal is kept down to the present level of about 20%. This is supported by additional price-induced energy efficiency investments. In the second half of the century, a further rise in the carbon tax is required because the marginal costs of efficiency improvements rise continuously and the carbon tax is now also needed to replace oil and gas with renewables. This latter process is rather difficult because there is still a lot of relatively cheap oil and gas. One reason these carbon tax profiles are high in comparison with those in most literature studies is, that in our simulation there is no feedback on the rate of economic growth and hence on end-use energy demand. Optimisation analyses with the TIME model give similar, high carbon tax profiles (Janssen, 1996). This is being investigated in more detail with the WorldScan model (CPB, 1996).

Can a carbon tax come too early or too late?

In another set of other experiments we simulated the introduction of a constant carbon tax from a given year onwards throughout the next century. The carbon tax policy is in all cases complemented with RD&D programmes for biofuels and non-thermal electric power sources to speed up their introduction. This leads to CO_2 emissions which first decrease, and then start increasing again, because the opportunities for energy efficiency improvements, substitution of coal by oil and gas, and the penetration of non-carbon based energy sources operate on different timescales. A stabilisation of the CO_2 concentration would require continuously decreasing emissions.

The results of these experiments are given in Table 18.4. For the 550 ppmv case it appears that waiting until 2000 would imply a carbon tax level of $ 300. If the tax is introduced 25 years later, it has to be twice as large. If introduction is postponed until 2050, it will have to be extremely high. In this case the resurgence of coal is not suppressed and emissions follow the hierarchist scenario until 2050 after which they have to be reduced at about twice the rate at which they fall in the two previous model experiments. This may turn out to be impossible and is to be labelled a risk strategy. These results suggest the existence of a window-in-time for policy instruments like a carbon tax. One explanation is that, in our model, the opportunities to reduce the energy intensity at low cost are gradually disappearing as the cheapest conservation options are implemented. Also, the new supply technologies tend to stabilise fuel costs and their learning diminishes (Chapter 13).

Target CO_2 concentration (ppmv)	450	550	650
Year of introduction			
2000	2000	300	250
2025	-	600	350
2050	-	1500	500

Table 18.4 Level of a constant carbon tax as a function of the target CO_2 concentration to be reached by 2100 and the introduction year. For the reference case the CO_2 concentration rises to about 730 ppmv.

Postponing population policy measures

We also performed a model experiment on the consequences of timing population policy measures. If the effort put into family planning policies is halted completely in 2000, there is a significantly slower fall in the average family size. The corresponding higher birth rate causes the world population to grow faster. By 2100 there are some 2×10^9 additional people. If the effort put into family planning stops by 2025, the effect is much less because in our model formulation the desired family size will decrease, also without policy measures, as a consequence of increasing welfare. If family planning policy efforts stop by 2050, there is hardly a noticeable effect on population. This shows the 'window-in-time' character of policy efforts like the one related to family planning, as has also been found for a carbon tax.

Overconsumption and the allocation of scarce investment resources

As explained in Chapters 11 and 17, it is assumed that the required investment goods to operate the food, water and energy systems are always and fully met by the

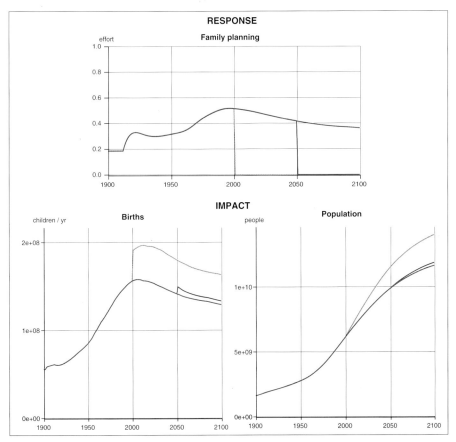

Figure 18.5 The effectiveness of population policy as a function of the introduction year.

economic system. This is in no way guaranteed in the future nor is it the case in the present-day situation. In a third set of experiments we explore the dystopian tendencies which result from overconsumption, i.e. from a consumption level (including non-health services) which exceeds the maximum permitted growth of consumption (Figure 17.7d). This is to be expected in a world with large differences in income and wealth. It causes a shortage in investment goods to build up the capital stocks required to satisfy the demand for food, water and energy. All model experiments are done with the hierarchist utopia as the reference.

We implement the shortage of investment resources in the form of a pulse of overconsumption in the hierarchist scenario which peaks at 2% in 2025, i.e. a 2% higher annual growth rate of per capita consumption than shown in Figure 17.7d. This creates a shortage of investment goods available for maintaining the agricultural, water and energy sectors. A key question, of course, is which of the nine investment categories distinguished in our model is to suffer from the shortage (see

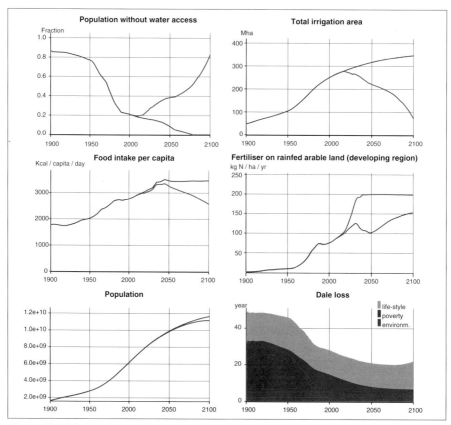

Figure 18.6 The hierarchist utopia (blue lines) if the shortage of investment goods affects drinking water, and sanitation and irrigation. The utopian transitions with regard to water and food fail (red lines).

Figure 17.11). *Figure 18.6* illustrates what happens if the overconsumption occurs at the expense of domestic water supply and irrigation. In real-world terms, this would mean that a proportion of the available investments is not available for drinking water and sanitation infrastructure and for new irrigation schemes, and is instead used for consumer goods. The direct result is twofold: the number of people without access to safe water start to increase again and the agricultural area under irrigation starts to decline. The latter results in increasing use of fertiliser in the developed region. Due to declining marginal productivity, this option is no longer available after some time and a food shortage emerges. The combined effect of less access to safe water and less food results in a lower life expectancy and a slightly lower population as it influences the health transition. Obviously, several of the aforementioned transitions may fail if no adequate investments are made. This may pose serious risks for parts of the world, especially if there are weak governments and financial instabilities.

18.5 Uncertainty and risk

In combination with the analyses in Chapters 12 to 16, the exploration of dystopias gives us an idea of the uncertainty and risk in the wider context of the TARGETS 1.0 model. One way to assess risks is to evaluate the probability that certain assumptions behind the utopian world view and/or management style will be invalid, in combination with the consequences of such undesired, dystopian courses of events. The experiments presented in the previous sections indicate that one obvious risk aspect is an overestimation of the capacity of the environmental system to absorb increasing amounts of greenhouse gases. Another one is associated with the chance that high expectations on technological or behavioural changes do not materialise.

For the *hierarchist utopia* a major risk is that the egalitarian view of the climate system is correct. This would make the 'Business-as-Usual' future as depicted in the IPCC-IS92a scenario, a high-risk future indeed. At the same time, it can be argued that this coal-based scenario has a rather low plausibility anyhow, in view of the options for higher energy use efficiency and non-carbon energy sources. The *individualist utopia* with its large population and economic growth faces a double risk with regard to climate change impacts: either that the egalitarian view of the climate system turns out to be right or that the continuous technical innovations in efficiency and renewables are not borne out. In both cases the chances are that the climate is seriously disturbed. Despite its high population and economic growth, the individualist utopia is the least vulnerable with respect to food and water shortages – provided it is right in its risk-taking attitude with regard to the environmental system. If this is not borne out by the facts, it may easily be a recipe for disaster – as the egalitarians fear will be the case. The individualist with her/his heavy reliance on high-tech solutions is also prone to technology and management failures, and public resistance against options considered as unethical or unsafe.

The *egalitarian utopia* with its low population and economic growth is obviously the one with the lowest environmental risks but the price for this is paid in the form of less material welfare and a higher risk of severe food shortages, and limited and high-cost water supplies. These in turn affect people's health. Like the other two scenarios, it also faces unquantifiable risks. The slow economic growth may slow down the pace of technical innovations; people's behaviour with regard to desired family size, diet and the like may change much more slowly than expected. This, in turn, in combination with low and for some stagnating welfare levels could lead to authoritarianism and civil war.

Table 18.5 evaluates the extent to which the three utopias are prone to risk with regard to four events: too optimistic a view of the resilience of the Earth's environmental systems, too high an estimate of cheap and high-quality natural resources, too much technological optimism, and overconfidence about the willingness of people to change their behaviour in response to policies. This table synthesises the results of the previous, dystopian model experiments. With regard to damaging the environment system, the hierarchist utopia poses the highest risk in view of its large greenhouse gas emissions. Both the individualist and the egalitarian utopias expect lower emission paths, though for various reasons and with different policy measures. We have not performed systematic experiments for overestimation of the amount of cheap, high quality resources, especially water and land resources and natural gas deposits. However, the submodel experiments in Chapters 12-16 indicate that this will form a strong disturbance of the individualist dream. The hierarchist, with her/his cautious estimate of available resources, is less dependent on resource optimism. The egalitarian, and even more so, the individualist utopias are the most vulnerable to disappointingly slow rates of technological innovation, but here too such failures would take different forms. For example, a stagnation in energy efficiency and renewable energy technology would affect both utopias, but a

Perspective :	Hierarchist	Egalitarian	Individualist
Prone to the risk of :			
Overestimating the resilience of global element cycles and the climate system	++++	----	++
Overestimating the amount of cheap, high quality resources	+-	--	+++
Overestimating the rate at which adequate technology is developed and made available	+-	++	++++
Overestimating the degree to which people's behaviour can be guided by policy measures	+-	++++	-----

Table 18.5 Risks associated with the three utopias.

stagnation in biotechnology would only pose a risk in the individualist utopia. The hierarchist utopia, with its middle-of-the-road perception of what technology can do, is less vulnerable to this kind of failure. Finally, the risk factors related to the feasibility and effectiveness of policy-induced change in people's behaviour are highest for the egalitarian. His or her dream may be cruelly disturbed by the stubbornness of human conventions and attitudes as they prove inflexible in the absence of major catastrophes or revolutions.

18.6 Conclusions

The utopias presented in this book are idealised images of the future. For a variety of reasons, the dominant style of management can be at odds with the corresponding world view, and dystopian tendencies will evolve. Without any response policies, this can lead to highly undesirable futures. One of the most catastrophic dystopias we have explored is the situation in which high growth in population and the economy is pursued while the climate system proves very sensitive to a continuing increase of greenhouse-gas emissions. Disappointing technological breakthroughs may turn the individualist cornucopia into a world which is overpopulated, ecologically seriously degraded and for many a place of hunger, despite the assumption of resource abundance and great resilience of natural systems. Failure to change people's lifestyle may threaten the egalitarian utopia because of food shortages and a flawed health transition.

As it turns out, there are several policy options to divert such dystopian catastrophies, but their implementation will not be easy. Moreover, postponing such policy measures may reduce their effectiveness, which stresses the need for careful analysis of future trends and warning signals. Another aspect of the utopias is that they assume an adequate channelling of industrial outputs into health, food, water and energy infrastructure. If this is not the case – which surely happens in today's world – developments will also be less utopian, or in some cases and places even catastrophic. From these model-based explorations it can be concluded that each of the three utopias are subject to risks inherent to the uncertainties and controversies about how the world functions. Future global change research should focus on improving our understanding of these uncertainties if we are to improve our abilities to anticipate doom and our capacity to implement adequate response policies to avert it.

Global change – fresh insights, no simple answers

"Now that man has developed consummate skill in technology – the art of how to do things – can he develop equal ability to choose wisely which things are worth doing?"
W. Harman, An incomplete guide to the future (1976)

19 GLOBAL CHANGE – FRESH INSIGHTS, NO SIMPLE ANSWERS

Bert J.M. de Vries, Jan Rotmans, Arthur H.W. Beusen, Michel G.J. den Elzen, Henk B.M. Hilderink, Arjen Y. Hoekstra, Marco A. Janssen, Louis W. Niessen, Bart J. Strengers and Marjolein B.A. van Asselt

19.1 Introduction

We know that the future is inherently uncertain, yet we are fascinated by insights into ways in which we may be influencing the planet. This interest is intensified because there is widespread perception that the world is changing at an unprecedented speed. Undeniably, many parts of the global system are accelerating or decelerating compared to previously observed, natural rates of change. For some people these processes of change may just look like more of the same. There are, however, underlying behavioural and structural changes at work which suggest deeper, more radical change in the longer term. Many of those long-term changes can be viewed as part of transition processes. Several of these are within the human system: from many to 1 or 2 children per family, twice as many older people per thousand compared to today, a factor of 3 to 5 less energy and water use per unit of economic activity, increasing pressure to cultivate more land and use it more intensively to feed the population. More gradual, but possibly of overriding importance, are the changes in the environmental system, such as the accelerating increase in the concentration of some atmospheric gases and increasing accumulation of pollutants in soils and water bodies which are the result of past and present practices. It is difficult to disentangle the human-induced, structural long-term changes from the natural changes, which makes it even harder to see where the world is heading.

For those involved, the changes are experienced and evaluated quite differently. In Chapter 10 the metaphor of a walk in a landscape is used to illustrate this situation. One may also use the metaphor of a huge ship which is moving at ever greater speed in an unknown direction. Some will find it an exciting experience, which offers all kinds of challenges and opportunities. Others get increasingly scared on this runaway trip. They try to slow things down, arguing that it will become more and more difficult to change course, should this prove necessary later on. Some focus on their responsibilities and use all available knowledge and controls to keep existing systems intact. Obviously there is no single, valid way in which one can interpret what's going on now and does formulate common goals for the future.

In this book we have presented a framework for analysing issues of global change and sustainable development. Using a systems orientation, the Pressure-State-

Impact-Response concept has been introduced to bring some coherence into this analysis. The notion of a family of interwoven transitions is used to put the many developments within the human and environmental system in a common context. The TARGETS1.0 model has been constructed with these concepts in mind. We have also attempted to deal with uncertainties in a novel and explicit way, recognising that many of the uncertainties in such a complex system cannot be resolved with the standard natural scientific method. These uncertainties have been framed as controversies and have been used to implement perspective-based model routes. This has led to the construction of utopian and dystopian possible futures, based on model experiments. In this chapter, we present a synthesis of the results of these experiments and some conclusions.

19.2 Synthesis of the results

Social, economic and ecological capital
Economic activities are, along with population, the major driving force in the global human-environment system. It is driven by re-investment of a proportion of industrial output into agricultural and manufacturing production facilities. This process of economic growth feeds on a stream of productivity-enhancing innovations and the gradual unfolding of an infrastructure of roads, schools, hospitals, etc. Together, these comprise the stocks of economic capital. Adequate functioning of this 'economic capital' has to be complemented by other, less tangible forms of investments so as to maintain and enhance what is called 'social capital'. This refers to characteristics of the human population, such as social coherence, institutional arrangements and capacities and skills. The economic system can only be sustained by a continuous influx of energy and materials which are withdrawn from and again dissipated into the natural environment. This natural resource base constitutes 'ecological capital', also called 'environmental' or 'natural' capital. History has shown time and again that a proper balance between the use of these forms of capital is an important condition for human aspirations towards a fulfilling and prosperous life.

Recently, it has been proposed to frame indicators of sustainable development around these forms of capital stocks (Serageldin, 1996)[1]. As explained in Chapter 9, such indicators can help to communicate major trends and insights. The TARGETS framework permits a quantitative indication of the three capital stocks: social, economic and ecological (Table 9.1). For the present synthesis of the results, we make some basic definitions. Economic capital is the sum of industrial and service capital and the capital stocks for the supply of food, water and energy. The latter is

---

[1] Serageldin (1996) distinguishes four forms of capital: man-made, natural, human and social. This is comparable to the categories used here, although there are slight differences in emphasis.

Definition of capital indicators

The economic capital indicators are only rough measures in view of our highly simplified description of the economic system. Social capital, defined here as the number of literate people in the age group of 15 to 65 years old, can be associated with the potential labour force. The ecological capital indicators are calculated for arable land, clean water and high-quality oil and gas resources and are referred to as Arable Land resource Index (ALI), Clean Water resource Index (CWI) and High-quality fossil Energy resource Index (HEI). Each of these is defined as the ratio of the unused resource base in terms of quality and size in year t and its value in year 1990. For arable land, we use the product of potential arable land and average soil quality (Q factor, Chapter 7) as a measure of the resource base. Hence, a decline in the ALI indicates that less of the potentially arable land is left unused which we feel is also a crude substitute for more complex quantities like forests and biodiversity. For water, we use the remaining fraction of the potentially usable, clean water flow (two highest classes, A and B; Chapter 6) as the indicator for the resource base. Hence, a decline in the CWI indicates that clean water sources are increasingly being tapped for human purposes. With regard to energy, we relate the resource base to the amount of oil and gas deposits which can be expoited at capital-output ratios less than 20 times the value in 1900 (depletion multipliers, Chapter 5). Hence, if the HEI falls, this implies that fewer cheap and accessible fossil fuel deposits remain. The composite indicator for Ecological Capital is a weighed average of the ALI, the CWI and the HEI.

---

dealt with explicitly and gives an indication of the relative importance of the food, water and energy-supply sectors discussed in Chapters 11, 17 and 18. As a first step, we associate social capital with the number of literate people in the age group from 15 to 65. Ecological capital is defined in terms of the size and quality of the remaining land, water and energy resource base. We confine ourselves to the 'source' side of the environment system; elsewhere indicators of the 'sink' side are presented (see, for example, the CYCLES state index in Figure 17. 11). One should also realise that changes on the sink side of the environment system such as a temperature rise do affect the land, water and energy resource indicators in our integrated model experiments.

Absolute amounts are not what matters here – it is the trends that are relevant. *Figure 19.1* shows the trajectories for the capital indicators for all three utopias for the years 1900, 1990, 2020, 2050 and 2100 normalised to the year 1990. Economic capital in the hierarchist and individualist utopia grows exponentially as a result of the assumption of exogenous GWP growth. The combination of a growing and structurally changing population and increasing literacy rates leads to an even faster growth in social capital. The capital stock needed to sustain the food, water and energy sectors grows much more slowly in the individualist utopia than in the hierarchist one. This reflects the individualist reliance on productivity-enhancing technology and a large, low-cost resource base. In both utopias, the ratio of economic to human capital keeps on growing during the second half of next century. This is most striking in the individualist utopia: it pictures a world which is inhabited by over 13×10^9 people, many of which are – and have to be – literate, skilled people who manage an increasingly man-made world. In the process, ecological capital falls roughly in a linear fashion over time. Here, too, the decline in the individualist utopia is slower because the resource base is assumed to be abundant.

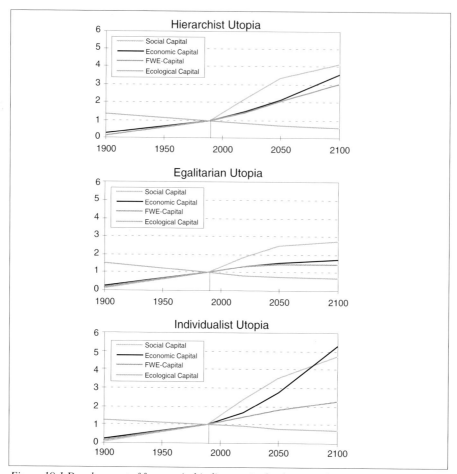

Figure 19.1 Development of four capital indicators in the three utopias.

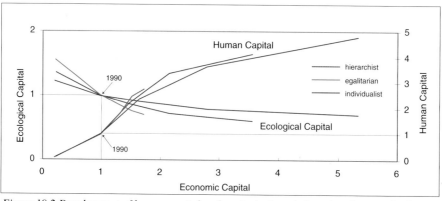

Figure 19.2 Development of human capital and ecological capital against the growth of economic capital in the three utopias.

The egalitarian utopia is rather different. Social capital increases the fastest, while growth of economic capital levels off. The resulting ratio of economic to social capital is only half the value of the individualist utopia by the year 2100, in a world with about half as many people and only one third as much machinery, buildings, roads, etc. Because the egalitarian world view has a conservative view of the possibilities of maintaining the resource base, ecological capital in this utopia also keeps declining, despite the much lower pressure. Indeed, it is at a lower level than in the individualist utopia.

A different way of displaying the same information is shown in *Figure 19.2*. The development of human capital and ecological capital against the growth of economic capital clearly indicates, at this high level of aggregation, the substitutability between ecological and economic capital on the one hand and the complementarity between human and economic capital on the other. The acceptability of such a trade-off is behind the difference in 'strong' and 'weak' definitions of sustainability. Our assessment suggests that substitutability and complementarity are strongest in a world which functions as the egalitarians suspect and which is managed according to their prudent style. As a result of emphasis on educating women, access to safe water and the like, modest economic growth goes with a significant increase in social capital. At the same, this modest increase in economic capital is still responsible for a noticeable decline in the stock of ecological capital, despite a reduction in the resource intensity of the economy. The individualist utopia presents the opposite picture. A large expansion of economic capital goes with a rise in human capital, albeit slightly less than in the less populated hierarchist utopia. This expansion creates major pressures on the environment, but less than in the hierarchist utopia which lacks the individualist high-tech orientation. Because of its underlying optimism about the natural system's resilience, it is still a world with plenty of good land, clean water, cheap oil and gas – and an abundance of the greatest resource of all: skilled humans.

In another attempt to synthesise the mass of results, we make use of the multi-dimensional star or 'amoeba' representation (*Figure 9.3*). The first step is to select sets of indicators for each of the four parts of the Pressure-State-Impact-Response (PSIR) chain. We have sought a compromise between the desired characteristics of indicators (Chapter 9) and what the model can provide. A total of 21 indicators have been chosen. The first six are considered representative of the pressure on the system, and are extensive: the larger the more. The second set of five indicators are a measure of the state of the human and the environmental system. These are closely related to the capital stock indicators presented in the previous paragraph. The third set represents a selection of indicators that are associated with impacts identifying aspects related to the quality of human life. They are intensive and their value ranges from some biological or technical lower limit to some upper limit, which in some cases is also related to biological or technical considerations. The last set is a – quite limited – selection of variables which can be associated with response actions in the form of prices, taxes and technology. Because the outer circle is normalised to the maximum value in the three perspectives, this representation is primarily a comparison. The globe in the middle is the unit circle representing the situation in 1990. The formulation of the indicators is such that outward expansion corresponds with 'more': the more pressure, the more depletion of resources, disturbance of the climate system, affluence, and the more drastic responses.

The multi-dimensional star, or amoeba, representation

The starting point are five concentric circles which are associated with the levels 1 (inner circle) to 5 (outer circle). The indicators used are defined as follows. We first calculate the ratio of the value at time t to the value in 1990. Next, we scale the resulting normalised indicator in such a way that 1990 is on the unit 1 circle and the maximum value of one of the three utopias is on the outer, unit 5 circle. Hence, the position along the five concentric circles represents the position of the indicator in between its value in 1990 and its value in 2020, 2050 or 2100. For instance, life expectancy is presented as the number of years ranging from 66 in 1990 (unit 1 circle) to 88 years, which is the maximum (unit 5 circle). The list of indicators and range of values is given in *Table 19.1*.

We have arranged the indicators according to the PSIR chain discussed in Chapter 2. Pressure refers, in general, to those quantities which are the driving forces behind the exploitation of natural resources. The state variables reflect changes in the size and quality of natural resources, usually related to source depletion and sink accumulation. The impact variables are important quality aspects of the reservoirs and fluxes in the human and environmental system. There are lower and upper limits for these quantities, beyond which they are considered to be inhuman or impossible. Finally, the response indicators represent a few of the endogenous response mechanisms within the TARGETS1.0 model. Many of these indicators are discussed, in varying degrees of detail, in Chapters 12 to 16.

Pressure	**State** [b]	**Impact**	**Response**
GWP	1/ALI (1-5)	GWP/cap (4030-41300 $/cap)	water price (4-220 ¢/ton)
Population POP (5.25-13.3×10^9)	1/CWI (1-5)	Life Expectancy LE (66-88 yr)	energy price (5-16 $/GJ)
Food Demand FD (X14.2-56.5 Gton CCE)	1/HEI (1-5)	Food/cap [c] (2720-3940 kcal/cap/day)	Health Services/GWP HS/GWP [d] (8.6-12.9%)
Water Demand WD (4.1-14.9 Tton)	CO_2 concentration (352-727 ppmv)	Water consumption /cap [c] (0.8-1.28 kton/cap)	GWP/emissions [e] (1-2.75)
Energy Demand ED [a] (177-843 EJ)	temperature Change (0.3-3.6 °C)	Energy /cap [c] (34-71 TJ/cap)	
CO_2, SO_2 and NO_x emissions [e] (1-4.4)	sea-level rise *slr* (0-120 cm)		

[a] use of commercial, secondary fuels
[b] the ALI, CWI and HEI are the environmental capital indicators discussed previously
[c] primary energy supply, water use (withdrawal plus re-use) and food intake per capita, respectively
[d] health service investments
[e] the sum of $0.5 \times$ indexed CO_2 and $0.5 \times$ indexed SO_2 and NO_x emissions

Table 19.1 Indicators used for amoebas

Let us first look at the utopias, indicated for the years 2020, 2050 and 2100 in *Figure 19.3a-c*. In 2020 the hierarchist and the individualist utopia differ less than 15% with respect to all but 6 indicators. These 6 have to do with a more optimistic assessment of water resources and climate sensitivity and technical progress in food and energy provision. The egalitarian utopia differs more than 15% from the hierarchist one for 13 out of 21 indicators. The difference is mainly in the form of lower pressure and lower affluence levels and yet less remaining resources and a larger climate impact. The low estimate of the potential arable land and the negative impact from climate change shows up in the state part; the high carbon tax in the response part. By the middle of the next century all these differences have become much larger, with the exception of emissions where technology in the individualist utopia does for emission reduction what the carbon tax and technology do in the egalitarian utopia.

By the end of the next century, the differences between the three utopias have become quite marked. Whereas the individualist utopia has the highest pressure and the highest impact values for the selected indicators, the climate impacts are least and water and energy prices are lowest. This is a world of abundance and resilience: , land, water and energy resources are huge and the environment system is rather insensitive to human disturbances, although the cheap oil and gas deposits are largely depleted. The egalitarian utopia shows the opposite picture: at much lower pressure and impact levels, and despite rather radical response measures, some aspects of the environment – notably land and climate – are in worse shape. The hierarchist world is somewhere in between these two extremes, with the exception of the energy-related indicators.

The three pictures give a useful visual impression of possible future shapes of the planet Earth. However, these utopias are the rosy part of the picture. There are numerous less utopian futures, a few of which have been explored with our model (see Chapter 18). To give an impression of such dystopias for the year 2100, *Figure 19.4* pictures the amoebas for the six dystopias which are generated if management style and world view clash[2]. If the medium growth 'Conventional Development' or 'Business-as-Usual' trends continue, but as it turns out that the world functions according to the egalitarian world view, humankind's ecological capital is in a disastrous state (*Figure 19.4, upper, green curve*). There are serious, negative feedbacks on human well-being; the lower population leads to a lower pressure, but technical progress falls far short of slowing down environmental degradation. In the high-growth world of the individualist, the confrontation with a small resource base and a vulnerable climate system would lead to a similar, catastrophic situation (*Figure 19.4, lower, green curve*). Already in 2050 the world would be confronted with significant climate change and, partly as a consequence, with serious food and water shortages. The major transitions are slowed down or reversed: life expectancy

---

2 The GWP trajectory always corresponds with the management style.

19 GLOBAL CHANGE - FRESH INSIGHTS, NO SIMPLE ANSWERS

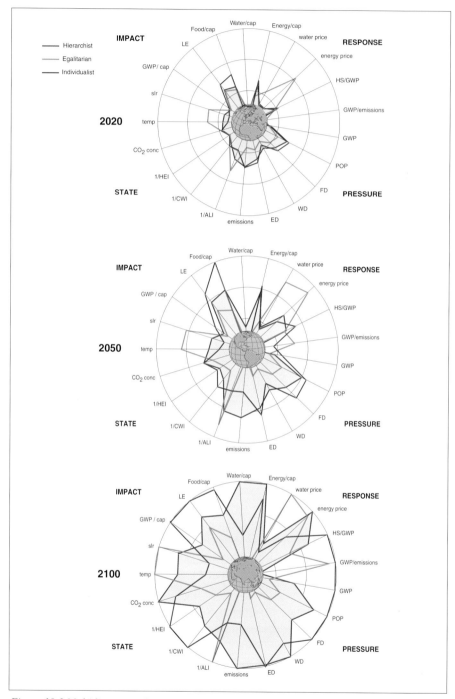

Figure 19.3 Multidimensional star, or amoeba, representation of the state of the world in 2020, 2050 and 2100 for the three utopias. The shaded blue area represents the state of the world according to the hierarchist utopia. The globe in the middle represents the situation in 1990.

19 GLOBAL CHANGE - FRESH INSIGHTS, NO SIMPLE ANSWERS

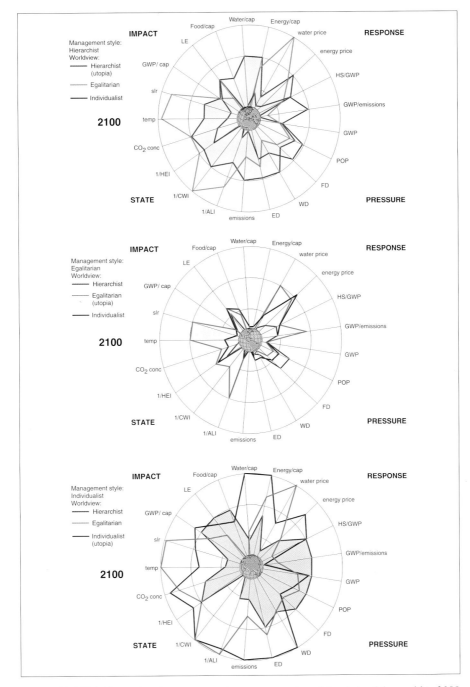

Figure 19.4 Multidimensional star, or amoeba, representation of the state of the world in 2100 for the dystopias which evolve when a certain management style is confronted with a world view which does not match. For reference, the utopias are indicated in the form of the shaded coloured areas. The outer circle is level 7.5, not 5 as in Figure 19.3.

increases much less, the resource base is squandered and welfare and health are even more unevenly distributed. These outcomes reflect the experiments described in Chapter 18 where it is also found that the most catastrophical future is one in which an egalitarian world view is combined with medium to high growth in population and economy and a hierarchist or individualist management style. Or, as egalitarians would put it: a future in which a vulnerable world is ruled by myopic materialists. The main conclusion here is that governments should heed the warnings of environmentalists if they expect – and, as is often the case, promote – an individualist style of government. Indeed, it is because of this risk that environmentalists advocate strong, not weak government.

The mirror image of the above disquieting dystopia is also interesting: the combination of an egalitarian management style with an individualist world view (*Figure 19.4, middle, red curve*). Or, as individualists would say, a low growth world run by anxious and frugal prophets of doom whose risk averseness has spoilt huge opportunities. Climate change is hardly noticeable, but there are more people and they are less affluent and have a lower life expectancy than in the egalitarian utopia. Because of the failure of technology, the resource base is in worse shape. If we call the previous dystopia an egalitarian nightmare, then this one might be called the individualist regret.

19.3 World in transition

In the previous chapters, we have explored population, health, energy, food, water, and environmental change issues on the basis of current controversies. The issues behind these controversies can be summarised in a single question: *Can we provide a future world population with enough food, clean water and energy to guarantee a healthy life, while safeguarding our natural resource basis?* The fundamental uncertainties in the functioning of the Earth system and human behaviour do not permit an unambiguous answer to this question. Outspoken and often controversial views on the above question are published regularly, fuelling the debate. Many assessments have either an apocalyptic or a sanguine character. They adhere to the (neo-)Malthusian view that mass starvation is unavoidable as the human population already exceeds the carrying capacity of the Earth (Pimentel *et al.*, 1997). Or they anticipate an overshoot and collapse future which is triggered by a continuous degradation of natural capital (Meadows *et al.*, 1972, 1991; Barney, 1980, 1993; Brown and Kane, 1995). In striking contrast to these warnings of doom are a growing number of global studies which propagate an optimistic future for humankind. Well-known is Simon's 'The Ultimate Resource' (1981). There are some more recent examples: 'The True State of the Planet' (Bailey, 1995), which is a positive response to the 'State of the World' report of the Worldwatch Institute, and 'A moment on Earth' (Easterbrook, 1995). A more balanced picture emerges from the recent 'Global Environmental Outlook' (UNEP, 1997). Many analyses of the longer term future do not even mention the natural system, at least not explicitly; the focus is on current liberalisation, globalisation and technological trends (Petrella et al., 1994; Shell Planning, 1996). Obviously, there are as many ways of viewing the need to interfere with current trends and formulate policies for a more sustainable future as there are in these views themselves. The picture is further obscured by the fact that many transition processes are occurring simultaneously and that some of these are scarcely discernible now but may become dominant factors of change as a consequence of emerging technologies, insights and attitudes.

In our global assessment we identify three interrelated transitions (Figure 17.6-9). The first is the health transition, which comprises the demographic and epidemiological transitions. The second is the economic transition which represents the shift from a largely agricultural to an industrial economy, followed by a shift to a service and information-oriented economy. Thirdly, there is the environmental or ecological transition during which material and energy intensities, after an initial upturn, start to fall. The associated emission intensities follow a similar pattern. These major transitions are often associated with minor ones which are sometimes a precondition and sometimes a consequence and have also a cultural component. For food, for instance, there is the trend towards a higher proportion of meat in the average diet and, in agriculture, towards more intensive farming. For water and energy, the trend of increasing intensity-of-use with the onset of industrialisation is reversed once service and information-oriented activities start to dominate. This reversal is supported by the wider use of more efficient equipment.

The combined dynamics of population and economic growth and these transitions determines the demand for food, water and energy (Figures 11.3 and 17.1-2). The resulting forward projections of food, water and energy supply provides an indication of the use of important natural capital stocks and flows: arable land, water reservoirs and fluxes, and energy resources and fluxes (Figure 19.1-2). This is mediated by all kinds of pressures on the environment such as the use of fertiliser in agriculture, which results in water pollution, and the emissions of various substances into the air. Will the Earth be able to sustainably meet humanity's demand for these resources? And if not, why not, and what should we do about it?

Focusing on the physical level (Figure 2.6), current trends suggest that there is reason for concern. The supply of food, both for humans and animals, will cause a further increase in the area of land use for cultivation and of consequent erosion. Use of inputs per hectare may steadily rise too, leading to higher yields. Effluents to land and water, especially nitrogen compounds, will increase. Great demands on water resources lead to an increase in the average costs of water supply and to a decline in the available groundwater resources. The supply of energy in various forms will cause a decline in the fossil fuel resource base. Supply costs are expected to increase as the high costs of new deposits are no longer compensated for by innovations.

Fortunately, new sources and technologies slowly penetrate the market which will, in combination with abatement measures, tend to slow down or reverse the trends in emissions of carbon, sulphur and nitrogen compounds. Some changes in environmental capital directly affect the functioning of human society; others have a more indirect impact in the sense that they necessitate changes in activity and investment patterns.

There is also a broader aspect which involves behavioural and cultural dynamics; the medium level in Figure 2.6. Throughout these transitions and as part of them, there will be human response behaviour. Declining productivity of land is counteracted by additional fertiliser use; rising marginal costs to supply water require additional investments; depletion of fossil fuel resources induces energy conservation and the exploitation of renewable energy sources. Much of what is called 'current trends' is actually rooted in human behaviour at the local scale and determined by factors such as traditions, habits, markets and prices, governance and trade structures. The three cultural perspectives we distinguish in this book express part of this behavioural richness.

The highest level along the vertical axis of Figure 2.6 refers to values, beliefs and ideas. At this level, belief systems about Nature and Man are constructed to be able to act coherently and meaningfully – such as the three world views formulated in this book. One of the basic differences in world views is whether human history is seen as a series of 'rise-and-fall' cycles or as steady progression (Kahn and Wiener, 1967). The dominant, western view has been that the future will be better than the past. This vision is increasingly challenged, not in the least because of the growing awareness that humankind faces unprecedented global threats to its well-being or even to its survival. It is in this context that the notion of sustainable development has emerged as a vision which may help development along desirable transition pathways.

Of particular importance is the observation that the less developed regions in the world aspire to follow a path similar to the one that the developed regions have followed over the last century. If this process unfolds without an acceleration of the health and ecological transitions described above, the world will most probably be confronted with increasing pressure on both local and global resources. This will lead to increasingly intense political conflicts about food, water and fuels. A leap towards advanced technologies is required, because humankind can no longer afford the luxury of 19th century inefficiencies. But technical fixes are not sufficient; they have to be complemented by sustained, long-term population, income and price policies. One of the most precious – and scarce – resources for this undertaking is that most intangible of the capital stocks: 'social capital'.

With regard to the controversy formulated above, there can be hardly any doubt that continuation of current trends will increasingly confront societies with physical limits to growth. The interesting question is whether we anticipate the approach of such limits and whether the responses, both precautionary and adaptive, will be timely enough to avoid major overshoot and collapse situations. A key factor is the rules which govern human behaviour – there may be social and psychological limits to the rate and extent to which these can be changed. This, in turn, is at least partly a matter of perception and vision. If leading groups in society develop a coherent and convincing image of a sustainable and fair future, the probability that it will be realised will increase significantly. In this respect, much still remains to be done. In the wordings of the recent Global Environmental Outlook: 'progress towards a global sustainable future is just too slow. A sense of urgency is lacking' (UNEP, 1997 p.3).

19.4 Epilogue

At the end of a book like this, one faces the question: what has been our contribution? We have presented a number of concepts and constructed a simulation model. Such a model is a tool, like a telescope or a microscope: you hope to see new things or to see things more clearly. Which new insights can be gained from our research? The past 25 years of global forecasting have shown that dogmatic predictions about the Earth's future are misleading and unreliable, even politically counterproductive. With hindsight, world developments have turned out to be more complicated and more surprising than anticipated. Many problems identified in earlier doom scenarios persist but they have not overwhelmed the planet. Some threats, such as fossil fuel depletion, have receded; others, such as industrial pollution, appear susceptible to determined policy intervention. Unfortunately, new and unexpected threats have emerged: depletion of stratospheric ozone, resurgence of infectious diseases, anticipated global climate change, increasing scarcity of fresh water, land degradation.

Our model-based analyses add at least three elements to the more speculative statements about the future of the globe. First, our quantifications of future trends are based on numerical consistency and on a variety of 'stylised facts' and insights which are at the core of the sciences. Although this offers no guarantee that the future does not hold surprises which may make such consistency and insights irrelevant, it nevertheless complements less quantitative, fiction-like analyses of the future. Moreover, several of the submodels have quite novel elements too. Secondly, we look at these future trends from a more integrated perspective than is usually the case. In particular, we investigate some of the most important feedbacks between the human and the environment system. This increases the plausibility of the resulting images of the future because it provides a consistency which is absent from many more narrowly-based analyses. Thirdly, the use of perspective-based model routes contains a clear invitation to others to participate in the search for more sustainable futures. We do not hide behind scientific expertise when such a position cannot be defended, given the range of views expressed in current controversies. This explicit inclusion of values and beliefs gives our endeavour an open, process-oriented flavour which we feel is essential in the context of an enquiry into nothing less than the future of the planet. In this sense, the TARGETS project extends an invitation to people with a broad spectrum of views to participate in the debate about that future.

Our experiments indicate that certain combinations of assumptions can lead to developments in the world system which are deemed acceptable or even desirable from the point of view of the value system which is supposed to be reflected in those assumptions. These are called utopias. Our quantitative framework suggests that such utopias are feasible in the sense that they do not violate the prevailing insights about the dynamics of the various subsystems and their interlinkages. The experiments also indicate that such utopias are particularly dependent on certain assumptions. If these turn out to be incorrect, dystopian trajectories evolve unless adaptive and, within the corresponding world view, sometimes undesirable policies are implemented. Another insight is that the human-environment system is characterised by changes that take place according to radically different time-scales. Despite the appearance of rapid change, many of the forces behind the transition processes within the human and environment system are, in fact, quite slow. Apart from geological and evolutionary processes, some changes in the environment system which are caused by human activities, occur on a time-scale of five to ten generations, as our model experiments up to the year 2200 clearly show (Figure 17.12). This appears very long from an individual's point-of-view: people's behaviour is generally influenced by much shorter-term considerations. This discrepancy is the cause of a great deal of inertia in the system with regard to policy actions designed to influence global trends.

Global catastrophe does not appear to be imminent. However, the projections presented in this book indicate the risks and uncertainties associated with perpetuating current trends, as presented in many official reports and plans. It seems unlikely that, on current trends, the route to sustainable development as pictured in Agenda 21 will be chosen. Yet, as our explorations confirm, there are ample opportunities to change track without denying the majority of the world population their legitimate aspirations for a better life. Many policy interventions have been identified, which have the potential to accelerate the transitions towards a more healthy life for all and towards a much more resource-efficient economy. They can, and should, focus on reversing negative long-term trends. This requires honest appraisal of the current situation as well as vision and courage, the more so because many of these interventions and measures will require one or even more generations before there are visible effects. Our analyses offer some guidance here: the perspectives of the hierarchist, the egalitarian and the individualist all contain part of the problems and part of the solutions. The best of all worlds might be one in which the stability and responsibility of institutions, the vigour and ingenuity of the entrepreneur, and the prudence and respect of critical citizens work in unison in the search for a more sustainable development path for humankind.

We consider this book as one more step in the integrative research cycle described in Chapter 1. We sincerely hope that it inspires other researchers, especially integrated assessment modellers, to participate in this exciting and necessary venture. We also hope that it helps policy makers and interested lay-people to see the larger picture and to formulate the vision and the strategies which are needed if humankind is not to harm the interests of our children out of pure habit, greed or ignorance.

REFERENCES

Achterhuis, H. (1995) *Nature between myth and technology* Ambo, Baarn, the Netherlands.
Ackoff, R.L. (1971) 'Towards a system of systems concepts' *Management Science* **July 1971**, no. 17.
Addiscot, T.M., A.P. Whitemore and D.S. Powlson (1991) *Farming, fertilizers and the nitrate problem* CAB International, Oxon, UK.
Adriaanse, A. (1993) *Environmental policy performance indicators: a study on the development of indicators for environmental policy in the Netherlands* SDU Publishers, The Hague, the Netherlands.
AES (1990) 'An overview of the Fossil2 Model: a dynamic long-term policy simulation model of US energy supply and demand', AES (Associated Energy Services) Corporation, Washington.
Alcamo, J. (ed.) (1994) *IMAGE 2.0: integrated modeling of global climate change* Kluwer Academic Publishers, Dordrecht, the Netherlands.
Alcamo, J., A. Bouwman, J. Edmonds, A. Grübler, T. Morita, A. Sugandhy (1995) 'An Evaluation of the IPCC IS92 Emission Scenarios (ch. 6)' In: *Climate Change 1994* J. T. Houghton *et al.* (ed.), 338, Cambridge University Press, Cambridge.
Alcamo, J., E. Kreileman, M. Berk, J. Bollen, M. Krol, R. Leemans and S. Toet (1996) 'The global climate system: near term action for long term protection' RIVM Report, 481508001, National Institute for Public Health and the Environment, Bilthoven, the Netherlands.
Allen, J.C. and F.B. Barnes (1985) 'The causes of tropical deforestation in developing countries' *Annals of the Association of American Geographers* **75**, no. 2, 163-184.
Alter, G. and J.C. Riley (1993) 'Frailty, sickness and death: models of morbidity and mortality in historical populations' *Population Studies*, 25-45.
Ambroggi, R.P. (1977) 'Underground reservoirs to control the water cycle' *Scientific American* **236**, no. 5, 21-27.
Ambroggi, R.P. (1980) 'Water' *Scientific American* **243**, no. 3, 90-105.
Anderson, K. (1995a) 'The political economy of coal subsidies in Europe' *Energy Policy* **23**, no. 6, 485-496.
Anderson, T.L. (1995b) 'Water options for the blue planet' In: *The true state of the planet* R. Bailey (ed.), The Free Press, New York, USA.
Arnold, D. (1993) 'Diseases of the modern period in South Asia' In: *The Cambridge world history of human disease* Kiple (ed.), Cambridge University Press.
Arnoldus, H.M.J. (1977) 'Methodology used to determine the maximum potential average annual soil loss due to sheet and rill erosion in Morocco' In: *Assessing soil degradation* FAO (ed.), Food and Agriculture Organization (FAO), Rome, Italy.
Arnoldus, H.M.J. (1978) 'An approximation of the rainfall factor in the universal soil loss equation' In: *Assessment of erosion* M. de Boodt and D. Gabriels (eds.), John Wiley & Son, New York, USA.
Arrhenius, S.A. (1896) 'On the influence of carbonic acid in the air upon the temperature of the ground' *Philosophical Magazine and Journal of Science*, **41**, 237-276.
Aselmann, I. and P.J. Crutzen (1989) 'Global distribution of natural freshwater wetlands and rice paddies, their net primary production, seasonality and possible methane emissions' *Journal of Atmospheric Chemistry* **8**, 307-358.
Bacon (1628) *The new Atlantis*.
Baes, C.F., A. Börkström and P.J. Mulholland (1985) 'Uptake of carbon dioxide by the oceans' In: *Atmospheric carbon dioxide and the global carbon cycle* J.R. Trabalka (ed.), US Department of Energy, Washington D.C., USA.
Bailey, R. (ed.) (1995) *The true state of the planet* The Free Press, New York, USA.
Bakkes, J.A., G.J. van den Born, J.C. Helder, R.J. Swart, C.W. Hope and J.D.E. Parker (1994) 'An overview of environmental indicators: state of the art and perspectives' Environmental Assessment Technical Report, Bilthoven.
Baldwin, R. (1995) 'Does sustainability require growth?' In: *The economics of sustainable development* I. Goldin and L. Winters (eds.), 312, Cambridge University Press, Cambridge.
Balling, R.C. (1995) 'Global warming: messy models, decent data, and pointless policy' In: *The true state of the planet* R. Bailey (ed.), The Free Press, New York, USA.

Barendregt, J.J. and L. Bonneux (1992) 'Desease-free survival: results of a dynamic model' *Tijdschrift Sociale Geneeskunde* **A2**, 28-33.

Barney, G.O. (ed.) (1980) *The global 2000 report to the president: entering the twenty-first century* US Council on Environmental Quality and the Department of State, Washington D.C., USA.

Barney, G.O., J. Blewett and K.R. Barney (1993) 'Global 2000: what shall we do? The critical issues of the 21th century', Millennium Institute, Arlington, Virginia, USA.

Baughman, M.L. (1972) 'Dynamic energy system modeling: interfuel competition' MIT School of Engineering, Energy Analysis and Planning Group.

Bazzaz, F.A. and E.D. Fajer (1992) 'Plant life in a CO_2-rich world' *Scientific American* **1992**, 18-21.

Becker, G.S. (1991) *A treatise on the family* Harvard University Press, Cambridge.

Benders, R.J.M. (1996) 'Interactive simulation of electricity demand and production' RUG, Groningen, the Netherlands.

Bergsma, E. (1981) 'Indices of rain erosivity: a review' *ITC Journal*, no. 4, 460-484.

Billharz, S. and B. Molda (1995) 'Scientific workshop on indicators of sustainable development' Report of SCOPE meeting, Wuppertal, Germany.

Bilsborrow, R.E. and H.W.O. Okoth Ogendo (1992) 'Population-driven changes in land use in developing countries' *Ambio* **21**, no. 1, 37-45.

Blok, K., E. Worrell, R. Culenaere and W. Turkenburg (1993) 'The cost effectiveness of CO_2 emission reduction achieved by energy conservation' *Energy Policy* **21**, no. 6, 656-667.Bobadilla, J.L., J. Frenk, R. Lozano, Frejka and C. Stern (1993) 'The epidemiological transition and health priorities' In: *Disease control priorities in developing countries* D.T. Jamison, W.H. Mosley, A.R. Measham and J.L. Bobadilla (eds.), Oxford Medical Publications, New York.

Bolin, B. and R.B. Cook (eds.) (1983) *The major biogeochemical cycles and their interactions* John Wiley & Sons, New York.

Bolin, B., E.T. Degens, S. Kempe and P. Ketner (eds.) (1979) *The global carbon cycle* John Wiley & Sons, New York, USA.

Bollen, J.C., A.M.C. Toet and H.J.M. de Vries (1996) 'Evaluating cost-effective strategies for meeting regional CO_2 targets' *Global Environmental Change* **6**.

Bollen, J.C., A.M.C. Toet, H.J.M. de Vries and R.A. van den Wijngaart (1995) 'Modelling regional energy use for evaluating global climate scenarios' RIVM report, 481507010, National Institute of Public Health and the Environment (RIVM), Bilthoven, the Netherlands.

Bongaarts, J. and R.G. Potter (1983) *Fertility, biology and behavior: an analysis of the proximate determinants* Academic Press, New York.

Bongaarts, J.P. (1994) 'Population policy options in the developing world' *Science* **263**, 771-776.

Bongaarts, J.P. (1996) 'Global and regional population projections to 2025' In: *Population and food in th early twenty-first century: meeting future food demand of an increasing population* N. Islam (ed.), International Food Policy Institute, Washington D.C., USA.

Bongaarts, J.P., O. Frank and Lesthaeghe (1984) 'The proximate determinants of fertility in Sub-Saharan Africa' *Population Development Review* **10**, no. 3, 511-537.

Bonneux, L., J.J. Barendregt, K. Meeter, G.J. Bonsel and P.J. van der Maas (1994) 'Estimating clinical morbidity due to ischemic heart disease and congestive heart failure: the future rise of heart failure' *American Journal of Public Health* **84**, 20-25.

Bonsel, G.J. and P.J. van der Maas (1994) *Aan de wieg van de toekomst: scenarios voor de menselijke voortplanting 1995-2010* Bohn Stafleu Van Loghum, Houten, the Netherlands.

Bos, E. (1995) 'Comments on chapter 2: how reliable are popution projections?' In: *Population and food in the early twenty-first century: meeting future food demand of an increasing population* N. Islam (ed.), International Food Policy Research Institute.

Bossel, H. (1987) *Systemdynamik: Grundwisen, Methoden und BASIC-programme zur Simulation dynamischer Systeme*, Braunschweig-Wiesbaden.

Bouwman, A.F., L. van Staalduinen and R.J. Swart (1992) 'The IMAGE land use model to analyze trends in land use related emissions' RIVM Report, 222901009, National Institute of Public Health and the Environment, (RIVM), Bilthoven, the Netherlands.

Braddock, R., J. Filar, J. Rotmans and M.G.J. den Elzen (1993) 'The IMAGE greenhouse model as a mathematical system' *Applied Mathematical Modelling* **18**, 234-254.

Braudel, F. (1981) *The structures of every day life: the limits of the possible* Harper & Row, New York, USA.
Brecke, P. (1993) 'Integrated global models that run on personal computers' *SIMULATION* **60**, no. 2.
Brewer, G.D. (1986) 'Methods for synthesis: policy exercises' In: *Sustainable development of the biosphere* W.C. Clark and R.E. Munn (eds.), 455-473, Cambridge University Press, Cambridge, UK.
Broekner, W.S. (1987) 'Unpleasant surprises in the greenhouse?' *Nature* **328**, 123-126.
Brown, L.R., J. Abramovitz, C. Bright, C. Flavin, G. Gardner, H. Kane, A. Platt, S. Postel, D. Roodman, A. Sachs and L. Starke (1996) *State of the world 1996: a Worldwatch Institute report on progress toward a sustainable society* W.W. Norton & Company, London, UK.
Brown, L.R., A. Durning, C. Flavin, H. French, J. Jacobson, N. Lenssen, M. Lowe, S. Postel, M. Renner, J. Ryan, L. Starke and J. Young (1991) *State of the world: 1991* W.W. Norton & Company, New York, USA.
Brown, L.R. and H. Kane (1995) *Full House: reassessing the earth's population carrying capacity* Earthscan, London, UK.
Brown, L.R., H. Kane and E. Ayres (1993) *Vital signs 1993-1994: the trends that are shaping our future* Earthscan Publications Ltd., London, UK.
Brückner, E. (1889) 'To what extent is the present climate constant?' In: *Verhandlungen des VIII. Deutschen Geographen* , Teubner, Leipzig, Germany.
Brulacich, M., R. Stewart, V. Kiruwood and R. Muma (1994) 'Effects of global climate change on wheat yields in the Canadian prairie' In: *Implications of climate change for international agriculture: crop modeling study* C. Rosenzweig and A. Iglesias (eds.), U.S. Environmental Protection Agency, Washington D.C.
Bulatao, R.A. (1993) 'Mortality by cause, 1970-2015' In: *The epidemiological transition; Policies and planning implications for developing countries* J.N. Gribble and S.H. Preston (eds.), National Academy Press, Washington D.C.
Buringh, P., H.D.J. van Heemst and G.J. Staring (1975) 'Computation of the absolute maximum food production of the world', Report 598, Wageningen Agricultural University. Dept. of tropical soil science, Wageningen, the Netherlands.
Butcher, S.S., R.J. Charlson, G.H. Orians and G.V. Wolfe (eds.) (1992) *Global biogeochemical cycles* Academic Press, London.
Calder, K. (1996) 'Asia's empty tank' *Foreign Affairs* **75**, no. 2, 55-69.
Caldwell, J.C. (1993) 'Health transition: the cultural, social and behavioural determinants of health in the third world' *Social Science and Medicine* **36**, no. 2, 125-135.
CBS (1990) *Negentig jaar statistiek in tijdreeksen 1899-1989* SDU, The Hague, the Netherlands.
Cernea, M.M. (1994) 'The sociologist's approach to sustainable development' In: *Making development sustainable* I. Serageldin and A. Steer (eds.), The World Bank, Washington, USA.
Charlson, R.J., S.E. Schwartz, J.M. Hales, R.D. Cess, J.A. Coakley, J.E. Hansen and D.J. Hofmann (1992) 'Climate forcing by anthropogenic aerosols' *Science* **255**, 423-429.
Chenje, M. and P. Johson (eds.) (1994) *State of the environment in southern Africa* The Penrose Press, Johannesburg, South Africa.
Christmas, J. and C. de Rooy (1991) 'The decade and beyond: at a glance' *Water International* **16**, no. 3, 127-134.
Clark, W.C. and C.S. Holling (1985) 'Sustainble development of the biosphere: human activities and global change' In: *Global change* T.F. Malone and J.G. Roederer (eds.), 474-490, Cambridge University Press, Cambridge, UK.
Clark, W.C. and R.E. Munn (eds.) (1986) *Sustainable development of the biosphere* Cambridge University Press / IIASA, Cambridge / Laxenburg.
Codispoti, L.A. (1995) 'Biogeochemical cycles - is the ocean losing nitrate?' *Nature* **376**, 724-728.
Coffin, T. (1991) 'Earth, the crowded planet' *The Washington Spectator* **17**.
Cohen, J.E. (1995) *How Many People Can the Earth Support?* W.W. Norton & Company, New York, USA.
Colglazier, E.W. (1991) 'Scientific uncertainties, public policy and global warming: how sure is sure enough?' *Policy Studies Journal* **19**, no. 2, 61-72.

Comins, H.N. and R.E. McMurtrie (1993) 'Long-term response of nutrient limited forests to CO_2 enrichment; equilibrium behaviour of plant-soil models' *Ecological Applications* **3**, no. 4, 666-681.
Commission on Global Governance (1995) *Our global neighbourhood* Oxford University Press, Oxford, UK.
Cowley, P. and R.J. Wyatt (1993) 'Schizophrenia and manic-depressive illness' In: *Disease control priorities in developing countries* D.T. Jamison, W.H. Mosley, A.R. Measham and J.L. Bobadilla (eds.), Oxford Medical Publications, New York.
CPB (1992) 'Scanning the future', CPB (Central Planning Bureau).
Cramer, W.P. and A.M. Solomon (1993) 'Climate classification and future global redistribution of agricultural land' *Climate Research* **3**, 97-110.
Crutzen, P.J. (1988) 'Tropospheric ozone: an overview' In: *Tropospheric ozone-regional and global scale interactions* I.S.A. Isaksen (ed.), 227, 3-32, Reidel, Boston, USA.
Cubasch, U., K. Hasselmann, H. Höck, E. Maier-Reimer, U. Mikolajewicz, B.D. Santer and R. Sausen (1992) 'Time-dependent greenhouse warming computations with a coupled ocean-atmosphere model' *Climate Dynamics* **8**, 55-69.
Cumper, G.E. (1984) 'Determinants of health levels in developing countries' In: *Tropical Medicine Series*, Research Studies Press Ltd., New York.
Dargay, J. and D. Gately (1994) 'Oil demand in the industrialized countries' *The Energy Journal*, The changing world petroleum market (Special Issue), 39-67.
Davidsen, P. (1988) 'A dynamic petroleum life-cycle model for the United States 1870-2050' MIT Sloan School of Management, MIT Report D-3974.
de Bruin, J., P. de Vink and J. van Wijk (1996) 'M - a visual simulation tool' *Simulation in the Medical Sciences*, 181-186.
de Vries, H.J.M. (1989) 'Sustainable resource use - an inquiry into modelling and planning' RUG-IVEM, Groningen, the Netherlands.
de Vries, H.J.M., D. Dijk and R. Benders (1991) 'PowerPlan: an interactive simulation model for electric power planning', IVEM/ RUG, Groningen, the Netherlands.
de Vries, H.J.M., T. Fiddaman and R. Janssen (1993) 'Strategic planning exercise about global warming' RIVM report, 461502002, National Institute of Public Health and the Environment (RIVM), Bilthoven, the Netherlands.
de Vries, H.J.M. and M.A. Janssen (1996) 'Global energy futures: an integrated perspective with the TIME-model' RIVM report, 461502017, National Institute of Public Health and the Environment (RIVM).
de Vries, H.J.M. and R.A. van den Wijngaart (1995) 'The Targets/IMage Energy (TIME) 1.0 model' RIVM report, 461502016, National Institute of Public Health and the Environment (RIVM), Bilthoven, the Netherlands.
de Wit, C.T. (1965) 'Photosynthesis of leaf canopies', Report 663, PUDOC, Wageningen, the Netherlands.
Dean, R.B. and E. Lund (1981) *Water reuse: problems and solutions* Academic Press, London, UK.
Delft Hydraulics and RIKZ (1993) 'Sea level rise: a global vulnerability assessment (second revised edition)', Delft Hydraulics, Delft, the Netherlands.
Demeny, P. (1990) 'Population' In: *The earth as transformed by global and regional change in the biosphere over the past 300 years* B.L. Turner II, W.C. Clark, R.W. Kates, J.F. Richards, J.T. Matthews and W.B. Meyer (eds.), Cambridge University Press.
den Elzen, M.G.J. (1993) 'Global environmental change: an integrated modelling approach' Rijksuniversiteit Limburg, Maastricht, International Books, Utrecht, the Netherlands.
den Elzen, M.G.J., A.H.W. Beusen and J. Rotmans (1995) 'Modelling global biogeochemical cycles: an integrated assessment approach' RIVM report, 461502007, National Institute of Public Health and the Environment (RIVM), Bilthoven, the Netherlands.
den Elzen, M.G.J., A.H.W. Beusen and J. Rotmans (1997) 'An integrated modelling approach to global carbon and nitrogen cycles: balancing their budgets' *Accepted by Global Biogeochemical Cycles*.
Descartes, R. (1637) *On the method*.
di Castri, F. (1989) 'Global crises and the environment' Study week on a modern approach to the protection of the environment, Vatican City, 7-39.
Doll, R. (1992) 'Health and the environment in the 1990s' *American Journal of Public Health* **82**, no. 7, 933-941.

Douglas, M. (ed.) (1982) *Essays in the sociology of perception* Routledge and Kegan Paul, London, UK.

Douglas, M. and A. Wildavsky (1982) *Risk and culture: essays on the selection of technical and environmental dangers* University of California Press, Berkley, USA.

Dovers, S. (1990) 'Sustainability in context: an Australian perspective' *Environmental Management* **14**, 297-305.

Dowlatabadi, H. and M.G. Morgan (1993a) 'Integrated assessment of climate change' *Science* **259**, 1813-1814.

Dowlatabadi, H. and M.G. Morgan (1993b) 'A model framework for integrated studies of the climate problem' *Energy Policy*, 209-221.

Drake (1993) 'Towards building a theory of population-environment dynamics: a family of transitions' In: *Population-environment dynamics: ideas and observations* G.D. Ness, W.D. Drake and S.R. Brechin (eds.), The University of Michigan Press, Michigan, USA.

Duce, R.A., P.S. Liss, J.T. Merril, E.L. Atlas, P. Buat-Menard, B.B. Hicks, J.M. Miller, J.M. Prospero, R. Arimoto, T.M. Church, W. Ellis, J.N. Galloway, L. Hansen, T.D. Jickells, A.H. Knap, K.H. Reinhardt, B. Schneider, A. Soudine, J.J. Tokos, S. Tsunogai, R. Wollast and M. Zhou (1991) 'The atmospheric input of trace species to the world ocean' *Global Biogeochemical Cycles* **5**, 193-259.

Duchin, F. and G.-M. Lange (1994) *The future of the environment* Oxford University Press, Oxford, UK.

Duinker, P.N., S. Nilsson and F.L. Toth (1993) 'Testing the policy exercise in studies of Europe's forest sector: methodological reflections on a bittersweet experience', WP-93-23, International Institute for Applied Systems Analysis (IIASA), Laxenburg, Austria.

Dunkerley, J. (1995) 'Financing the energy sector in developing countries: context and overview' *Energy Policy* **23**, no. 11, 929-939.

Dürrenberger, G., U. Dahinden, J. Behringer, M.B.A. van Asselt and B. Kasemir (1996) 'A preliminary manual for IA-juries: recommendations for ULYSSES and CLEAR' ULYSSES working paper, 2, Swiss Institute for Environment and Technology (EAWAG), Zürich, Switzerland.

Easterbrook, G. (1995) *'A moment on earth': the coming age of environmental optimism* Penguin Books, USA.

Eberstadt, N. (1995) 'Population, food, and income: global trends in the twentieth century' In: *The True State of the Planet* R. Bailey (ed.), The Free Press, New York, USA.

EC (1995) 'For a European energy policy', Green Paper COM(94)659, European Commission.

Edmonds, J. and J.M. Reilly (1985) *Global energy: assessing the future* Oxford University Press, Oxford, UK.

Ehrlich, P.R. (1968) *The population bomb* Ballantine, New York, USA.

Ellerman, A. (1995) 'The world price of coal' *Energy Policy* **23**, no. 6, 499-506.

EMF (1996) 'Markets for energy efficiency', EMF Report 13, Energy Modeling Forum, Standford University, Stanford, California, USA.

Enting, I.G., T.M.L. Wigley and M. Heimann (1994) 'Future emissions and concentrations of carbon dioxide' Technical Paper, 0 643 05256 9, Mordialloc, Australia.

Esrey, S.A., R.G. Feacham and J.M. Hughes (1985) 'Interventions for the control of diarrhoeal disease among young children: improving water supplies and excreta disposal facilities' .

Esrey, S.A., J.B. Pothas, L. Roberts and C. Shiff (1991) 'Effects of improved water supply and sanitation on ascariasis, diarrhea, racunculiasis, hookworm infection, schistosomiasis and trachoma' .

Euroconsult (1989) *Agricultural compendium: for rural development in the tropics and subtropics (third revised edition)* Elsevier Science Publishers, Amsterdam, the Netherlands.

Falkenmark, M. (1989) 'The massive water scarcity now threatening Africa: why isn't it being addressed?' *Ambio* **18**, no. 2, 112-118.

Falkenmark, M., L. da Cunha and L. David (1987) 'New water management strategies needed for the 21st century' *Water International* **12**, 94-101.

Falkenmark, M. and G. Lindh (1974) 'How can we cope with the water resources situation by the year 2015?' *Ambio* **3**, no. 3-4, 114-122.

FAO (1978-1981) *Report on the agro-ecological zones project* Food and Agricultural Organization of the United Nations (FAO), Rome, Italy.

FAO (1987) 'Report on the agro-ecological zones project. Vol. 3. Methodology and results for South and Central America.', World Soil Report, Food and Agriculture Organization of the United Nations (FAO), Rome, Italy.

REFERENCES

FAO (1992) 'AGROSTAT-PC, Computerized information series 1 (1/1 User manual, 1/2 population, 1/3 Land use, 1/4 Production, 1/5 Trade, 1/6 Food balance sheets, 1/7 Forest products)', Food and Agricultural Organization of the United Nations (FAO), Rome, Italy.

FAO (1993a) 'Agriculture towards 2010', C93/24, Food and Agricultural Organization of the United Nations (FAO), Rome, Italy.

FAO (1993b) *Forest resources assessment 1990: tropical countries* Food and Agricultural Organization of the United Nations (FAO), Rome, Italy.

FAO (1994) *Readings in sustainable forest management, FAO forestry papers.*

Feacham, R.G., D.T. Jamison and E.R. Bos (1991) 'Changing patterns of disease and mortality in Sub-Saharan Africa' In: *Disease and mortality in Sub-Saharan Africa* R.G. Feacham and D.T. Jamison (eds.), Oxford University Press, Oxford, UK.

Feinstein, A.R. (1994) 'Clinical judgement revisited: the distraction of quantitative models' *Annual International Medicine* **120**, 799-805.

Fischer, A. (1988) 'One model to fit all' *MOSAIC* **19**, no. 34, 53 - 59.

Fischer, G., K. Froberg, M.A. Keyzer and K.S. Pahrik (1988) *Linked national models: a tool for international food policy analysis* Kluwer, Dordrecht, the Netherlands.

Flavin, C. (1991) 'The heat is on: the greenhouse effect' In: *Worldwatch Reader* L. Brown (ed.), Norton, New York, USA.

Forrester, J.W. (1961) *Industrial dynamics* MIT Press, Cambridge, USA.

Forrester, J.W. (1968) *Principles of systems* Wright-Allen Press Inc., Cambridge, USA.

Freeman, C. (1973) 'Malthus with a computer' *Futures* **5**, no. 1, 5-13.

Frenk, J., J.L. Bobadilla, C. Stern, T. Frejka and R. Lozano (1993) 'Elements for a theory of the health transition' In: *Health and social change in international perspective* L.C. Chen, A. Kleinman and N.C. Ware (eds.), Harvard University Press, Cambridge, Massachusetts, USA.

Frey, H.C. (1992) 'Quantitative analysis of uncertainty and variability in environmental policy making', Directorate for science and policy programs, American association for the advancement of science, Washington, USA.

Friedman, H. (1985) 'The science of global change: an overview' In: *Global change* T.F. Malone and J.G. Roederer (eds.), Cambridge University Press, Cambridge, UK.

Funtowicz, S.O. and J. Ravetz (1989) 'Managing uncertainty in policy-related research' Les experts sont formels: controverse scientifques et decisions politiques dans le domaine de l'environment, Arc et Sanas, France.

Funtowicz, S.O. and J.R. Ravetz (1993) 'Science for the post-normal age' *Futures* **25**, no. 7, 739-755.

Gallopin, G.C. (1996) 'Environmental and sustainability indicators and the concept of situational indicators as a cost-effective approach' *Environmental Modeling and Assessment* **2**.

Galloway, J.N., W.H. Schlesinger, H. Levy, A. Michaels and J.L. Schnoor (1995) 'Nitrogen fixation: anthropogenic enhancement-environmental response' *Global Biogeochemical Cycles* **9**, no. 2, 235-252.

Ghana Health Assessment Project Team (1981) 'A quantitative method of assessing the health impact of different diseases in less developed countries' *International Journal of Epidemiology* **10**, no. 1, 7380.

Gifford, R.M. (1980) 'Carbon storage by the biosphere' In: *Carbon Dioxide and Climate* G.I. Pearman (ed.), 167-181, Australian Academy of Science, Canberra, Australia.

Gleick, P.H. (ed.) (1993) *Water in crisis: a guide to the world's fresh water resources* Oxford University Press, New York, USA.

Goodman, M.R. (1974) *Study notes in system dynamics* Wright-Allen Press Inc., Cambridge, USA.

Gordon, R.L. (1970) *The evolution of energy policy in Western Europe: the reluctant retreat from coal* Praeger Publishers, New York.

Gore, A. (1992) *Earth in the balance: ecology and human spirit* Plume, New York, USA.

Goudriaan, J. (1989) 'Modeling biospheric control of carbon fluxes between atmosphere, ocean and land in view of climatic change' In: *Climate and geo-sciences: a challenge for science and society in the 21th century* A. Berger, S. Schneider and J.C. Duplessy (eds.), 285, 481-499, Kluwer, Dordrecht, the Netherlands.

Goudriaan, J. and P. Ketner (1984) 'A simulation study for the global carbon cycle, including man's impact on the biosphere' *Climatic Change* **6**, 167-192.

Graham, W.J. (1991) 'Maternal mortality: levels, trends and data deficiencies' In: *Disease and mortality in Sub-Saharan Africa* R.G. Feacham and D.T. Jamison (eds.), Oxford University Press, New York, U.S.A..

Gregory, R.L. (1981) *Mind in science: a history of explanation of psychology and physics* Penguin Books Ltd, Middlesex, UK.

Grendstad, G. (1994) 'Classifying cultures' LOS report, 9502, Norwegian Research Center on Organization and Management and Department of Comparative Politics, Bergen, Norway.

Grendstad, G. and P. Selle (1995) 'Ecology, environmentalism and political cultures' Risk, policy and complexity, University of Bergen, Norway.

Grigg, D.B. (1974) *The agricultural systems of the world: an evolutionary approach* Cambridge University Press, London.

Grigg, D.B. (1982) *The dynamics of agricultural change: the historical experience* Hutchinson & Co. Ltd, London.

Gross, R.N. (1980) 'Interrelation between health and population: observations derived from field experiences' *Social Science and Medicine* **14c**, 99-120.

Grossman, G. (1995) 'Population and growth: what do we know?' In: *The economics of sustainable development* I. Goldin and L. Winters (eds.), 312, Cambridge University Press, Cambridge.

Grübler, A. and H. Nowotny (1990) 'Towards the fifth Kondratiev upswing: elements of an emerging new growth phase and possible development trajectories', RR-90-7, IIASA, Laxenburg, Austria.

Gunning-Schepers, L.J. (1989) 'The health benefits of prevention' *Health Policy* **12**, 93A-129SA.

Haines, A. (1991) 'Improving health: a key factor in sustainable development' In: *The world at the crossroads* Study group of the 41st Pugwash Conference on Science and World Affairs (ed.), Beijing, China.

Hall, D. and J. House (1994) 'Biomass a modern and environmentally acceptable fuel' *Journal of Solar Energy Materials and Solar Cells*.

Hann, J. (1903) *Handbook of climatology* Macmillan, New York, USA.

Hansen, J., I. Fung, A. Lacis, D. Rind, S. Lebedeff, F. Ruedy and G. Russel (1988) 'Global climate change as forecasted by Goddard Institute for space studies three dimensional model' *Journal of Geophysical Research* **93**, 9341-9363.

Hardin, G. (1968) 'The tragedy of the commons' *Science* **162**.

Harris, J.M. (1990) *World agriculture and the environment* Garland Publishing, Inc., New York, USA.

Harrison, P. (1992) *The third revolution: environment, population and a sustainable world* I.B. Tauris & Co. Ltd., London.

Harvey, L.D. (1989) 'Effect of model structure on the response of terrestrial biosphere models to CO_2 and temperature increases' *Global Biogeochemical Cycles* **3**, no. 2, 137-153.

HDP (1995) 'Global change, local challenge' HDP Third Scientifc Symposium, Geneva, Switzerland.

Hoekstra, A.Y. (1995) 'Aqua: a framework for integrated water policy analysis' RIVM report, 461502006, National Institute of Public Health and the Environment (RIVM), Bilthoven, the Netherlands.

Hoekstra, A.Y. (1997) 'Water in crisis? An inquiry into possible futures' Delft University of Technology, Delft, the Netherlands.

Holling, C.S. (1979) 'Myths of ecological stability' In: *Studies in crisis management* G. Smart and W. Stansbury (eds.), Butterworth, Montreal, Canada.

Holling, C.S. (1986) 'The resilience of terrestrial ecosystems' In: *Sustainable development of the biosphere* W.C. Clark and R.E. Munn (eds.), Cambridge University Press, Cambridge, UK.

Holling, C.S. (1994) 'An ecologist view of the Malthusian conflict' In: *Population, economic development, and the environment* K. Lindahl-Kiessling and H. Landberg (eds.), Oxford University Press, New York, USA.

Hope, C. and J. Parker (1993) 'Forum: sharpening the environmental debate' *Energy Policy* **20**, 1075-1076.

Hope, C., J. Parker and S. Peake (1992) 'A pilot environmental index for the UK in the 1980s' *Energy Policy* **12**, no. 4, 335-342.

Hordijk, L. (1991) 'An integrated assessment model for acidification in Europe' Free University of Amsterdam, Amsterdam, the Netherlands.

Houghton, J.T., B.A. Callander and S.K. Varney (eds.) (1992) *Climate change 1992: the supplementary report to the IPCC scientific assessment* Cambridge University Press, Cambridge, UK.

Houghton, R.A. (1994) 'Emission of carbon from land-use' In: *The global carbon cycle* T.M.L. Wigley and D. Schimel (eds.), Cambridge University Press, Stanford, USA.

Houghton, R.A., R.D. Boone, J.R. Fruci, J.E. Hobbie, J.M. Melillo, C.A. Palm, B.J. Peterson, G.R. Shaver, G.M. Woodwell, B. Moore, D.L. Skole and N. Myers (1987) 'The flux of carbon from terrestrial ecosystems to the atmosphere in 1980 due to changes in land use: geographic distribution of the global flux' *Tellus* **39B**, 122-139.

Houghton, R.A., J.E. Hobbie, J.M. Melillo, B. Moore, B.J. Peterson, G.R. Shaver and G.M. Woodwell (1983) 'Changes in the carbon content of terrestrial biota and soils between 1860 and 1980: a net release of CO_2 to the atmosphere' *Ecological Monographs* **53**, no. 3, 235-262.

Houghton, R.A. and G.M. Woodwell (1989) 'Global climatic change' *Scientific American* **260**, 36-44.

Hourcade, J.C., K. Halsnaes, M. Jaccard, W.D. Montgomery, R. Richels, J. Robinson, P.R. Shukla, P. Sturm, W. Chandler, O. Davidson, J. Edmonds, D. Finon, K. Hogan, F. Krause, A. Kolesov, E. La Rovere, P. Nastari, A. Pegov, K. Richards, L. Schrattenholzer, R. Shackleton, Y. Sokona, A. Tudini and J. Weyant (1995) 'A review of mitigation cost studies' In: *Economic and social dimensions of climate change: contribution of working group III to the second assessment report of the intergovernmental panel on climate change* J. Bruce, Hoesung Lee and E. Haites (eds.), Cambridge University Press, Cambridge, UK.

Houwaart, E.S. (1991) *De hygienisten: artsen, staat and volksgezondheid in Nederland 1840-1890* Historische Uitgeverij, Groningen, the Netherlands.

Howe, D.W. (1974) *Political cultures of the American whigs* University of Chicago Press, Chicago, USA.

Hudson, R.J.M., S.A. Gherini and R.A. Goldstein (1994) 'Modelling the global carbon cycle: nitrogen fertilization of the terrestrial biosphere and the "missing" CO_2 sink' *Global Biogeochemical Cycles* **8**, no. 3, 307-333.

Hulme, M., S.C.B. Raper and T.M.L. Wigley (1994) 'An integrated framework to address climate change (ESCAPE) and further developments of the global and regional climate modules (MAGICC)' In: *Integrative assessment of mitigation, impacts and adaptation to climate change* N. Nakícenovíc, W.D. Nordhaus, R. Richels and F.L. Toth (eds.), 289-308, IIASA, Laxenburg.

Humphrey, C.R. and F.H. Buttel (1982) *Environment, energy and society* Wadsworth, Belmont, USA.

Hurowitz, J.C. (1993) 'Toward a social policy for health' *N Engl J Med* **329**, 130-133.

Hutter, I., F.J. Willekens, H.B.M. Hilderink and L.W. Niessen (1996) 'Fertility change in India' RIVM report, 461502013, National Institute of Public Health and the Environment (RIVM) and Population Research Centre of University of Groningen, Bilthoven and Groningen, the Netherlands.

ICWE (1992) 'Report of the conference', International Conference on Water and the Environment, Dublin, Ireland.

IFPRI (1995) *A 2020 vision for food, agriculture and the environment; the challenge and recommended action* International Food Policy Research Institute, Washington, D.C., USA.

IGBP (1988) 'The International Geosphere-Biosphere Programme: a study of global changes - A plan for action', International Geosphere - Biosphere Programme, Stockholm, Sweden.

IGBP (1990) 'A study of global change: the initial core programmes', International Geosphere-Biosphere Programme, The Royal Swedish Academy of Sciences, Stockholm, Sweden.

IGBP (1992) 'Global change: reducing uncertainties', The Royal Swedish Academy of Sciences, Stockholm, Sweden.

IGBP (1993) 'Biospheric aspects of the hydrological cycle: the operational plan', International Geosphere - Biosphere Programme, Berlin, Germany.

IIASA/WEC (1995) 'Global energy perspectives to 2050 and beyond' IIASA report, WP-95-127, IIASA, Laxenburg, Austria.

International Energy Agency IEA (1990) 'World Energy statistics and balances 1985-1988', IEA / OECD.

IPCC (1990) *Climate change: the IPCC scientific assessment* Cambridge University Press, Cambridge, UK.

IPCC (1992) *Climate change 1992: the supplementary report to the IPPC scientific assessment* Cambridge University Press, Cambridge, UK.

IPCC (1994) *Climate change 1994: radiative forcing of climate change and an evaluation of the IPCC 1992 emission scenario* Cambridge University Press, Cambridge, UK

IPCC (1995a) *Impacts, adaptions and mitigation of climate change: scientific - technical analyses: contribution of working group II to the second assessment report of the intergovernmental panel on climate change* Cambridge University Press, New York, USA.
IPCC (1995b) *IPCC guidelines for national greenhouse gas inventories* Cambridge University Press, Cambridge, UK.
IPCC (1995c) *The science of climate change: contribution of working group I to the second assessment report of the intergovernmental panel on climate change* Cambridge University Press, Cambridge, UK.
IUCN, UNEP and WWF (1980) *World conservation strategy: living resource conservation for sustainable development*, Gland, Switzerland.
IUCN, UNEP and WWF (1991) *Caring for the earth: a strategy for sustainable living*, Gland, Switzerland.
Jaeger, C.C. (1995) 'ULYSSES: urban lifestyles, sustainability and integrated environmental assessment', EC program, Zürich, Switzerland.
Jaeger, J. (ed.) (1988) *Developing policies for responding to climatic change: a summary of discussions and recommendations of the workshops in Villach and Bellagio* WMO.
Jaeger, J., N. Sonntag, D. Bernard and W. Kurz (1990) 'The challenge of sustainable development in a greenhouse world: some visions for the future' Policy Exercise, Bad Bleiberg, Austria.
Jager, W., M.B.A. van Asselt, J. Rotmans, C.A.J. Vlek and P. Costerman-Boodt (1997) 'Consumer behaviour: a modelling perspective in the context of integrated assessment of global change' RIVM report, 461502017, National Institute of Public Health and the Environment (RIVM), Bilthoven, the Netherlands.
Janssen, M.A. (1996) 'Meeting targets: tools to support integrated assessment modelling of global change' Maastricht University, Maastricht, the Netherlands.
Janssen, M.A. and J. Rotmans (1994) 'Allocation of fossil CO_2-emissions rights: quantifying cultural perspectives' *Ecological Economics* **13**, 65-79.
Janssen, P.H.M., P.S.C. Heuberger and R. Sanders (1992) 'UNCSAM 1.1: a software package for sensitivity and uncertainty analysis manual' RIVM-report, 959101004, National Institute of Public Health and the Environment (RIVM), Bilthoven, the Netherlands.
Janssen, P.H.M., W. Slob and J. Rotmans (1990) 'Uncertainty analysis and sensitivity analysis: an inventory of ideas, methods and techniques from the literature' RIVM-report, 958805001, National Institute of Public Health and the Environment (RIVM), Bilthoven, the Netherlands.
Jensen, J. (1994) 'Gas supplies for the world market' *The Energy Journal* **15**, no. The changing world petroleum market (Special Issue), 237-250.
Johansson, T., B. Bodlund and R. Williams (eds.) (1989) *Electricity - efficient end-use and new generation technologies, and their planning implications* Lund University Press, Lund.
Johansson, T., H. Kelly, A. Reddy and R. Williams (eds.) (1993) *Renewable energy* Island Press, Washington, USA.
Joos, F., U. Siegenthaler and J.L. Sarmiento (1991) 'Possible effects of iron fertilisation in the southern ocean on atmospheric CO_2 concentration' *Global Biogeochemical Cycles* **5**, 135-150.
Jörgensen, S.E. and H. Meijer (1976) 'Modeling the global cycle of carbon, nitrogen and phosphorus and their influence on the global heat balance' *Ecological Modeling* **2**, 19-31.
Kahn, H. and A. Wiener (1967) *The year 2000 - A framework for speculation on the next thity-three years* MacMillan, New York.
Kane, R.L., G.J. Evans and D. Macfadyen (1990) 'Improving the health of older people - a world view' In: WHO (ed.), Oxford University Press, Oxford, UK.
Kane, S. and J. Reilly (1992) 'An empirical study of the economics effects of climate change on world agriculture' *Climate Change* **21**, 277-279.
Kassam, A.H., H.T. van Velthuizen, G.W. Fischer and M.M. Shah (1991) 'Agro-ecological land resources assessment for agricultural development planning. A case study of Kenya. Resources database and land productivity. Technical Annex 6. Fuelwood productivity.' World Soil Resources Reports, Rome.
Kassler, P. (1994) 'Energy for development', Shell Selected Paper, Shell Int'l Petr Company.
Kasun, J. (1988) *The unjust war against population* Ignatius, San Francisco, USA.

Kates, R.W., B.L. Turner and W.C. Clark (1990) 'The great transformation' In: *The earth as transformed by human action* B.L. Turner, W.C. Clark, R.W. Kates, J.F. Richards and J.T. Mathews (eds.), Cambridge University Press, Cambridge, UK.

Keepin, B. and B. Wynne (1984) 'Technical analysis of the IIASA energy scenarios' *Nature* **312**, 691-695.

Keller, A.A. and R.A. Goldstein (1994) 'The human effect on the global carbon cycle: response functions to analyze management strategies' *World Resource Review* **6**, no. 1, 63-87.

Keller, W.J. and J. van Driel (1985) 'Differential consumer demand systems' *European Economic Review* **27**, 375-390.

Keyfitz, N. (1993) 'Are there ecological limits to population?' IIASA working paper, WP-93-16, International Institute for Applied Systems Analysis (IIASA), Laxenburg, Austria.

Kiessling, K.L. and H. Landberg (1994) *Population, economic development and the environment - the making of our common future* Oxford University Press, Oxford, UK.

King, M. (1990) 'Health is a sustainable state' *The Lancet* **226**, 664-667.

King, M.H. and C.M. Elliott (1994) 'Cairo: damp squib or roman candle?' *The Lancet* **344**, no. August, 528.

Klein Goldewijk, C.G.M. and J.J. Battjes (1995) 'The IMAGE 2 hundred year (1890-1900) data base of the global environment (HYDE)' RIVM Report, 481507008, National Institute of Public Health and the Environment (RIVM), Bilthoven, the Netherlands.

Klein Goldewijk, C.G.M., J.G. van Minnen, G.J.J. Kreileman, M. Vloedbeld and R. Leemans (1994) 'Simulating the carbon flux between the terrestrial environment and the atmosphere' *Journal of Water Air Soil Pollution* **76**, no. 1-2, 199-230.

Kohlmaier, G.H., A. Janecek and J. Kindermann (1990) 'Positive and negative feedback loops within the vegetation/soil system in response to a CO_2 greenhouse warming' In: *Soils and the greenhouse effect* A.F. Bouwman (ed.), Wiley & Sons, London, UK.

Kooreman, P. (1993) 'De prijsgevoeligheid van huishoudelijk waterverbruik' *Economisch Statistische Berichten* **78**, 181-183.

Kramer, N.J.T.A. and J.d. Smit (1991) *System Thinking* Stenfert Kroese Uitgevers, Leiden, the Netherlands.

Krapivin, V.F. (1993) 'Mathematical Model for Global Ecological Investigations' *Ecological Modelling*, no. 67, 103-127.

Kuik, O.J. and H. Verbruggen (eds.) (1991) *In search of indicators of sustainable development* Kluwer Academic Publishers, Dordrecht, the Netherlands.

Kulshreshtha, S.N. (1993) 'World water resources and regional vulnerability: impact of future changes' Research report, RR-93-10, International Institute for Applied Systems Analysis, Laxenburg, Austria.

Kuznets, S. (1966) *Modern Economic Growth* Yale University Press, New Haven, USA.

Kwa, C. (1987) 'Representations of nature mediating between ecology and science policy: the case of the international biological programme' *Social Studies of Science* **17**, 413-442.

Lal, R. (ed.) (1994) *Soil erosion research methods* St. Lucie Press, Delray Beach (Fl), USA.

Latour, B. (1987) *Science in action: how to follow scientists and engineers through society* Harvard University Press, Cambridge, USA.

Latour, B. (1993) *We have never been modern* Harvard University Press, Cambridge, USA.

Lazarus, M. (1993) 'Towards a fossil free energy future: the next energy transition', Stockholm Environment Institute (SEI) Boston, for Greenpeace International.

Leach, G. (1995) 'Global land and food in the 21st century: trends and issues for sustainability' SEI Report, Stockholm Environment Institute, Stockholm.

Lee, T., H. Linden, D. Dreyfus and T. Vasko (ed.) (1988) *The Methane Age* Kluwer Academic Publishers, Dordrecht, the Netherlands.

Leemans, R. and W. Cramer (1991) 'The IIASA database for mean monthly values of temperature, precipitation and cloudiness on a global terrestrial grid' Research Report, RR-91-18, International Institute for Applied Systems Analysis (IIASA), Laxenburg, Austria.

Leemans, R. and A.M. Solomon (1993) 'Modeling the potential change in yield and distribution of the earth's crops under a warmed climate' *Climate Research* **3**, 79-96.

Leemans, R. and G.J. van den Born (1994) 'Determining the potential distribution of vegetation, crops and agricultural productivity' In: *IMAGE 2.0: integrated modeling of global climate change* J. Alcamo (ed.), Kluwer Academic Publishers, Dordrecht, the Netherlands.

Leggett, J. (ed.) (1990) *Global Warming: The Greenpeace Report* Oxford University Press, Oxford, UK.
Leggett, J., W. Pepper and R. Swart (1992) 'Emissions scenarios for the IPCC : an update' In: *Climate Change 1992* J.T. Houghton, B.A. Callendar and S.K. Varney (eds.), 69-95, Cambridge University Press, Cambridge, UK.
Lele, S.M. (1991) 'Sustainable development: a critical review' *World Development* **19**, no. 6, 607-621.
Lenssen, N. and C. Flavin (1996) 'Sustainable energy for tomorrow's world: the case for an optimistic view of the future' *Energy Policy* **24**, no. 9, 769-781.
Leopold, A. (1949) *The land ethic* Oxford University Press, Oxford, UK.
Lindblom, C.E. (1959) 'The science of "muddling through"' American Society for Public Administration 19, 79-88.
Lindzen, R.S. (1990) 'Some coolness concerning global warming' *Bulletin American Meteorological Society* **71**, no. 3.
Liverman, D.M., Hanson, B. Brown and R.W. Merideth (1988) 'Global sustainability: towards measurement' *Environmental Management* **12**, no. 2, 133-143.
Lopez, A. (1990) 'Who dies of what? A comparative analysis of mortality conditions in developed countries around 1987' *World Health Statistics Quarterly* **43**, 105-114.
Lovelock, J.E. (1988) *The ages of Gaia: a biography of our living earth* Norton and Company, New York, USA.
Lovins, A.B., and H.L. Lovins (1991) 'Least-cost climatic stabilization' *Ann. Rev. of Energy Environment* **16**, 433-531.
Lutz, W. (1991) *Future demographic trends in Europe and North America - What can we assume today?* Academic Press, London.
Lutz, W. (1994) 'The future of world population: What can we assume today?', IIASA.
Luyten, J.C. (1995) 'Sustainable world food production and environment' AB-DLO report, 37, Research Institute for Agrobiology and Soil fertility, Wageningen.
L'vovich, M.I. (1977) 'World water resources: present and future' *Ambio* **6**, no. 1, 13-21.
L'vovich, M.I. and G.F. White (1990) 'Use and transformation of terrestrial water systems' In: *The earth as transformed by human action: global and regional changes in the biosphere over the past 300 years* B.L. Turner II, W.C. Clarck, R.W. Kates, J.F. Richards, J.T. Mathews and W.B. Meyer (eds.), Cambridge University Press, Cambridge, UK.
Mackenbach, J.P. (1991) 'Mortality and medical care: studies of mortality by cause of death in the Netherlands and other European countries' Erasmus University Rotterdam, Rotterdam.
MacKenzie, F.T., L.M. Ver, C. Sabine, M. Lane and A. Lerman (1992) 'C, N, P, S Global Biogeochemical Cycles and modeling of global change' In: *Interactions of C, N, P, and S Biogeochemical Cycles and Global Change* R. Wollast, F.T. Mackenzie and L. Chou (eds.), 4, 1-64, Springer Verlag.
Malone, T.F. (1993) 'Perspectives on science, technology and society in 2050' Third Stockholm Water Symposium, Stockholm, Sweden.
Malthus, T. (1789) *Essays on population*.
Margat, J. (1994) 'Les utilisations d'eau dans le monde, etat présent et essai de prospective', Contribution au Project M-1-3 du Programme Hydrologique International, PHI-IV, UNESCO, Paris, France.
Marmot, M. and P. Elliott (1994) *Coronary health disease epidemiology - from aetiology to public health* Oxford University Press, Oxford, UK.
Marsh, G.P. (1864) *Man and nature or physical geography as modified by human action* Scribner, New York, USA.
Martens, W.J.M., T.H. Jetten, J. Rotmans and L.W. Niessen (1995a) 'Climate change and vector-borne diseases: A global modelling perspective' *Global Environmental Change* **5**, no. 3, 195-209.
Martens, W.J.M., L.W. Niessen, J. Rotmans, T.H. Jetten and A.J. McMichael (1995b) 'Potential impacts of global climate change on malaria risk' *Environmental Health Perspectives* **103**, no. 5, 458-464.
Matsuoka, Y., M. Kainuma and T. Morita (1995) 'Scenario analysis of global warming using the Asian-Pacific Integrated Model (AIM)' *Energy Policy* **23**, no. 4/5, 357-371.
Mazur, A. (1996) 'The inevitablility - and limits - of dissent' *21stC* **1**, no. 3, 5-6.
McCaffrey, S.C. (1993) 'Water, politics, and international law' In: *Water in crisis: a guide to the world's fresh water resources* P.H. Gleick (ed.), Oxford University Press, New York, USA.

McGuire, A.D., J.M. Melillo, L.A. Joyce, D.W. Kicklighter, G. A.L., B. Moore III and C.J. Vorosmarty (1992) 'Interactions between carbon and nitrogen dynamics in estimating net primary productivity in North America' *Global Biogeochemical Cycles* **6**, no. 2, 101-124.
McLaren, D. and B. Skinner (eds.) (1987) *Resources and world development* John Wiley & Sons Ltd., Berlin, Germany.
McMichael, A.J. (1993) 'Global environmental change and human population health: a conceptual and scientific challenge for epidemiology' *International Journal on Epidemiology* **22**, 1-8.
McMichael, A.J. and W.J.M. Martens (1995) 'The health impacts of global climate change: grappling with scenarios, predictive models and multiple uncertainties' *Ecosystem Health* **1**, no. 1, 23-33.
Meadows, D.H., D.L. Meadows and J. Randers (1991) *Beyond the limits: confronting global collapse, envisioning a sustainable future* Earthscan Publications Ltd., London, UK.
Meadows, D.H., D.L. Meadows, J. Randers and W.W. Behrens (1972) *The limits to growth* Universe Books, New York, USA.
Meadows, D.L., W.W. Behrens III, D.H. Meadows, R.F. Nail, J. Randers and E.K.O. Zahn (1974) *Dynamics of growth in a finite world* Wright-Allen Press Inc., Cambridge, Massachusetts, USA.
Meier, M.F. (1984) 'Contribution of small glaciers to global sea level' *Science* **226**, 1418-1421.
Melillo, J.M., A.D. McGuire, D.W. Kicklighter, B. Moore, III, C.J. Vorosmarty and A.L. Schloss (1993) 'Global climate change and terrestrial net primary production' *Nature* **363**, no. 6426, 234-239.
Meybeck, M. (1982) 'Carbon, nitrogen, and phosphorus transport by world rivers' *American Journal of Science* **282**, 401-450.
Meybeck, M. and R. Helmer (1989) 'The quality of rivers: from pristine stage to global pollution' *Palaeogeography, Palaeoclimatology, Palaeoecology* **75**, 283-309.
Millard, A.V. (1994) 'A 'causal' model of high rates of child mortality' *Social Science and Medicine* **38**, no. 2, 253-268.
Ministry of Public Works (1985) 'De waterhuishouding van Nederland', Staatsuitgeverij, The Hague, the Netherlands.
Mitchell, G.J. (1991) *World on fire: saving an endangered earth* Charles Scribner's Sons, New York, USA.
Molina, M.J. and F.S. Rowland (1974) 'Stratospheric sink for chlorofluoromethanes: chlorine atomic catalysed destruction of ozone' *Nature* **249**, 810-814.
Montgomery, M.R. and J.B. Casterline (1993) 'The diffusion of fertility control in Taiwan: evidence from pooled cross-section time series model' *Population Studies* **47**, 457-479.
Morgan, G.M. and M. Henrion (1990) *Uncertainty: a guide to dealing with uncertainty in quantitative risk and policy analysis* Cambridge University Press, New York, USA.
Morgan, M.G. and D.W. Keith (1995) 'Subjective judgments by climate experts' *Environmental Science and Technology* **29**, no. 4/5, 468-476.
Morita, T., Y. Matsuoka and M. Kainuma (1995) 'Long term global scenarios based on the AIM model', AIM Interim Paper IP-95-03, AIM Project Team.
Morrison, D.E. (1976) 'Growth, environment, equity and scarcity' *Social Science Quarterly* **57**, 292-306.
Mosley, W.H. and P. Cowley (1991) 'The challenge of world health' *Population Bulletin* **46**, no. 4, 2-39.
Moxnes, E. (1989) 'Interfuel substitution in OECD-European electricity production', 89/03, Chr. Michelsens Institutt (CMI), Bergen, Norway.
Munasinghe, M. (1990) 'The pricing of water services in developing countries' *Natural Resources Forum*.
Munasinghe, M. (1993) 'Environmental Economics and Sustainable Development' World Bank Environment Paper, 3, The World Bank, Washington, USA.
Murdoch, W.W. and A. Oaten (1975) 'Population and food: a critique of lifeboat ethics' *Bioscience* **25**.
Murray, C.J.L. and A.D. Lopez (1994) 'Quantifying disability: data, methods and results' *WHO bulletin* **72**, 447-480.
Myers, N. and J.L. Simon (1994) *Scarity or abundance? A debate on the environment* W.W. Norton & Company, New York, London.
Naill, R. (1977) *Managing the energy transition: a systems dynamics search for alternatives to oil and gas* Ballinger, Cambridge, MA, USA.
Naisbitt, J. and P. Aburdene (1990) *Megatrends 2000: ten new directions for the 1990's* Avon Books, New York, USA.

Najera, J.A., B.H. Liese and J. Hammer (1993) 'Malaria' In: *Disease control priorities in developing countries* D.T. Jamison, W.H. Mosley, A.R. Measham and J.L. Bobadilla (eds.), World Bank, New York.

Nakicenovic, N. (1989) 'Technological progress, structural change and efficient energy use: trends worldwide and in Austria', 700/76.716/9, IIASA, Laxenburg, Austria.

Nakicenovic, N. and H.H. Rogner (1995) 'Global financing needs for long-term energy perspectives', WP-95-101, IIASA, Laxenburg, Austria.

Nathwani, J.S., E. Siddall and N.C. Lind (1992) *Energy for 300 years: benefits and risks* Institute for Risk Research, University of Waterloo, Waterloo, Ontario, Canada.

Ness, G.D., W.D. Drake and S.R. Brechin (eds.) (1993) *Population-environment dynamics: ideas and observations* The University of Michigan Press, Michigan, USA.

Niessen, L. and H. Hilderink (1997) 'Roads to health: modelling the health transition' GLOBO report series, National Institute of Public Health and the Environment (RIVM), Bilthoven, the Netherlands.

Niessen, L.W. and J. Rotmans (1993) 'Sustaining health: towards an integrated global health model' RIVM-report, 461502001, National Institute of Public Health and the Environment (RIVM), Bilthoven, the Netherlands.

Nieswiadomy, M.L. (1992) 'Estimating urban residual water demand: effects of price structure, conservation and education' *Water Resources Research* **28**, no. 3, 609-615.

Nordhaus, W.D. (1991) 'To slow or not to slow: the economics of the greenhouse effect' *The Economic Journal* **101**, 920-937.

Nordhaus, W.D. (1992) 'An optimal transition path for controlling greenhouse gases' *Science* **258**, 1315-1319.

Nordhaus, W.D. (1994) *Managing the global commons: the economics of climate change* The MIT Press, Cambridge, Massachusetts and London, UK.

O'Riordan, T. and S. Rayner (1991) 'Risk management for global environmental change' *Global Environmental Change* **1**, no. 2, 91-108.

OECD (1993) 'Environmental indicators: basic concepts and terminology' Indicators for use in environmental performance reviews, Paris, France.

Oerlemans, J. (1989) 'A projection of future sea level' *Climatic Change* **15**, 151-174.

Oldeman, L.R., R.T.A. Hakkeling and W.G. Sombroek (1991) 'World map of the status of human-induced soil degradation: an explanatory note (second revised edition)', Wageningen, the Netherlands.

Oldeman, L.R. and H.T. van Velthuyzen (1991) 'Aspects and criteria of the agro-ecological zoning approach of FAO', International Soil Reference and Information Centre (ISRIC), Wageningen, the Netherlands.

Olshansky, S.J., B.A. Carnes and C. Cassel (1990) 'In search of Methuselah: estimating the upper limits to human longevity' *Science* **25**, 634-640.

Olshansky, S.J., M.A. Rudberg, B.A. Carnes, C. Cassel and J.A. Brody (1991) 'Trading off longer life for worsening health: the expansion of morbidity hypothesis' *Journal Aging Health* **3**, 194-216.

Opschoor, J.B. and L. Reijnders (1991) 'Towards sustainable development indicators' In: *In search of indicators of sustainable development* O.J. Kuik and H. Verbruggen (eds.), Kluwer Academic Publishers, Dordrecht, the Netherlands.

Oreskes, N., K. Schrader-Frechette and K. Belitz (1994) 'Verification, validation and conformation of numerical models in the earth sciences' *Science* **263**, 641-646.

Ott, W. (1978) 'Environmental indices: theory and practice', Ann Arbor Science, Michigan, U.S.A.

Pahl-Wostl, C., C. Jaeger, S. Rayner, C. Schaer, M.B.A. van Asselt, D. Imboden and A. Vckovski (1996) 'Regional integrated assessment and the problem of indeterminacy' In: *Climate and environment in the Alpine region: an interdisciplinary view* P. Cebon, H. Davies, D. Imboden and C. Jaeger (eds.).

Parkin, D.M. (1994) 'Cancer in developing countries' In: *Trends in cancer incidence and mortality* R. Doll, J.F. Fraumeni jr. and C.S. Muir (eds.), Cold Spring Harbor Lab Press.

Parson, E.A. (1996) 'Integrated assessment and environmental policy making: in the pursuit of usefulness' *Energy Policy* **23**, no. 4/5, 463-475.

Peccei, A. (1982) 'Global modelling for humanity' *Futures* **14**, no. 2, 91-94.

Pelletier, D.L., E.A. Frongillo, D.G. Schroeder and J.P. Habicht (1995) 'The effects of malnutrition on child mortality in developing countries' *WHO bulletin* **73**, 443-448.

Penning de Vries, F.W.T., H. van Keulen and R. Rabbinge (1995) 'Natural resources and limits of food production in 2040' In: *Eco-regional approaches for sustainable land use and food production* J. Bouma, A. Kuyvenhoven, B.A.M. Bouman and J.C. Luyten (eds.), Kluwer Academic Publishers, Dordrecht, the Netherlands.

Peterson, B.J. and J.M. Melillo (1985) 'The potential storage of carbon caused by eutrophication of the biosphere' *Tellus* **37B**, 117-127.

Peto, R., A.D. Lopez, J. Boreham, M. Thun and C. Heath (1992) 'Mortality from tobacco in developing countries: indirect estimation from national vital statistics' *The Lancet* **339**, 1268-1278.

Petrella, R., J.C. Burgelman, B. de Schutter, S. Gutwirth and Groep van Lissabon (1994) *Grenzen aan de concurrentie* VUBPRESS, Brussel, Belgium.

Petty, W. (1899) 'Another essay in political arithmetic' In: *The economic writings of Sir William Petty* C.H. Hull (ed.), Cambridge University Press, Cambridge, UK.

Pieri, C. (1989) *Fertilité des terres de savannes: bilan de trente ans de recherche et dévelopement agricole au sud du Sahara* Ministère de la Cooperation CIRIAD-IRAT, Montpellier, France.

Pimentel D. et al. (1997) 'Impact of population growth on food supplies and environment', College of Agriculture and Life Sciences, Cornell University, New York.

Pimentel, D., L.E. Hurd, A.C. Belotti, M.J. Foster, I.N. Oka, O.D. Sholes and R.J. Whitman (1973) 'Food production and the energy crisis' *Science* **182**, 443-449.

Pojman, L.P. (ed.) (1994) *Environmental ethics: readings in theory and application* Jones and Bartlett Publishers, Boston, USA.

Ponting, C. (1993) *A green history of the world: the environment and the collapse of great civilizations* Perguin, New York, USA.

Postel, S. (1992) *Last oasis: facing water scarcity* W.W. Norton & Company, New York, USA.

Postel, S. (1996) 'Divind the waters: food security, ecosystem health, and the new politics of scarity', Worldwatch Paper 132, Worldwatch Institute, Washington D.C., USA.

Postel, S.L., G.C. Daily and P.R. Ehrlich (1996) 'Human appropriation of renewable fresh water' *Science* **271**, 785-788.

PPPP and Delft Hydraulics (1989) 'Cisadane - Cimanuk integrated water resources development', BTA-155, Pusat Penelitian dan Pengembangan Pengairan, Ministry of Public Works, Bandung, Indonesia.

Prather, M., R. Derwent, D. Ehhalt, P. Fraser, E. Sanhueza and X. Zhou (1995) 'Other trace gases and atmospheric chemistry' In: *Radiative forcing of climate change and an evaluation of the IPCC IS92 emission scenarios* J.T. Houghton, L.G. Meira Filho, J. Bruce, H. Lee, B.A. Callander, E. Haites, N. Harris and K. Maskell (eds.), Cambridge University Press, Cambridge, UK.

Preston, S.H. (1975) 'The Changing Relation between Mortality and Level of Economic Development' *Population Studies* **29**, no. 2, 231-248.

Preston, S.H. (1976) *Mortality patterns in national populations - with special reference to recorded causes of death* Academic Press, New York, USA.

Preston, S.H. (1980) 'Causes and consequences of mortality declines in less developed countries during the 20th century' In: *Population and Economic Change in Developing Countries,* R. Easterlin (ed.).

Price, M.F. (1989) 'Global Change: Defining the Ill-defined' *Environment* **31**, no. 8, 18-20, 42, 44.

Price, M.F. (1990) 'Humankind in the Biosphere: The Evolution of International Interdisciplinary Research' *Global Environmental Change* **1**, no. December, 3-13.

Pritchett, H. (1994) 'Desired Fertility and the Impact of Population Policies' *Population and Development Review* **20**, 1-55.

Raich, J.W., E.B. Rastetter, J.M. Melillo, D.W. Kicklighter, P.A. Steudler, B.J. Peterson, A.L. Grace, B. Moore III and C.J. Vörösmarty (1991) 'Potential net primary productivity in South America: application of a global model' *Ecological Applications* **1**, no. 4, 399-429.

Randers, J. (ed.) (1980) *Elements of the systems dynamics method* Productivity Press, Cambridge, Mass.

Raskin, P., E. Hansen and R. Margolis (1995) 'Water and sustainability: a global outlook', Polestar Report Series No. 4, Stockholm Environment Institute, Boston, USA.

Raskin, P.D., E. Hansen and R.M. Margolis (1996) 'Water and sustainability: global patterns and long-range problems' *Natural Resources Forum* **20**, no. 1, 1-5.

Rastetter, E.B., M.G. Ryan, G.R. Shaver, J.M. Melillo, K.J. Nadelhoffer, J.E. Hobbie and J.D. Aber (1991) 'A general biogeochemical model describing the responses of the C and N cycles in terrestrial ecosystems to changes in CO_2, climate and N deposition' *Tree Physiology* **9**, 101-126.

Ravetz, J. (1996) 'Integrated environmental assessment: developing guidelines for "good practise"' Conference on Integrated Assessment of Climate Change, Toulouse, France.
Ray, D.L. and L. Guzzo (1990) 'The Greenhouse Effect: Hype and Hysteria' In: *Trashing the Planet*, Regenery Gateway, Washington D.C., USA.
Rayner, S. (1984) 'Disagreeing about risk: the institutional cultures of risk management and planning for future generations' In: *Risk analysis, institution and public policy* S.G. Hadden (ed.), Associated Faculty Press, Port Washington, USA.
Rayner, S. (1991) 'A cultural perspective on the structure and implementation of global environmental agreements' *Evaluation Review* **15**, no. 1, 75-102.
Rayner, S. (1992) 'Cultural theory and risk analysis' In: *Social theory of risk* G.D. Preagor (ed.), Westport, USA.
Rayner, S. (1994) 'Governance and the global commons' Discussion Paper, 8, The Centre for the Study of Global Governance, London School of Economics, London, UK.
Rayner, S. and E. Malone (eds.) (1996) *Human Choice and Climate Change: An International Social Science Assessment*.
Redclift, M. (1987) *Sustainable Development: Exploring the Contradictions* Methuen, London, UK.
Rees, C. (1994) 'The Ecologist's Approach to Sustainable Development' In: *Making Development Sustainable* I. Serageldin and A. Steer (eds.), The World Bank, Washington, USA.
Reich, C. (1970) *The greening of America* Random House, New York, USA.
Reijntjes, C., B. Haverkort and A. Waters-Bayer (1992) *Farming for the future: an introduction to low-external-input-sustainable-agriculture* Macmillan Press Ltd., London. UK.
Renzetti, S. (1992) 'Evaluating the welfare effects of reforming municipal water prices' *Journal of Environmental Economics and Management* **22**, no. 2, 147-163.
Resource Analysis (1994) 'Socio-economic impacts study of international environmental problems, Phase 3, Part I: software implementation activities, Part II: data gathering for the world version, Part III: data collection for coastal defense model', RA/94-150, 151 & 159, Resource Analysis, Delft, the Netherlands.
Riley, J.C. (1990) 'The risk of being sick: morbidity trends in four countries' *Population Development Review* **3**, 403-431.
RIVM (1991) *National Environmental Outlook 1990-2010* National Institute of Public Health and the Environment (RIVM), Bilthoven, the Netherlands.
RIVM (1994) *National Environmental Outlook 3: 1993-2015* National Institute of Public Health and the Environment (RIVM), Bilthoven, the Netherlands.
Robey, B., S.O. Rutstein, L. Morris and R. Blackburn (1992) 'The reproductive revolution: new survey findings' *Population Reports Series* **11**, no. Special Issue.
Rogers, H.H. and R.C. Dahlmann (1993) 'Crop responses to CO_2 enrichment' *Vegetatio* **104/105**, 117-131.
Root, T.L. and S.H. Schneider (1995) 'Ecology and climate: research strategies and implications' *Science* **269**, no. 5222, 334-341.
Rosegrant, M.W., M. Agcaoili-Sombilla and N.D. Perez (1995) 'Global food projections to 2020: implications for investment' *International food policy institute*.
Rosen, R. (1985) *Anticipatory systems: philosophical, mathematical and methodlogical foundations* Pergamon Press, New York.
Rosenzweig, C., B. Curry, J.T. Richie, J.W. Jones, T.Y. Chou, R. Goldberg and A. Iglesias (1994) 'The effects of potential climate change on simulated grain crops in the United States' In: *Implications of climate change for international agriculture: crop modeling study* C. Rosenzweig and A. Iglesias (eds.), 1-24, U.S. Environmental Protection Agency, Washington D.C.
Rosenzweig, C. and M.L. Parry (1994) 'Potential impact of climate change on world food supply' *Nature* **367**, 133-138.
Rosero-Bixby, L. and J.B. Casterline (1993) 'Modelling diffusion effects in fertility transition' *Population Studies* **47**, 147-167.
Rotmans, J. (1990) *IMAGE: an integrated model to assess the greenhouse effect* Kluwer Academics, Dordrecht, the Netherlands.
Rotmans, J. and M.G.J. den Elzen (1993a) 'Halting global warming: should fossil fuels be phased out?' In: *Global warming: concern for tomorrow* M. Lal (ed.), Tata McGraw-Hill, New Delhi.

Rotmans, J. and M.G.J. den Elzen (1993b) 'Modelling feedback mechanisms in the carbon cycle: balancing the carbon budget' *Tellus* **45B**, no. 4, 301-320.

Rotmans, J., H. Dowlatabadi and E.A. Parson (1996) 'Integrated assessment of climate change: evaluation of methods and strategies' In: *Human choice and climate change: an international social science assessment* S. Rayner and E. Malone (eds.), Cambridge University Press, New York.

Rotmans, J. and R.J. Swart (1991) 'Modelling tropical deforestation and its consequenses for global climate' *Ecological Modelling* **58**, 217-247.

Rotmans, J. and M.B.A. van Asselt (1996) 'Integrated assessment: a growing child on its way to maturity' *Climate Change* **34**, 327-336.

Rotmans, J., M.B.A. van Asselt, A.J. de Bruin, M.G.J. den Elzen, J. de Greef, H. Hilderink, A.Y. Hoekstra, M.A. Janssen, H.W. Köster, W.J.M. Martens, L.W. Niessen and H.J.M. de Vries (1994) 'Global change and sustainable development: a modelling perspective for the next decade' RIVM report, 461502004, National Institute of Public Health and the Environment (RIVM), Bilthoven, the Netherlands.

Rozanow, B.G., V. Targulian and D.S. Orlov (1990) 'Soils' In: *The earth as transformed by human action. Global and regional changes in the biosphere over the past 300 years* B.L. Turner II, W.C. Clark, R.W. Kates, J.F. Richards, J.T. Mathews and W.B. Meyer (eds.), Cambridge University Press, New York, USA.

Ruwaard, D., R.T. Hoogeveen, H. Verkleij, D. Kromhout, A.F. Casparie and E.A. van der Veen (1993) 'Forecasting the number of diabetic patients in the Netherlands in 2005' *American Journal of Public Health* **83**, 989-995.

Sage, R.F., T.D. Sharkey and J.R. Seeman (1989) 'Acclimation of photosynthesis to elevated CO_2 in five C_3 species' *Plant Physiology* **89**, 590-596.

Sahagian, D.L., F.W. Schwartz and D.K. Jacobs (1994) 'Direct anthropogenic contributions to sea level rise in the twentieth century' *Nature* **367**, 54-57.

Santer, B.D., T.M.L. Wigley, M.E. Schlesinger and J.F.B. Mitchell (1990) 'Developing climate scenarios from equilibrium GCM results' report no., 47, Max Planck Institute for Meteorology, Hamburg, Germany.

Schimel, D., I.G. Enting, M. Heimann, T.M.L. Wigley, D. Raynaud, D. Alves and U. Siegenthaler (1995) 'CO_2 and the carbon cycle' In: *Radiative forcing of climate change and an evaluation of the IPCC IS92 emission scenarios* J.T. Houghton, L.G. Meira Filho, J. Bruce, H. Lee, B.A. Callander, E. Haites, N. Harris and K. Maskell (eds.), 35-71, Cambridge University Press, Cambridge, UK.

Schindler, D.W. and S.E. Bayley (1993) 'The biosphere as an increasing sink for atmospheric carbon: estimates from increased nitrogen deposition' *Global Biogeochemical Cycles* **7**, no. 4, 717-733.

Schipper, L. and S. Meyers (1992) *Energy efficiency and human activity: past trends, future prospects* Cambridge University Press, Cambridge, UK.

Schlesinger, W.H. (1991) *Biogeochemistry: an analysis of global change* Academic Press, London.

Schlesinger, W.H. (1993) 'Response of the terrestrial biosphere to global climate change and human perturbation' *Vegetatio* **104/105**, 295-305.

Schlesinger, W.H. and A.E. Hartley (1992) 'A global budget for atmospheric NH_3' *Biogeochemistry* **15**, 191-211.

Schmidheiny et al., S. (1992) *Changing course: a global business perspective on development and the environment* MIT Press, Cambridge, MA, USA.

Schneider, S. (1989) 'The changing global climate' *Scientific American* **261**, no. 3, 38-47.

Schofield, R., D. Reher and Bideau (1991) *The decline of mortality in Europe* Clarendon Press, Oxford, UK.

Schwarz, M. and M. Thompson (1990) *Divided we stand: redefining politics, technology and social choice* Harvester Wheatsheaf, New York, USA.

Scrabanek, P. and McCormick (1992) *Follies and fallacies in medicine* Tarragon Press, 2nd edition.

Scrimshaw, N.S. and L. Taylor (1980) 'Food' *Scientific American* **243**.

Serageldin, I. (1995) 'Water resources management: a new policy for a sustainable future' *Water International* **20**, 15-21.

Serageldin, I. (1996) 'Sustainability and the wealth of nations: first steps in an ongoing journey', Environmentally Sustainable Development Studies and Monographs Series No. 5, World Bank, Washington, D.C.

Serageldin, I. and A. Steer (eds.) (1994) *Making development sustainable* The World Bank, Washington, USA.
Shaw, E.M. (1994) *Hydrology in practice (third edition)* Chapman & Hall, London, UK.
Shaw, R., G. Gallopin, P. Weaver and S. Öberg (1992) 'Sustainable development: a systems approach' Status Report, SR-92-06, International Institute for Applied Systems Analysis, Laxenburg, Austria.
Shell Planning (1996) 'Global scenarios 1995-2020', Shell, London.
Shiklomanov, I.A. (1993) 'World fresh water resources' In: *Water in crisis: a guide to the world's fresh water resources* P.H. Gleick (ed.), Oxford University Press, New York, USA.
Shiklomanov, I.A. (1995) 'Assessment of water resources and water availability in the world: scientific and technical report', State Hydrological Institute, St. Petersburg, Russia.
Siegenthaler, U. and H. Oeschger (1987) 'Biospheric CO_2 emissions during the past 200 years reconstructed by deconvolution of ice core data' *Tellus* **39B**, 140-154.
Simon, J.L. (1980) 'Resources, population, environment: an oversupply of false bad news' *Science* **208**, 1431-1437.
Simon, J.L. (1981) 'Against the Doomsdayers' In: *The ultimate resource*, Princeton University Press, Princeton, USA.
Simonnot, P. (1978) *Les nucléocrates* Presses universitaires de grenoble, Grenoble.
Speidel, D.H. and A.F. Agnew (1988) 'The world water budget' In: *Perspectives on water: uses and abuses* D.H. Speidel, L.C. Ruedisili and A.F. Agnew (eds.), Oxford University Press, Oxford, UK.
Statoil and Energy Studies Program (1995) 'Global Transport Sector Energy Demand towards 2020' WEC Tokyo Congress, Tokyo, October 8-13, 1995, 151.
Stehr, N., H.v. Storch and M. Flugel (1995) 'The 19th Century Discussion of Climate Variability and Climate Change: Analogies for Present Debate?', 157, Max Planck Institute for Meteorology, Hamburg, Germany.
Sterman, J.D. (1981) 'The energy transition and the economy: a system dynamics approach' MIT, Sloan School of Management.
Sterman, J.D. and G.P.R. Richardson (1983) 'An experiment to evaluate methods for estimating fossil fuel resources', D-3432-1, MIT, Sloan School of Management.
Stevens, R.D. and C.L. Jabara (1988) *Agricultural development principles: economic theory and empirical evidence* The John Hopkins University Press, Baltimore, Maryland, USA.
Stoffers, M.J. (1990) 'A tentative world oil price model and some possible future developments of crude oil prices', CPB (Central Planning Bureau), The Hague, the Netherlands.
Subroto (1993) 'The road from Rio' Seminar on energy for sustainable development post-Rio challenges and Dutch responses, Amsterdam, the Netherlands.
Suess, E. (1980) 'Particulate organic carbon flux in the oceans - surface productivity and oxygen utilization' *Nature* **288**, 260-263.
Swart, R. and J. Bakkes (1995) 'Scanning the global environment : a framework and methlodgy for integrated environmental reporting and assessment', UNEP/EAP.TR/95-01 RIVM/402001002, UNEP/RIVM.
Tans, P.P., I.Y. Fung and T. Takahashi (1990) 'Observational constraints on the global atmospheric CO_2 budget' *Science* **247**, 1431-1438.
Teramura, A.H., J.H. Sullivan and L.H. Ziska (1990) 'Interaction of elevated ultraviolet-B radiation and CO_2 on productivity and photosynthetic characteristics in wheat, rice, and soybean' *Plant Physiology* **94**, 470-475.
Thom, T.J. and F.H. Epstein (1994) 'Heart disease, cancer, and stroke mortality trends and their interrelations' *Circulation* **90**, 574-582.
Thomas, W.L. (ed.) (1956) *Man's Role in Changing the Face of the Earth* University of Chicago, Chicago, USA.
Thompson, M. (1982) 'Among the energy tribes: the anthropology of the current energy debate' Working Paper, 82-59, International Institute for Applied Systems Analysis (IIASA), Laxenburg, Austria.
Thompson, M., R. Ellis and A. Wildavsky (1990) *Cultural theory* Westview Press, Boulder, USA.
Thornthwaite, C.W. (1948) 'An approach toward a rational classification of climate' *The Geographical Review* **38**, 55-94.

Thornthwaite, C.W. and J.R. Mather (1957) *Instructions and tables for computing potential evapotranspiration and the water balance* Drexel Institute of Technology, Laboratory of Climatology, Centerton, New Jersey,USA.

Timmerman, P. (1986) 'Myths and paradigms of intersections between development and environment' In: *Sustainable development of the biosphere* W.C. Clark and R.E. Munn (eds.), Cambridge University Press, Cambridge, UK.

Timmerman, P. (ed.) (1989) *The human dimensions of global change: an international programme on human interactions with the earth* HDGCP Secretariat, Toronto, Canada.

Toet, A.M.C., H.J.M. de Vries and R.A. van den Wijngaart (1994) 'Background report for the Image2.0 energy-economy model' RIVM report, 481507006, National Institute of Public Health and the Environment (RIVM), Bilthoven, the Netherlands.

Tolba, M.K. and O.A. El-Kholy (eds.) (1992) *The world environment 1972 - 1992. Two decades of challenge* Chapman & Hall, London.

Toth, F.L. (1988) 'Policy Exercises' *Simulation and Games* **19**, no. September, 235-276.

Toth, F.L. (1992) 'Global change and the cross-cultural transfer of policy games' In: *Global interdependence* D. Crookall and K. Arai (eds.), Springer Verlag, Tokyo, Japan.

Toth, F.L. (1993) 'Practice and progress in integrated assessment of climate change: a review' Integrative assessment of mitigation, impacts and adaptation to climate change, IIASA, Laxenburg, Austria.

Toth, F.L., E. Hizsnyk and W.C. Clark (1989) 'Scenarios of socio-economic development for studies of global environmental change: a critical review', RR-89-4, IIASA, Laxenburg, Austria.

Trainer, F. (1995) 'Can renewable energy sources sustain affluent societies ?' *Energy Policy* **23**, no. 12, 1009-1026.

Troeh, F.R., J.A. Hobbs and R.L. Donahue (1991) *Soil and water conservation* Prentice Hall Inc., EDnglewood Cliffs, New Jersey, USA.

Tunstall, H., K. Kuulasmaa, P. Amouyei, D. Arveiler, A.M. Rjakangas and A. Pajak (1994) 'Myocardial infarction and coronary deaths in the WHO MONICA project - registration procedures, event rates, and case-fatality rates in 38 populations from 21 countries in four continents' *Circulation* **90**, 583-612.

Turner II, B.L., W.C. Clark, R.W. Kates, J.F. Richards, J.T. Mathews and W.B. Meyer (eds.) (1990a) *The earth as transformed by human action: global and regional changes in the biosphere over the past 300 years* Cambridge University Press, New York, USA.

Turner II, B.L., R.E. Kasperson, W.B. Meyer, K.M. Dow, D. Golding, J.X. Kasperson, R.C. Mitchell and S.J. Ratick (1990b) 'Two types of global environmental change' *Global Environmental Change* **1**, no. 4, 14 - 22.

Turner, R.K. and W.R. Dubourg (1994) 'Water resource scarcity: an economic perspective' Proceedings of Third Stockholm Water Symposium, Stockholm, Sweden.

Tversky, A. and D. Kahneman (1974) 'Judgement under uncertainty: heuristics and biases' *Science* **185**, 1124-1131.

Tversky, A. and D. Kahneman (1980) 'Causal schemes in judgements under uncertainty' In: *Progress in Social Psychology* M. Fishbein (ed.), Lawrence Erlbaum Associates, Hillsdale, USA.

Tversky, A. and D. Kahneman (1981) 'The framing of decisions and the psychology of choice' *Science* **211**, 453-458.

Tylecote, A. (1992) *The long wave in the world economy: the present crisis in historical perspective* Routledge, London.

Uemura, K. and Z. Pisa (1988) 'Trends in cardiovascular disease mortality in industrialized countries since 1950' *World Health Statistics Quarterly* **41**, 155-178.

Ueshima, H.et al. (1987) 'Declining mortality from is ischemic heart disease and change in coronary risk factors in Japan - 1956-1980' *American Journal of Epidemiology* **125**, no. 1, 37-39.

UN (1958) *Integrated river basin development* United Nations, New York, USA.

UN (1970) *Integrated river basin development: second edition* United Nations, New York, USA.

UN (1990) 'Population prospects 1990', United Nations, New York, USA.

UN (1992) *Agenda 21: the United Nations programme of action from Rio* United Nations Publications, New York, USA.

UN (1993) 'World population prospects: the 1992 revision', ST/ESA/SER.A/135, United Nations, Department for Economic and Social Information and Policy Analysis, New York, USA.

UN (1995) 'World population prospects: the 1994 revision', ST/ESA/SER.A/145, United Nations, Department for Economic and Social Information and Policy Analysis, New York, USA.
UNCED (1992) 'Aganda 21', United Nations Conference on Environment and Development, Conches, Switzerland.
UNDP (1994) 'Human development report', New York, USA.
UNDP (1995) 'Human development report', New York, USA.
UN-DPCSD (1995) 'Work programme on indicators of sustainable development of the Commission on Sustainable Development', UN-DPCSD, New York, U.S.A.
UN-ECE/FAO (1992) 'The forest resources of the temperate zones: 1990 Forest Resource Assessment', ECE/TIM/80, UN-ECE/FAO, Rome, Italy.
UNEP (1984) 'General assessment of progress in the implementation of the plan of action to combat desertification 1978-1985', United Nations Environmental Programme, Nairobi, Kenya.
UNEP (1991) *Status of desertification and implementation of the United Nations plan of action to combat desertification* UNEP, Nairobi. Kenya.
UNEP (1997) *Global environmental outlook* Oxford University Press, Oxford, UK.
UNEP/RIVM (1997) 'The future of the global environment: a model-based analysis supporting UNEP's first global environment outlook', Environment Information and Assessment Technical Report, RIVM/402001007 and UNEP/DEIA/TR.97.1-1, Bilthoven, the Netherlands and Nariobi, Kenya.
UNFPA (1994) 'World population report', New York, USA.
UNFPA (1995) 'World population report', New York, USA.
UNFPA (1996) 'World population report', New York, USA.
UNICEF (1994) 'The state of the world's children'.
UNICEF (1995) 'The state of the world's children'.
Vallin, J. (1992) 'Theories of mortality decline and the African situation' In: *Mortality and society in Sub-Saharan Africa* E. van der Walle, G. Pison and M. Sala-Diakanda (eds.), Clarendon Press, Oxford.
van Asselt, M.B.A., A.H.W. Beusen and H.B.M. Hilderink (1996) 'Uncertainty in integrated assessment: a social scientific approach' *Environmental Modelling and Assessment* **1**, no. 1/2, 71-90.
van Asselt, M.B.A. and J. Rotmans (1995) 'Uncertainty in integrated assessment modelling: a cultural perspective-based approach' RIVM report, 461502009, National Institute of Public Health and the Environment (RIVM), Bilthoven, the Netherlands.
van Asselt, M.B.A. and J. Rotmans (1996) 'Uncertainty in perspective' *Global Environmental Change* **6**, no. 2, 121-157.
van den Berg, H. (1994) 'Calibration and evaluation of a global energy model', National Institute of Public Health and the Environment (RIVM) and University of Groningen (IVEM-RUG), the Netherlands.
van der Pol, F. (1993) 'Analysis and evaluation of options for sustainable agriculture, with special reference to southern Mali' In: *The role of plant nutrients for sustainable crop production in Sub-Saharan Africa* H. van Reuler and W.H. Prins (eds.), Verening van Kunstmest Producenten, Leidschendam, the Netherlands.
van der Walle, E., G. Pison and M. Sala-Diakanda (1992) *Mortality and society in Sub-Saharan Africa* Clarendon Press, Oxford, UK.
van Deursen, W.P.A. and J.C.J. Kwadijk (1994) 'The impacts of climate change on the water balance of the Ganges-Brahmaputra and Yangtze basin', RA/94-160, Resource Analysis, Delft. the Netherlands.
van Vianen, H., F.J. Willekens, H.B.M. Hilderink and L.W. Niessen (1997) 'Fertility change in Mexico', National Institute of Public Health and the Environment (RIVM) and Population Research Centre of University of Groningen, Bilthoven and Groningen, the Netherlands.
van Vianen, H.A.W., F.J. Willekens, I. Hutter, M.B.A. van Asselt, H.B.M. Hilderink, L.W. Niessen and J. Rotmans (1994) 'Fertility change: a global integrated perspective' RIVM-Report, 461502008, National Institute of Public Health and the Environment (RIVM), Bilthoven, the Netherlands.
Verbrugge, L.M. (1989) 'Recent, Present and Future Health of American Adults' *Annual Review of Public Health* **10**, 33-61.
Vernadsky, V.I. (1945) 'The Biosphere and the Noosphere' *American Scientist* **33**, no. 1, 1-12.
Vis, J.W.D. (1996) 'Experiments with and further development of the AQUA model for the Zambezi basin', Thesis for the study Science and Policy, Utrecht University, Utrecht, the Netherlands.

Vitousek, P.M. and R.W. Howarth (1991) 'Nitrogen limitation on land and in the sea: how can it occur?' *Biogeochemistry* **2**, 86-115.

von Storch, H. and K. Hasselman (1994) 'Climate variability and change' European Conference on Grand Challenges in Ocean and Polar Sciences, Bremen, Germany.

VROM (1994) 'Towards a single indicator for emissions - an exercise in aggregating environmental effects', 1994/12, Ministry of Public Housing and the Environment, The Hague, the Netherlands.

Walsh, J. and K. Warren (1980) 'Selective Primary Health Care: An Interim Strategy for Disease Control in Developing Countries' *Social Science and Medicine* **14C**, 145-164.

Walter, S.D. (1976) 'The estimation and interpretation of attributive risk in health research' *Biometrics* **32**, 829-849.

Walters, N.R. (1988) 'Central states forest management guides as applied in Stems, St.Paul: U.S.D.A.', General Technical Report, NC-119, United States Department of Agriculture, North Central Forest Experiment Station.

Warrick, R.A., C. Le Provost, M.F. Meier, J. Oerlemans and P.L. Woodworth (1995) 'Changes in sea level' In: *Climate change 1995: the science of climate change* J.T. Houghton, L.G. Meira Filho, B.A. Callander, N. Harris, A. Kattenberg and K. Maskell (eds.), Cambridge University Press, Cambridge, UK.

WCED (1987) *Our common future* World Commission on Environment and Development, Oxford University Press, Oxford,UK.

Weier, K.L., J.W. Doran, J.F. Power and D.T. Walters (1993) 'Denitrification and dinitrogen/nitrous oxide ratio as affected by soil water, available carbon, and nitrate' *Soil Science Society American Journal* **57**, 66-72.

Weinstein, M.C., P.G. Coxson, L.W. Williams, T.M. Pass, W.B. Stason and L. Goldman (1987) 'Forecasting coronary heart disease incidence, mortality and cost: the coronary heart policy model' *American Journal of Public Health* **77**, no. 1417-1426.

White, G.W. (1983) 'Water resources adequacy: illusion and reality' *Natural Resources Forum* **7**, no. 1, 11-21.

WHO (1978) 'Declaration of Alma Ata' International Conference of Primary Health Care, Alma Ata, USSR.

WHO (1984) 'The International Drinking Water Supply and Sanitation Decade: review of national baseline data (as at 31 December 1980)', Publication 85, World Health Organization, Geneva, Switzerland.

WHO (1991) 'The International Drinking Water Supply and Sanitation Decade: review of decade progress (as at December 1990)', World Health Organization, Geneva, Switzerland.

WHO (1994) 'Health future research' *World Health Statistics Quarterly* **47**, no. 3/4.

WHO (1996) *The World Health Report 1996, fighting disease fostering development* WHO, Geneva.

WHO/UNEP (1991) 'Water quality: progress in the implementation of the Mar del Plata Action Plan and a strategy for the 1990s', World Health Organization (WHO) and United Nations Environment Programme (UNEP), Geneva, Switzerland and Nairobi, Kenya.

Wigley, T.M.L. and S.C.B. Raper (1992) 'Implications for climate and sea-level of revised IPCC emissions scenarios' *Nature* **357**, 293-300.

Wigley, T.M.L., R. Richels and J.A. Edmonds (1996) 'Economic and environmental choices in the stabilisation of CO_2 concentrations: choosing the "right" emissions pathway' *Nature* **379**, 240-243.

Wigley, T.M.L. and M.E. Schlesinger (1985) 'Analytical solution for the effect on increasing CO_2 on global mean temperature' *Nature* **315**, 649-652.

Wild, A. (ed.) (1988) *Russell's soil conditions and plant growth* Longman Scientific and Technical, Harlow (Essex), UK.

Williams, R.H. (1995) 'Variants of a low CO_2-emitting energy supply system (less) for the world', PNL-10851, Pacific Northwest Laboratories, Richland, Washington, USA.

Wischmeier, W.H. and D.D. Smith (1978) 'Predicting rainfall erosion losses: a guide to conservation planning' Agricultural Handbook, 537, Washington D.C., USA.

Wisserhof, J. (1994) *Matching research and policy in integrated water management* Delft University, Delft, the Netherlands.

WMO (1992) 'Scientific Assessment of Ozone Depletion: 1991', 25, WMO/UNEP.

Wollast, R., F.T. Mackenzie and L. Chou (1993) *Interactions of C, N, P and S biogeochemical cycles and global change* Springer Verlag, Berlin.
Wolter, H.W. and A. Kandiah (1996) 'Harnessing water to feed a hungry world' Sixteenth congress on irrigation and drainage: sustainability of irrigated agriculture, Cairo, Egypt, Vol. 1H.
Woolgar, S. (1981) 'Interest and explanations in the social study of science' *Social Studies of Science* **11**, 365-397.
Woolgar, S. (1988) *Science: the very idea* Tavistock, London, UK.
World Bank (1991) *World development report 1991* Oxford University Press, New York, USA.
World Bank (1993) *World development report 1993 - Investing in Health* Oxford University Press, New York, USA.
World Bank (1994) 'A review of World Bank experience in irrigation' Report, 13676, The World Bank operations department, Washington DC, USA.
World Bank (1995) *World development report 1995* Oxford University Press, New York, USA.
World Bank (1996) 'World devolpment report 1996: from plan to market', World Bank, Washington D.C., USA.
Woytinski, W.S. and E.S. Woytinski (1953) *World population and production: trends and outlook* The Twentieth Century Fund, New York, USA.
WRI (1992) *World Resources 1992-93* Oxford University Press, New York, USA.
WRI (1994) *World Resources 1994-95* Oxford University Press, New York, USA.
WRI (1996) *World Resources 1996-97* Oxford University Press, New York, USA.
WRR (1992) 'Ground for choices; four perspectives for the ruaral areas in the European community', 42, Netherlands Scientific Council for Government Policy (WRR), The Hague, the Netherlands.
WRR (1994) *Sustained Risks: A Lasting Phenomenon* SDU Uitgeverij, The Hague, the Netherlands.
Wynne, B. (1982) *Rationality and ritual: the windscale inquiry and nuclear decisions in Brittain* British Society for the History of Science, Chalfont, St. Glies.
Wynne, B. and S. Shackley (1994) 'Environmental models: truth machines or social heuristics' *The Globe* **21**, 6-8.
Yergin, D. (1991) *The prize: the epic quest for oil, money and power* Simon & Schuster, London, UK.
Young, G.J., J.C.I. Dooge and J.C. Rodda (1994) *Global water resources issues* Cambridge University Press, Cambridge, UK.
Young, O.R. (1964) 'A survey of general systems theory' *General Systems* **9**.
Zeng Yi (1991) *Family dynamics in China: a life table analysis* The University of Wisconsin Press.
Zeng Yi, F.J. Willekens and L.W. Niessen (1997) 'Fertility change in China', RIVM report 461502016, Institute of Population Research, Beijing University, Univeristy of Groningen, RIVM (in preparation).
Zube, E.H. (1982) 'Increasing the Effective Participation of Social Scientists in Environmental Research and Planning' *International Social Science Journal* **34**, 481-492.
Zuidema, G., G.J. van den Born, J. Alcamo and G.J.J. Kreileman (1994) 'Simulating changes in global land cover as affected by economic and climatic factors' *Water, Air and Soil Pollution* **76**, no. 1-2, 163-198.
Zweers, W. and J.J. Boersema (1994) *Ecology, technology and culture* White Horse, Cambridge, UK.

ACRONYMS, UNITS, AND SYMBOLS

Acronyms: institutions and organisations

AGGG	Advisory Group on Greenhouse gases
ASCEND	Agenda of Science for Environment and Development into the 21st Century
CBS	Central Bureau of Statistics
CEC	Commission of European Communities
CPB	Central Planning Bureau
CSD	Commission on Sustainable Development
EPA	Environmantal Protection Agency, Washington, USA
FAO	Food and Agriculture Organisation of the United Nations
ICS	International Council of Scientific Unions
IEA	International Energy Agency
IGBP	International Geosphere-Biosphere Programme
IIASA	International Institute for Applied Systems Analysis
IMF	International Monetary Fund
IPCC	International Panel on Climate Change
ISSC	International Social Science Council
IUCN	World Conservation Union
OECD	Organisation for Economic Co-operation and Development
RIVM	National Institute of Public Health and the Environment, The Netherlands
SCOPE	Scientific Committee On Problems of the Environment
UN	United Nations
UNCED	United Nations Conference for Environment and Development, Rio de Janeiro, 1992
UNDP	United Nations Development Programme
UNEP	United Nations Environment Programme
UNGA	United Nations General Assembly
USDOE	Department of the Energy, USA
VROM	Dutch Ministry of Housing, Physical Planning and Environment
WCED	World Commission on Environment and Development
WCRP	World Climate Research Programme
WHO	World Health Organisation
WMO	World Meteorological Organisation
WRI	World Resources Institute

Acronyms: models and variables

AEEI	Autonomous Energy Efficiency Improvement
AQUA	Water submodel in TARGETS
BGF	BioGaseousFuels
BLF	BioLiquidFuels
CBD	Crude Birth Rate
CCE	Cereal Consumption Equivalent
CDR	Crude Death Rate
CYCLES	Biogeochemical cycles submodel in TARGETS
DALE	Disability Adjusted Life Expectancy
GCM	General Circulation Model
GF	Gaseous Fuels
GPP	Gross Primary Production
GWP	Gross World Product
HDI	Human Development Index

HDP	Human Dimensions Programme
HLE	Healthy Life Expectancy
HLF	Heavy Liquid Fuels
HYDE	Hundred Year Database of the global Environment
IA	Integrated Assessment
IMAGE	an Integrated Model to Assess the Greenhouse Effect
IS92a-f	IPCC-1992 scenarios a-f
LE	Life Expectancy
LF	Liquid Fuels
LLF	Light Liquid Fuels
NPP	Net Primary Production
NTE	Non Thermal Electric[ity]
PIEEI	Price-Induced Energy Efficiency Improvement
PSIR	Pressure-State-Impact-Response
Q	Soil reduction factor
RD&D	Research, Development & Demonstration
SCS	Strategic Cyclical Scaling
SF	Solid Fuel
SOM	Soil Organic Matter
TARGETS	Tool to Assess Regional and Global Environmental and health Targets for Sustainability
TE	Thermal Electric[ity]
TERRA	Land and Food submodel in TARGETS
TFR	Total Fertility Rate
TIME	Targets IMage Energy model, Energy submodel in TARGETS
UNCSAM	UNCertainty analysis by Monte Carlo SAMpling techniques

Units

°C	degrees Celsius
cap	capita
CCE	Cereal Consumption Equivalent
EJ	10^{18} joule
GJ	10^9 joule
Gt	Gigatonnes (10^{15} gram)
GtC	Gigatonnes of carbon
K	degrees Kelvin
kcal	kilocalory (= 4.1855×10^3 joule)
kg	kilogram
km	kilometre
l	litre
m	metre
mg	milligram
mm	millimetre
Mha	Million of hectares
mole	mol
ppmv	parts per million (10^6) by volume
ppbv	parts per billion (10^9) by volume
s	second
$	1990 US $
tonne	metric ton
Tg	Teragram (10^{12} gram)
ha	hectare
yr	year
Wm^{-2}	watts per square metre

Chemical symbols

C	carbon
CH_4	methane
CO	carbon monoxide
CO_2	carbon dioxide
DIN	dissolved inorganic nitrogen
DMS	dimethyl sulphide
DOC	dissolved organic carbon
DON	dissolved organic nitrogen
H_2O	water
N	nitrogen
N_2	molecular nitrogen
NH_3	ammonia
N_2O	nitrous oxide
NMHC	nonmethane hydrocarbons
NO	nitric oxide
NO_x	NO and NO_2
O_3	ozone
OH	hydroxyl
O	atomic oxygen
O_2	molecular oxygen
P	phosphorus
PO_4^{3-}	phosphate
POC	particulate organic carbon
PON	particulate organic nitrogen
S	sulphur
SO_2	sulphur dioxide

INDEX

A

Abandonment of rainfed arable land	145, 333
Age groups	62-63, 68, 73-74, 78, 80
Ageing	57, 61-62, 68, 73, 249-250, 256, 385
Aggregation	
spatial	18, 39-42
temporal	41
Agricultural systems	138, 158, 323, 325
AQUA submodel	109-134
calibration	130-134
dystopias	314-315
perspectives	300-304
model structure	114-115
policy	128-129, 315-317
position within TARGETS	115
simulation result	130-134
uncertainty analysis	311
uncertainties	295-296
utopias	304-310

B

Base Load Factor (BLF)	98-100, 274-276, 281
Biodiversity	5-6, 8, 50, 152
Biofuels	85, 95, 98-100, 102-103, 268, 274-277
BioGaseousFuels (BGF)	98-100, 274-275, 279, 457
BioLiquidFuels (BLF)	98-100, 274-276, 279, 457
Biogeochemical climate feedback	
CO_2 fertilisation	151, 164, 167, 172-173, 177, 179, 181, 323-326, 350-353
N fertilisation	164, 167, 173-177, 184-185, 350-354
temperature feedback	166-167, 173, 177, 179, 181, 350-355
Blood pressure	70, 79, 249

C

Capital-labour ratio	95, 97, 99-100
Capital-output ratio	95-96, 98-99, 230-231
Carbon tax	271-275, 282-289, 400, 407-411, 425
Carbon budget	163, 180-182, 184-185, 349-350, 353, 357, 360
Carbon cycle	167
model	170-175
Cause-effect chain	17, 20, 26, 37-38, 193
Cereal Consumption Equivalent (CCE)	147-148
Clearing costs	148, 329, 331-334, 336
Climate system	351-352
CO_2 fertilisation, see biogeochemical climate feedback	
Coastal defence	112, 129
Controversy	
biogeochemical cycles	348-349
energy	265-270
land and food	322
population and health	241-243
water	294
Crude Birth Rate (CBR)	74, 81, 251, 384-386
Crude Death Rate (CDR)	74, 81, 249-251, 384-386
Cultural theory	212, 215-217, 383
Curation	243
CYCLES submodel	161-185
calibration	179-185
dystopias	364-370
perspectives	353-356
model structure	168-169
position within TARGETS	169
uncertainty analysis	367
uncertainties	348-352
utopias	357-364

D

DALE	74-76, 384
Deforestation	137, 144-145, 154, 326, 329, 332, 335-336, 343
Depletion of fossil fuels resources	88, 268
Desalination	294-295, 298, 301, 303-304
Disease categories	71
Dystopia	252-253, 288, 364, 384, 405, 407, 428
biogeochemical cycles	364-370
energy	284-288
land and food	337-343
population and health	252-256
water	314-315

E

Economic scenario generator	38, 46, 49, 51-52, 231-232
Economic world regions	141
Electricity	
demand	94, 100-101
supply	49, 100-102
Emissions from fuel combustion	88, 269-270
End-use energy demand	92-93, 103, 266, 281, 285

Energy
    efficiency   85, 88, 90, 92-94, 103-104, 267, 269, 272, 275
    AEEI   92-94, 103, 267, 274-275, 281-285
    intensity   87-89, 92-93, 104, 266-267, 269, 274-277, 281, 283
    prices   87, 93, 104, 236, 267, 381
    PIEEI   92-94, 267, 274, 281-282
Energy submodel (TIME)   85-106
    calibration   102-106
    dystopias   284-289
    perspectives   270-273
    model structure   90
    position within TARGETS   89-91
    uncertainty analysis   284-285
    uncertainties   266-270
    utopias   274-284
Erosion   140-141, 145, 149-150, 174
Exposure categories   70-71, 74-75
Extensification   137

F

Fertiliser
    use   149, 156, 158, 165, 168-169, 176, 178-180, 329, 331, 333-336, 384-385
    response curve   149
Fertility   58, 66-68, 256-260
Food
    demand   139-144, 149, 153-158, 238, 321-326, 328-335, 337-343
    security   251, 322, 324, 341, 343
    supply   139, 146-149, 153, 155, 237, 251, 293, 331-340, 376-378
    surplus   340, 342-343
    shortage   338-342, 376, 403, 406, 413
    trade   137, 238, 321

G

Gaseous fuels (GF)   43, 89, 94, 98-99, 104, 268-269, 274, 279, 457
Glaciers   123, 132
Global change   3-14
Global environmental change   5-6
Greenland   112, 121, 123, 132, 305

H

Health
    determinants   62-71, 80, 197, 242-245, 251, 259
    services   57, 62, 65, 75-76, 78, 80, 230-231, 243-251, 254
Healthy Life Expectancy (HLE)
    46, 252, 258-260, 385, 394, 458

Heat demand   94
Heat stress   53, 66, 140, 151, 332
Human Development Index (HDI)
    65, 67, 80, 189, 192, 198, 230, 250
Hydrological cycle   47, 53, 109-110, 114-115, 119-120

I

Ice sheets   119-121
Indicator   30, 190-198, 200-202, 389-390, 420-422
Integrated Assessment (IA)   11-14, 18, 22, 27, 35-38, 189-193, 202, 207-211, 220-223, 262
Integrated experiments   225-226, 229, 256, 317, 347, 374-375, 394, 397
Integration   12, 17-18, 22, 25-27, 30
    horizontal   25-26
    vertical   25-27
Intensification   137-140, 149, 234, 335, 343
Investments   52, 60, 93-102, 231, 237-238
Irrigation costs   146, 153, 158, 299, 328, 331, 334, 336

L

Land use and land dynamics   145
Length of Growing period (LGP)   42, 127, 142-143, 147-152
Liquid fuels   94, 98-99, 104, 458
Life Expectancy (LE)   57, 61, 74, 79-81, 199, 244-248, 253-261, 405

M

Malnutrition   65, 69-70, 78-79, 152, 230, 235, 249, 321-322, 325, 401
Management style   216-220
Maximum total fertility   66-67
Metamodel   22, 41
Model route   209-211, 220, 231
Mortality
    maternal   71
    base   72
    disease-specific   57, 63, 78, 251

N

Natural gas   85-86, 88, 95, 102-106, 230, 268, 276-277, 283-284, 393
Net primary production (NPP)   167, 171-173, 177-181, 184-185, 352-356
N fertilisation,
    see biogeochemical climate feedback

Nitrogen budget	178, 182-185, 359, 364	Salinisation	204, 322-323, 331
Nitrogen cycle	164-167	Sanitation	65, 115, 126, 128, 203, 238, 245, 294, 300-304, 404, 412-413, 439, 454
model	176	Saturation level	144, 326, 328, 333-339
Non-thermal electricity	85, 90, 101, 281	Sea-level rise	438, 449
NTE	100-104, 274-277, 281-285, 458	Secondary fuels	89, 92, 94, 106, 275, 424
		Smoking	70, 79, 249

O

Oil products 86, 99

P

Peak Load Factor	101-102
Perspective	
biogeochemical cycles	353-356
energy	270-273
land and food	324-328
population and health	243-247
water	300-304
PLF	101-102
Potential arable land	146, 148, 326, 331-336
Population and Health submodel	
calibration	76-78
dystopias	252-256
perspectives	243-247
model structure	65-69
position within TARGETS	63-65
uncertainty analysis	251
uncertainties	242-243
utopias	247-250
Pressure-State-Impact-Response (PSIR)	17, 22-23
Prevention	
primary	75
secondary	46, 75
Proximate determinants	62, 66-67, 436
Public water supply	70, 117, 126, 128, 235, 257, 293-304, 316, 398

R

Reforestation	140, 145, 153-154, 305, 321, 324, 326, 328, 332, 336, 385
anthropogenic	145
natural	145
Renewable energy	87, 203, 233, 268, 271, 273, 288, 414
Reserve Product Ratio (RPR)	98-99
Reservoir	21, 42-43
Risk Assessment	220, 252, 314, 341, 364
Runoff	
stable	110, 124-125, 295-304, 314
total	124, 131, 295-304, 314

S

Safe drinking water 63-66, 69-71, 78-80, 115, 126, 249, 301, 374, 385

Socio-Economic Status (SES) 65, 68, 75-76, 78-79, 249, 257

Soil	
conservation	48, 140, 327
fertility	140, 169, 173, 328, 335, 337, 341-343, 445
moisture	119-123, 127, 151, 350, 353
organic matter (SOM)	140, 173, 182, 328, 335, 389
productivity reduction factor	143, 147
Solid fuels	91, 95
Sustainable development	3-14, 17-18, 20, 35-36, 40, 49

T

TARGETS	35-54
Temperature feedback, see biogeochemical climate feedback	
TERRA submodel	
calibration	153-158
dystopias	337-340
learning coefficient	102, 275, 281, 285
perspectives	324-328
model structure	139-140
policy measures	153
position within TARGETS	140
uncertainty analysis	339
uncertainties	322-324
utopias	328-336
Thermal electricity	276
Total Fertility Rate (TFR)	58, 67, 76, 199, 247, 252, 458
Transition	30-31, 384-389
agricultural	142, 384
energy	87, 270, 281, 384, 388
epidemiological	31, 58, 61, 198, 251, 384
fertility	58, 251, 259
health	58-59, 68, 80, 241-244, 384-386, 401, 413, 415
water	387-388

U

Utopia	
biogeochemical cycles	357-364
energy	274-284
land and food	328-336
population and health	247-250
water	304-310

V

Vegetable production   144, 148-149, 153-158, 234, 339
Vegetable supply   149, 155

W

Waste water   129, 377
Water
   availability   63, 109, 116, 127, 137-141, 146, 247, 256
   demand   111-119, 126-134, 141
   policy   109-114, 293, 295, 315-317, 398, 441
   pollution   109-110, 117, 170, 196, 200, 293, 296
   pricing   51, 111, 114, 130, 303, 316
   quality   111-114, 124, 128, 178, 233-236, 246, 296
   reuse   316, 438
   scarcity   109, 111, 153, 294-296, 299-305, 308, 312-315
   supply   62, 70-71, 78, 110-119, 125-134, 167, 235, 237
Wood demand   336
World view   211-212

Y

Yield per hectare   147, 404
   actual   147, 149, 339, 388
   potential   140-152, 328, 332-333, 336